Internet of Things

Technology, Communications and Computing

Series editors

Giancarlo Fortino, Rende (CS), Italy
Antonio Liotta, Eindhoven, The Netherlands

More information about this series at http://www.springer.com/series/11636

Franco Cicirelli · Antonio Guerrieri
Carlo Mastroianni · Giandomenico Spezzano
Andrea Vinci
Editors

The Internet of Things for Smart Urban Ecosystems

Springer

Editors
Franco Cicirelli
Institute for High Performance
Computing and Networking (ICAR)
National Research Council (CNR)
Rende (CS), Italy

Giandomenico Spezzano
Institute for High Performance
Computing and Networking (ICAR)
National Research Council (CNR)
Rende (CS), Italy

Antonio Guerrieri
Institute for High Performance
Computing and Networking (ICAR)
National Research Council (CNR)
Rende (CS), Italy

Andrea Vinci
Institute for High Performance
Computing and Networking (ICAR)
National Research Council (CNR)
Rende (CS), Italy

Carlo Mastroianni
Institute for High Performance
Computing and Networking (ICAR)
National Research Council (CNR)
Rende (CS), Italy

ISSN 2199-1073 ISSN 2199-1081 (electronic)
Internet of Things
ISBN 978-3-030-07212-4 ISBN 978-3-319-96550-5 (eBook)
https://doi.org/10.1007/978-3-319-96550-5

This Springer imprint is published by the registered company Springer Nature Switzerland AG
The registered company address is: Gewerbestrasse 11, 6330 Cham, Switzerland

Preface

Nowadays, the exploitation of ICT technologies in urban environments is enabling the realization of smart infrastructures that provide enhanced digital services improving the quality-of-life of citizens and the cities' efficiency.

In this field, significant smart applications include but are not limited to smart parking, traffic management, smart lighting, structural health monitoring, air quality and pollution monitoring, health monitoring, waste management, smart grids and city energy optimization, smart urban drainage networks, building automation, and emergency detection and management.

Since all these applications share a common urban environment and common goals, they need to interact with each other and with citizens, creating a Smart Urban Ecosystem (SUE).

An SUE is a people-centric system of systems needing a tight integration between cyber and physical components for sensing, reasoning, and controlling the urban environment. In this context, The Internet of Things (IoT) constitutes an enabling technology as it bridges the gap between physical things and software components and empowers the cooperation among distributed, pervasive, and heterogeneous entities.

The development of an SUE introduces several challenges, which include heterogeneous system integration, interoperability among different technologies, fault tolerance, scalability, system maintenance, geographical and functional extensibility, social networking, mobile computing, context-aware applications and services, human-in-the-loop modeling and simulation, big data analysis, cloud- and edge-based IoT frameworks and environments, field experiments and testbeds. Moreover, the realization of an SUE requires cross-domain knowledge expertise spanning from computer science to electronic, electrical, civil, hydraulic, and energy engineering.

The main objective of this book is to provide a multidisciplinary overview of methodological approaches, architectures, platforms, algorithms, and applications for the realization of Smart Urban Ecosystems.

The book includes 15 chapters covering three main topics: (i) software architectures and platforms for SUEs, (ii) development approaches and algorithms for SUEs implementation, and (iii) applications and case studies related to specific smart infrastructures and smart cities. A short introduction to the chapters is provided below.

The Chapter "A Social and Pervasive IoT Platform for Developing Smart Environments", by Orazio Briante, Franco Cicirelli, Antonio Guerrieri, Antonio Iera, Alessandro Mercuri, Giuseppe Ruggeri, Giandomenico Spezzano, and Andrea Vinci, gives an overview of the iSapiens platform, which is a Java-based platform specifically designed for the development and implementation of Smart Environments (SEs). iSapiens exploits the Social Internet of Things paradigm that allows to dynamically discover "things" without requiring intervention from humans. iSapiens provides tools for the realization of pervasive SEs and relies on the edge computing paradigm. Moreover, the chapter reviews some SE applications built by leveraging iSapiens features.

The Chapter "Smart City Platform Specification: A Modular Approach to Achieve Interoperability in Smart Cities", by Arianna Brutti, Piero De Sabbata, Angelo Frascella, Nicola Gessa, Raffaele Ianniello, Cristiano Novelli, Stefano Pizzuti, Giovanni Ponti, proposes a development methodology and a modular and scalable multilayered ICT platform to address the problem of cross-domain interoperability in the context of smart city applications. As a distinct feature, the chapter tackles issues about interoperability on the Information and Semantic levels and provides an approach for finding the correct balance between prescriptive and elastic specifications.

The Chapter "Integrated Cyber Physical Assessment and Response for Improved Resiliency", by P. Sivils, C. Rieger, K. Amarasinghe, and M. Manic, provides a summary and analysis of crucial concepts and challenges in assessing cyber-physical degradation, heterogeneous data fusion, and visualization under a smart city IoT architecture. The main objective is to provide a basis for enhancing the effectiveness of human response to physical and cyber events.

The Chapter "On the Integration of Information Centric Networking and Fog Computing for Smart Home Services", by Marica Amadeo, Andrea Giordano, Carlo Mastroianni and Antonella Molinaro, focuses on the role that can be assumed by the Information Centric Networking (ICN) paradigm to support future Internet communications and data delivery in smart urban ecosystems, including smart home/building services. The integration of ICN with Cloud/Fog resources is also discussed in the chapter, and a reference architecture is presented as a proof-of-concept, together with a preliminary testbed.

The Chapter "Optimal Placement of Security Resources for the Internet of Things", by Antonino Rullo, Edoardo Serra, Elisa Bertino, and Jorge Lobo, deals with the problem of efficiently and effectively securing IoT networks by carefully allocating security resources in the networked area. The problem is modeled according to the game theory, and a Pareto-optimal solution is provided, in which the cost of the security infrastructure and the probability of a successful attack are minimized. In the chapter, authors make a distinction between static and mobile networks, and address the problem of computing the best allocation plan with different approaches.

The Chapter "Embedding Internet-of-Things in Large-Scale Socio-technical Systems: A Community-Oriented Design in Future Smart Grids", by Yilin Huang, Giacomo Poderi, Sanja Šćepanović, Hanna Hasselqvist, Martijn Warnier, and Frances Brazier, focuses on the design of large-scale Socio-technical Systems (STS) relying on Internet of Things technologies. The design of such systems, especially in a complex social context like that of Urban Ecosystems, requires the use of suitable methodological approaches that have not yet entered into the mainstream of design practice. The chapter reviews the literature and presents some lessons learned in adopting an STS-based approach for the design of a community-oriented smart grid application.

The Chapter "Aggregation Techniques for the Internet of Things: An Overview", by Barbara Guidi and Laura Ricci, starts from the consideration that an Internet of Things environment can connect a very large number of sensors, so generating a huge amount of data. Aggregation techniques are required to reduce the size of data to be transmitted and stored, while maintaining a reasonable level of approximation. The chapter offers an overview of a set of aggregation techniques that can be exploited in the IoT, ranging from Space-Filling Curves to Q-digest, Wavelets, Gossip aggregation, and Compressive Sensing.

The Chapter "Swarm Intelligence and IoT-Based Smart Cities: A Review", by Ouarda Zedadra, Antonio Guerrieri, Nicolas Jouandeau, Giandomenico Spezzano, Hamid Seridi, and Giancarlo Fortino, reviews swarm intelligence-based algorithms and related smart city solutions. A swarm intelligence-based framework for smart cities is presented that uses a decentralized control over its components in order to build scalable and flexible smart cities. In addition, a set of trends on how to use swarm intelligence in smart cities, in order to make them flexible and scalable, is investigated.

The Chapter "Cost Saving and Ancillary Service Provisioning in Green Mobile Networks", by Muhammad Ali, Michela Meo, and Daniela Renga, discusses how mobile network operators are required to face huge operational costs, due to the staggering increase in mobile traffic and substantial bandwidth reliability requirements in Smart Urban Ecosystems. The chapter analyzes, for a real scenario, the notable benefits that can be achieved by combining the WiFi offloading approach, the techniques for dynamic adaptation load in a Demand Response context, and the usage of renewable energy sources.

The Chapter "Structural Health Monitoring (SHM)", by Raffaele Zinno, Serena Artese, Gabriele Clausi, Floriana Magarò, Sebastiano Meduri, Angela Miceli, and Assunta Venneri, aims at fitting the structural health monitoring into the Internet of Things. For this purpose, the chapter extensively details, through the description of a real application scenario, all the phases required for developing an effective online and real-time assessment of the structural health of a building.

The Chapter "A Smart Air-Conditioning Plant for Efficient Energy Buildings", by Roberto Bruno, Natale Arcuri, and Giorgio Cuconati, focuses on the correct management of energy fluxes in the context of prosumer systems. The goal is to make effective the process of producing, storing and consuming energy, and increasing users' remuneration. The chapter describes a smart air conditioning system and the correspondent control strategies adopted for its management. The system is based on the employment of photovoltaic-driven heat pumps with thermal storage connected to a radiant emission system.

The Chapter "A Comprehensive Approach to Stormwater Management Problems in the Next Generation Drainage Networks", by Patrizia Piro, Michele Turco, Stefania Anna Palermo, Francesca Principato and Giuseppe Brunetti, shows how the next generation of urban drainage networks can benefit from Internet of Things and ICT technologies. The chapter describes two innovative approaches for managing drainage networks, exploiting a decentralized real-time control system and low-impact development techniques.

The Chapter "Cooperative Video-Surveillance Framework in *Internet of Things* (IoT) Domain", by A. F. Santamaria, P. Raimondo, N. Palmieri, M. Tropea, and F. De Rango, presents the main issues related to the design of an architecture for a smart cooperative video-surveillance system. The chapter shows a surveillance system based on a cooperative tracking among cameras and involves advanced techniques of detection and tracking. In addition, the chapter presents a significant use case showing how an anomaly can be detected, followed by a set of cameras, and managed by generating alerts and messages without human intervention.

The Chapter "Personal Connected Devices for Healthcare", by Adina Riposan-Taylor and Ian J. Taylor, offers a comprehensive survey on personal connected health technologies, which are fast becoming integral in a person's daily life to help improve their health and well-being. The chapter provides a broad overview of a range of devices that allow people to monitor their own health and/or provide self-therapeutic benefits using biofeedback or neurofeedback to control their own physiologic functions. Specific focus is given to the techniques adopted to provide access to data, to the software used for data integration, to connectivity issues, and to user interfaces.

The Chapter "Evacuation and Smart Exit Sign System", by V. Ferraro and J. Settino, exploits the Internet of Things for the realization of emergency evacuation systems. The chapter proposes a smart system that relies on a set of sensors and smart exit signs which are coordinated by a reliable and dynamic evacuation algorithm, which is capable of fast adapting to changing conditions during the evolution of an emergency.

We would like to thank all the book contributors, the anonymous reviewers, and Prasanna Kumar Narayanasamy from Springer for his precious support during the publication process.

Rende (CS), Italy

Franco Cicirelli
Antonio Guerrieri
Carlo Mastroianni
Giandomenico Spezzano
Andrea Vinci

Contents

A Social and Pervasive IoT Platform for Developing Smart Environments

Orazio Briante, Franco Cicirelli, Antonio Guerrieri, Antonio Iera, Alessandro Mercuri, Giuseppe Ruggeri, Giandomenico Spezzano and Andrea Vinci

Abstract Nowadays, the increasing in the use of Internet of Things (IoT) devices is growing the realization of pervasive Smart Environments (SEs) and Smart Urban Ecosystems, where all the data gathered by the "Things" can be elaborated and used to improve the livability, the safety and the security of the environment, and to make inhabitants lives easier. Many efforts have been already done in the direction of SEs development and in the implementation of platforms specifically designed for SE realization. Anyway, such efforts miss of solutions regarding the interoperability among the realized SEs and other third-part "Things". This chapter gives an overview of iSapiens, which is a Java-based platform specifically designed for the development and implementation of SEs. iSapiens tries to overcome the interoperability issue by leveraging the Social Internet of Things (SIoT) paradigm that allows to dynamically include in an SE the new "Things" that can appear in an environment without requiring interventions from humans. iSapiens provides tools for the

---

F. Cicirelli · A. Guerrieri (✉) · A. Mercuri · G. Spezzano · A. Vinci
CNR - National Research Council of Italy - Institute for High Performance Computing and Networking (ICAR), Rende (CS), Italy
e-mail: guerrieri@icar.cnr.it

F. Cicirelli
e-mail: cicirelli@icar.cnr.it

A. Mercuri
e-mail: mercuri@icar.cnr.it

G. Spezzano
e-mail: spezzano@icar.cnr.it

A. Vinci
e-mail: vinci@icar.cnr.it

O. Briante · A. Iera · G. Ruggeri
University Mediterranea of Reggio Calabria - DIIES Department, Reggio Calabria, Italy
e-mail: orazio.briante@unirc.it

A. Iera
e-mail: antonio.iera@unirc.it

G. Ruggeri
e-mail: giuseppe.ruggeri@unirc.it

F. Cicirelli et al. (eds.), *The Internet of Things for Smart Urban Ecosystems*, Internet of Things, https://doi.org/10.1007/978-3-319-96550-5_1

realization of pervasive SEs and relies on the edge computing paradigm. Such paradigm is extremely important in a distributed system since it allows to use distributed storage and computation at the edge of a network, so reducing latencies with respect to move all the executions and storages in the cloud. Moreover, the chapter will review some SE applications realized by exploiting iSapiens concepts.

Keywords Smart environments · Social internet of things
Edge computing · Multi agent systems · IoT development platforms · IoT-based applications

1 Introduction

Smart Environments (SEs) are pervasive and highly dynamic systems involving a set (possibly huge) of devices which can interact each other, and which can be dynamically added to and removed from the system itself. SEs usually exploit the IoT paradigm for orchestrating heterogeneous devices [1, 2]. SEs final aim is to furnish advanced cyber-physical services to their inhabitants and administrators. Besides being dynamical in terms of physical things, such systems are required to be also dynamic regarding their functionalities, that should evolve according to changes in the SEs finalities. As a consequence, an SE system should have the capability of adding, updating, and removing functionalities depending on the goals and on the updates in the available devices. Among SEs, Smart Urban Ecosystems are gaining importance as a new research topic, involving the same issues presented for the SEs, but typically on a larger geographical scale, involving a wider number of users, and having a greater impact on people lifestyle.

Moreover, Smart Urban Ecosystem development requires to address issues related to dynamic and open scenarios, such as entity discovery, trustworthiness, interoperability, data processing and management. Furthermore, it is of paramount importance to have methodological guidelines and tools to foster the development of such systems by dealing with their complexity [3, 4].

This chapter is devoted to the review of iSapiens [5], which is an agent-based platform taking into account the above mentioned issues. This platform addresses data management and processing by coupling the cloud computing paradigm [6] with the edge computing one [7, 8], leveraging the agent metaphor [9], thus providing systems with both high computational power, exploitable for high-demanding tasks, and distributed elaboration capabilities supporting reactive and location-dependent processing. Entity discovery, trustworthiness, and interoperability are realized by integrating the Social Internet of Things (SIoT) computing paradigm [4, 10] with the iSapiens platform. Finally, device heterogeneity is reached by the introduction of specifically designed abstractions, namely Virtual Objects.

Besides providing details about the platform, the chapter also reviews a set of applications and case studies which involve the exploitation of the concepts, the features and the capabilities that are at the basis of the iSapiens platform.

The chapter is organized as follows. Section 2 introduces some related work, Sect. 3 provides details about the whole iSapiens platform, comprehending its architecture, abstractions and SIOT component. Section 4 reviews a set of application built on top of the iSapiens concepts. Finally, some conclusions are given to the reader.

2 Related Work

Smart Environments, IoT-based approaches, and the recent Social IoT para-digm are all based on the pervasive presence of a possibly huge number of heterogeneous smart objects, which are capable to interact and cooperate each other so as to create novel application or services for the user, and reach shared goals. In this direction, the research and industry communities are putting strong efforts, in particular for both the creation of IoT-based platforms for SEs [11–14], and the development and integration of social capabilities in the IoT [15, 16], which can help in the discovery/use of new devices that can enter, move inside, or leave an SE. Furthermore, such technologies and efforts have given a boost in the development of Smart City and, in general, Smart Urban Ecosystems applications.

2.1 IoT and Smart Environments Platforms

Nowadays, several SEs [17] and platforms/middlewares for their creation [18–21] have been introduced with the aim of tackling specific issues of these complex systems. Most of these platforms lack in interoperability and scalability which are important for the realization of comprehensive Smart Urban Ecosystems which use several enabling technologies such as cloud [6] and/or edge computing, and social capabilities [10, 22].

Actually, literature presents many approaches for implementing SEs. A selection of them is presented in the following. In [23] interactions among several bluetooth mobile devices are at the basis of a framework designed for implementing SEs capable of dynamically discover and exploit available bluetooth devices.

In [24], authors focused on system extensibility by introducing a framework which has been created for the support in the creation of context-aware applications.

The authors of [25] introduced a distributed middleware, based on the concept of *active spaces*, aimed at managing resources placed in physical environments. Active spaces are programmable environments where the involved SE devices are connected.

The development of context-sensing and context-aware applications is the aim of the work presented in [26], where they defined a framework providing a set of abstractions suitable for modeling different kinds of SEs.

A middleware for context-aware smart homes has been presented in [19], with the aim of foster the development of smart homes based on the sharing of contextual information among the smart home entities.

The well known *Syndesi* framework [18] has been developed for gathering data to profile users and make available to them customized services which exploit wireless sensor networks for both sensing and actuation purposes.

2.2 Smart Cities and Smart Urban Ecosystems

Regarding smart cities and smart urban ecosystems, which are hot topics for the research community, several works have been presented so far and several platforms implemented [27].

In [28] authors have presented an open source IoT platform for the pervasive use of actuators and sensors in smart cities. Such platform, called *Sentilo*, exploits both Big Data tools and cloud computing to gather, save, and analyze data coming from distributed sensors.

The work in [29] shows an IoT middleware that was developed for the EPIC (European Platform for Intelligent Cities) project. Issues such as heterogeneity, interoperability, extensibility, and (re)configurability were at the basis of the design and development of the middleware.

A platform for the realization of smart city is proposed in [30]. Such platform relies on Big Data analysis for achieving extensibility. Its architecture presents three layers that have been designed for (i) collecting and analyzing information, (ii) aggregating data in order to infer some knowledge, (iii) offering to the users instruments to take computed data.

Authors of [31] have introduced Smart Connected Communities in order to extend the concept of smart city considering also livability, preservation, revitalization, and sustainability of the areas in the city to improve. Smart Connected Community is thought as comprehending not only a specific city area but all its neighborhood. Smart Connected Community is based on a multi-layer architecture in which is specifically taken in consideration the importance of exploiting social capabilities.

2.3 The Social IoT

Recently, the Social Internet of Things (SIoT) paradigm risen to the attention of the the IoT research community by fostering the convergence between typical technologies and solutions of the Social Networks domain and the IoT [10]. The idea behind this paradigm is as simple as tremendously powerful: similarly to what happens in the human communities, the IoT objects are enhanced so to become Social IoT objects capable of establishing mutual social relationships in an autonomous way according to rules set by their owner. The insurgence of such a social network [32] has proven to facilitate resource visibility, service discovery, object reputation assessment, source crowding, and service composition in the IoT [10]. Let's consider as an example, the case of an object dealing with the likely overwhelming task of seeking an information provided by the multitude of its peers, might use its social connection to drive the research process by limiting the scope only to those nodes with mutual social relations, exactly in the same way as humans search over their Social Network platforms, then the complexity and the time duration of the search could be drastically reduced [10]. Also, social objects might asses their reciprocal trustworthiness and reputation based on their mutual social relationships just as human beings mostly do [33–36]. It has been proven that such an approach allows isolating almost any malicious device in a network [37].

So far, five types of relationships have been defined for this new paradigm [10]:

1. *Ownership Object Relationship* (*OOR*): that is created among objects belonging to the same user;
2. *Parental Objects Relationship* (*POR*): that is created among objects produced by the same manufacturer and belonging to the same production batch;
3. *Co-Work Objects Relationship* (*C-WOR*): that is created among objects cooperating towards the provision of the same IoT application;
4. *Co-Location Objects Relationship* (*C-LOR*): that bounds objects that are always used in the same place;
5. *Social Object Relationship* (*SOR*): that connects objects that come into contact, either sporadically or continuously, because their owners come in touch each other.

It is also possible to go beyond the types of relationships so far defined by specifically define special groups to include devices bounded by some common features or finalities. Following the interest the SIoT paradigm has attracted, an open source platform called *SIoT Platform*[1] was developed.

3 iSapiens: A Platform for Smart Environments

The iSapiens platform has been developed with the aim of providing a useful tool to developers and researcher for developing and implementing distributed cyber-physical systems [38–40] and smart environments [20, 41–44], even at large scale, thus enabling the creation of Smart Urban Ecosystems.

iSapiens is an IoT [15, 45] enabling, agent-based platform which exploits the edge computing paradigm [7] for its main operations, and permits the exploitation of the cloud for operations that are high-demanding in terms of storage, memory and computation requirement or that cannot be affordably executed in a distributed environment. The platform tackles with the heterogeneity of physical devices and

[1] Social Internet of Things (SIoT), https://www.social-iot.org.

protocols, which is a common issue of IoT application, by providing specific abstractions, namely Virtual Objects, so as to permit to focus on the functionalities/services that need to be realized in order to enhance the behavior of a specific environment.

Agents are used to implement the functionalities of the system under development. Agents, which suits well with the distributed and pervasive nature of smart environments, are deployed and run on a set of networked computational nodes, located close to the physical devices they have to manage and control. The network of computational nodes implements the edge computing paradigm and enables real-time analysis and a fast sensing-decision-actuation cycle. In addition, agents can also exploit out-of-the-edge services, purposely developed or third-party provided. Such out-of-the-edge services can be cloud-based and include predictive analysis, machine learning, data mining and other services that can take advantage of features and benefits of the cloud. Interoperability issues are dealt with by specific components devoted to integrating agents and virtual objects with the Social Internet of Things.

Figure 1 depicts the three layer of the iSapiens architecture, which are: (i) the physical device layer, (ii) the in-network computing layer, and (iii) the off-network computing layer.

The physical devices layer is populated by physical objects. As stated before, such objects can be heterogeneous in terms of communication protocols and computational capabilities, and span from simple sensors/actuators to more complex smart objects. A physical object offers a one or more physical functionalities, which regard on how it interacts with the environment, and can expose a *Social Senser*, that enables the object to integrate and exploit the social Internet of Things.

The In-network computation layer is composed of a set of computing nodes spread into the environment, each running an iSapiens server consisting of a Virtual Object Container and an Agent Server. The Virtual Object Container is devoted to the management of Virtual Objects, which abstract the physical object in order to hide devices heterogeneity. The functionalities offered by the Virtual Objects are exploited by the agents running on the Agent Server for doing the application business. Agents can directly interact with the Virtual Objects which are co-located in the same iSapiens server. A computational node hosting an iSapiens server is called *iSapiens node*. Social Object Containers, hosted by an iSapiens node or other computational nodes, provide functionalities enabling the integration of physical objects with the SIoT. In particular, the Social Object Container manages Social Objects, which are the social counterpart of the Virtual Objects.

The Off-network computation layer hosts cloud/internet services, which can provide storage, computation or specific information, e.g., weather forecasting. This layer also hosts the SIoT platform, which is the core of the social infrastructure. Through such component, iSapiens agents and iSapiens physical objects can interoperate with entities belonging to other platforms, provided that these entities are SIoT compliant.

All the iSapiens components, i.e., devices, VOs, agents or computing nodes, can be dynamically removed from, added into or updated to a system at runtime.

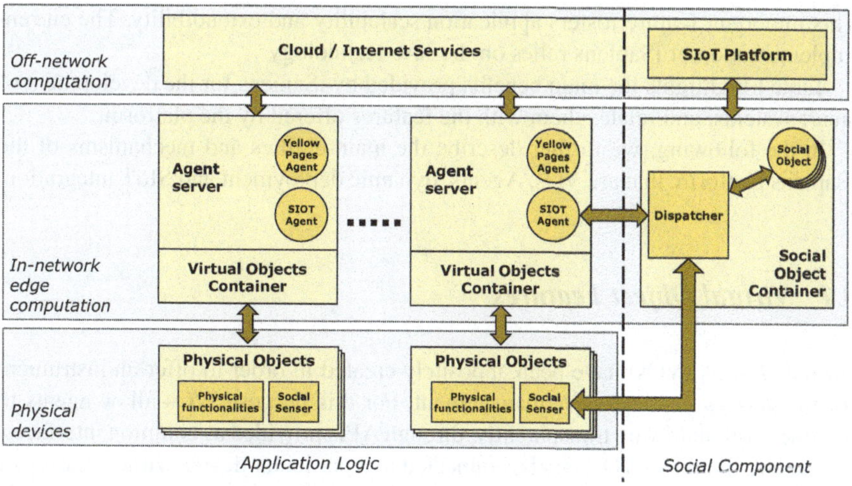

Fig. 1 The *iSapiens* platform and its computational layers

Table 1 The iSapiens platform: benefits and features

Benefits \ Features	In-network/Edge Computing	Agents	Virtual Objects	Dynamic Deployment	Yellow Pages / Acquaintances	Off-network/Cloud Computing	SIoT Integration
Functional Extendibility	+	++	++	++	++	+	+
Geographycal Extendibility	+	++	++	++	++	+	+
Online Analytics Support	++	/	/	/	/	+	/
Offline Analytics Support	+	/	/	/	/	++	/
Interoperability	/	/	/	/	/	/	++
Fault Tolerance	++	/	/	/	/	+	+
Heterogeneity Management	/	/	++	/	/	/	+
Software Maintenance	/	++	/	++	++	/	/
Hardware Maintenance	/	/	++	++	/	/	+

This important feature fosters application scalability and extendibility. The current implementation of iSapiens relies on the Java technology.

Table 1 highlights the main benefits provided by iSapiens for the development of smart systems, and relates them with the features offered by the platform.

In the following, we further describe the main entities and mechanisms of the iSapiens platform, that are VOs, Agents, dynamic deployment and SIoT integration.

3.1 Virtual Object Features

Virtual Objects (VOs) have been appositely created in order to offer an instrument for the *devices heterogeneity* management. For this purpose, VOs allow agents to interface with hardware transparently, through APIs provided as common interfaces. Any agent, thanks to VOs, can be connected to a particular device without caring of specific drivers or tackling specific technical issues. VO aim is that of decoupling (both for sensing and actuation) the specific devices from the functionalities offered by such devices. By using VOs, an update or a change of the devices or of the communication protocol used, will concern only the VOs themselves. The *Virtual Object Container* is the entity devoted to the VOs management. It allows to dynamically deploy new VOs and permits them to be addressed by the agents from the agent layer. Asynchronous and synchronous reading of data is supported by VOs. In particular, the publish/subscribe pattern is at the basis of the asynchronous one. Summarizing, VOs expose the features of the physical objects deployed in the environment. Such features are exported into iSapiens and named as *virtual object functionalities* and are connected to actuation and sensing. As an example, through a VO, a smart desk can expose a functionality called *isUsed* that allows agents to understand if someone is sit at the desk.

3.2 Agent Services and Features

An *Agent* [46] is defined ad an entity which autonomously is executed on an Agent Server and which runs its tasks also interacting with other agents that can be both local or remote. An agent residing on an iSapiens Agent Server can use all the functionalities that are exposed by the VOs on the same server. The Agent Server is devoted to support the life cycle of all the agents running on the server itself. Moreover, it permits, also at runtime, the creation of agents.

Agents can communicate through asynchronous messages, that can be both *untimed* or *timed*. These last timed messages hold an information on the time that specifies the time (absolute) in which the information must be given to the recipient agent. Direct acquaintance relationship can be set between agents by using *acquaintance messages*. An acquaintance message can provide information on the role and the identity of a specific agent. These messages are important since they allow to dynamically configure/reconfigure the relations among agents when a fault occurs or if new agents are introduced in the system or if agents are deleted. In order to dynamically discover agents, iSapiens allows the use of a distributed service of *yellow pages* which favors relationships establishment. Each agent can accede to yellow pages and register itself by entering its roles and its properties (i.e., a set of couples formed as <*key,value*>). Once an agent is registered, it can be found/discovered by other agents through the yellow pages (which can be queried).

3.3 SIoT Features

iSapiens implements the Social Internet of Things paradigm by leveraging the features offered by the SIoT Platform that is a cloud-based system that allows objects to autonomously establish social relationships [47]. Social relations between objects are based either on intrinsic properties of the objects such as their location, their producer and their owner, or on number and duration of mutual contacts. Like most human social networking platforms, the SIoT Platform offers the opportunity to define special groups including devices bounded by some common features or finalities. Its main features include: (i) the "Relationship management" to start, update, and terminate relationships among objects, (ii) the "Service discovery" that is finalized to find which objects can provide the required service, (iii) the "Service composition" that enables the interaction among objects and, (iv) the "Trustworthiness management" to understand how the information provided by other members has to be processed. Whenever an object is associated to iSapiens, it is also registered to the SIoT platform and its social profile is created.

The devices connected to an iSapiens node are abstracted as the union of *Virtual Objects* (VOs) (see Sect. 3.1) and *Social Objects* (SOs). All the SOs residing in a specific node are managed by a *Social Object Container*. The SO is responsible to manage all the interactions with the SIoT Platform on behalf of its real counterpart. More in detail, a SO is responsible to constantly update the object profile on SIoT by periodically updating all the relevant information about the object it represents. To carry out its tasks, a SO exploits the information provided by the *Social Senser*, which is installed on the real device and periodically overhears the transmission over the Wi-Fi and Bluetooth interfaces to search for other nearby social-enabled devices. This information, once forwarded by the social object to the SIoT Platform, is used to determine mutual contacts between social objects and, eventually, to create a social tie between those objects [48]. A SO is also responsible to search for a given data by analyzing all the information published by friend objects in the social object graph and to send queries to the SIoT platform or answering specific data requests. A SO can be requested to accomplish those functions by an external entity through a set of APIs called the *Social Hook*.

3.4 Application Deployment

As introduced above, the iSapiens platform allows the dynamic deployment of VOs and Agents during system runtime. Given a running set of iSapiens servers, an application can be deployed in all its software parts, by exploiting a *Deployer* entity, which is capable of sending agents and VOs code on the involved set of nodes and of instantiating and starting them. The Deployer is also responsible for the starting configuration of the application, made by establishing acquaintance relationships among agents and sending application-dependant configuration messages.

4 Application Domains and Examples

This section summarizes and overviews some carried out research activities and some significant applications realized by exploiting the concepts, the features and the capabilities that are at the basis of the iSapiens platform. All the reported applications fall in the domain of Smart Urban Ecosystems. They range from smart city applications [5] and smart offices [44] to smart drainage networks [49] and power cloud applications [50].

4.1 The Smart Street Cosenza Project

The project [5] was developed in the city of Cosenza (Italy) and it was devoted to the realization of a Smart Environment named *Smart Street Cosenza*. The target city area involved in the project is shown in Fig. 2a. Such area is located in the centre of Cosenza and cover: (i) a Bus Station, (ii) a square, (iii) the main commercial street of the city, (iv) part of one of the main driveways and (v) the area around a commercial centre. The goal is that of furnishing an IT infrastructure which provides services devoted to the real-time monitoring of the status of the involved areas. The infrastructure was designed so as to be further extended over time, either in terms of covered area or in term of provided features and offered functionalities.

The chosen area was instrumented by deploying in it a set of thirteen computational nodes (Fig. 2b) and seventy wireless sensors nodes (Fig. 2c). Each computational node is mains powered and consists of a Raspberry Pi mod.2 board hosting the iSapiens platform. Each sensor node is equipped with a battery embedded and is powered by a solar panel. Two types of sensors exist namely *type A* and *type B*. The former comprises noise, temperature, relative humidity and luminosity sensors. The latter comprise air quality sensors, measuring the concentration of CO, CO_2, NO and O_3. Sensors and computational nodes are connected by using a wireless network infrastructure.

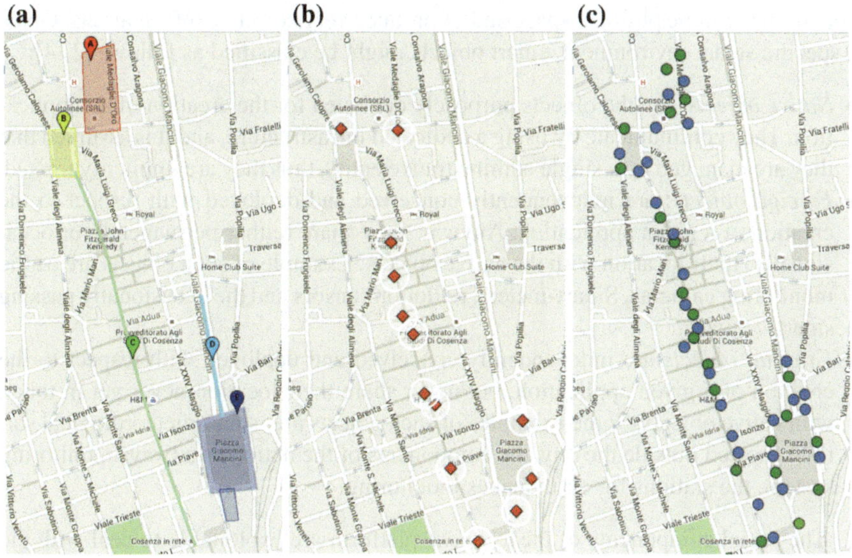

Fig. 2 Smart Street Cosenza. **a** Target areas; **b** Computational Nodes; **c** Sensors nodes. Green locations host one *type A* node and one *type B* node, blue locations hosts only a *type A* node

Raw measures are collected by the computational nodes, then filtered and aggregated. Final data are sent to a remote server that stores them in a DBMS. The geo-referenced sensor measures as well as the aggregated information indexes such as Simmer Summer Index (SSI), Humidex and Air Quality Index (AQI) are made available to citizens through a web-based application. Such information can be used to determine if a given monitored area is comfortable and healthy. Moreover, the city administrators can monitor critical climatic or pollution events, and the research community, can use gathered data for further investigations and analysis. The realized system are up and running since December 2015, with only a few sensors failures.

4.2 Enhancing Smart Objects with Social Capabilities

The use and availability of Smart Objects within our cities is ever increasing. Such objects are often the basic building blocks used to realize Smart Environments devoted to improving our quality of life by offering advanced services and innovative applications. However, we are still far from permitting a fully exploitation of the potentialities and features of such objects. In fact, usually they have different roles and are usually deployed and managed by a multitude of different players. They fulfill specific goals in isolation and, very rarely, devices belonging to different administrative entities are able to cooperate and/or share information even if they coexist in the same physical space and/or produce very correlate informations. Given a specific smart environment, smart objects might be classified as follows [2, 4]:

- *Native objects*: are the objects purposely deployed for the creation of an application. They communicate by using a dedicated infrastructure, and it is assumed that they are managed by a single administrative entity, thus they are implicitly trusted.
- *Foreign objects*: are independently conceived and deployed with respect to the creation of a given application. Anyway, they share (either permanently or occasionally) their location with the native objects. Example of such objects are traffic monitoring cameras, Smart-meters, pollution sensors and the occasionally passing smart-vehicles.
- *External objects* are independently conceived and deployed with respect to the creation of a given application, but never share their location never get in touch with native objects. Some examples of such objects are: smart meteorological stations located outside the city, the components of the national railways monitoring system, the stations for earthquakes monitoring.

The social components of the *iSapiens* platform are exploitable to deal with the segregation among those categories of objects. More in particular, *Native objects* become able to opportunistically discover and access the *Foreign* and *External objects*. The social components of *iSapiens* are inspired to the Social Internet of Things (SIoT) which permits to build a social network connecting smart-objects. Social relations are created according the characteristics of the objects such as their model, location, movement pattern and purpose (see Sect. 2.3). The *iSapiens* platforms leverages that social network by enabling: (i) information sharing between *Native, Foreign* and *External objects*, (ii) service discovery for finding objects providing some given services, (iii) service composition and, (iv) trustworthiness management.

In order to give evidence about the effectiveness of the social components of *iSapiens* we consider a city block of 2 Km^2 and a reference application which is based on 4 smart objects placed at fixed positions in the area. The SWIM software [51] was used to simulate N, $N \in [100, 300]$, people moving within the considered area. Each person bring with him his own personal mobile device. All together, the devices represent the *foreign objects*. While people move, their carried personal devices enter in contact (we assumed a sensing range of 10 [m]) and begin to create their SIoT network. By post-processing the output of SWIM through custom software written in *MATLAB*®, we asses how the social network between the *Native*, the *External*, and *Foreign objects* evolved after [10, 12, 24, 36] hours from the simulation beginning. Specifically, according to [32] we consider that two objects establish a *Social Object Relationship* (*SOR*) [10] after having experienced one or more contacts for a cumulative contact time of at least 10 [min]. We further considered all the *Native objects* connected by a *Co-work objects relationship* (*C-WOR*) [10]. Figure 3 shows the cumulative distribution function of the distance, in terms of hops, between *native* and *non native* objects in the social network.

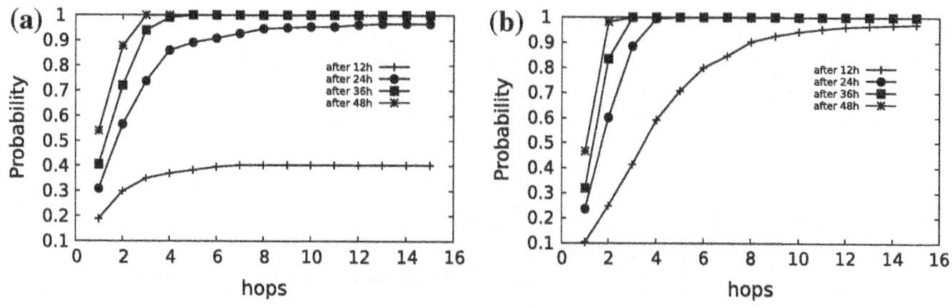

Fig. 3 Cumulative distribution function of the distance between *Native* and *Foreign* objects in the social network: **a** 100 nodes; **b** 300 nodes

From Fig. 3 it emerges that all the *Foreign* objects are included in the social network of the *Native* objects. As the time passes, the average distance in terms of hops between the *Natives* and the *Foreigners* get smaller, and, a shorter average distance in the social networks corresponds to a faster propagation and to a quick interaction between the objects. By comparing Fig. 3a, b it follows that the higher is the number of *Foreign* objects the faster is the inclusion process. Moreover, within 24*h*, 36*h*, 48*h* the inclusion process completed despite the number of the considered objects. As a final remark, we conclude observing that the *iSapiens* platform is really able in breaking down the segregation between smart objects belonging to different categories. More extensive results can be found in [4].

4.3 Urban Drainage Network

Urban floods are becoming more frequent as a result of the increase of impervious areas and of the occurrence of extreme weather conditions due to climate change. Moreover, obstructions and blockages, due to a poor maintenance of pipes and catch basins, may impair the hydraulic efficiency of an urban drainage system. Centralized stormwater measures [52], like the use of detention tanks and retention basins, can be exploited for managing urban flooding. However, those solutions are not easily applicable due to the lack of space, especially in densely populated areas.

The solution here considered [49], relies on a CPS which exploits the iSapiens architecture and its related concepts. The proposal consist in spreading in the drainage network a set of water level sensors along with some smart gates. Such devices are directly linked with a set of interconnected computing nodes. Computing nodes host a distributed and decentralized application based on cooperating agents whose aim is that of adjusting the gates in order to dynamically optimize the water load in all the conduits of the drainage network. The optimization of the water load makes the network able to fully exploit the storage capacity of the pipeline. In such a way, the

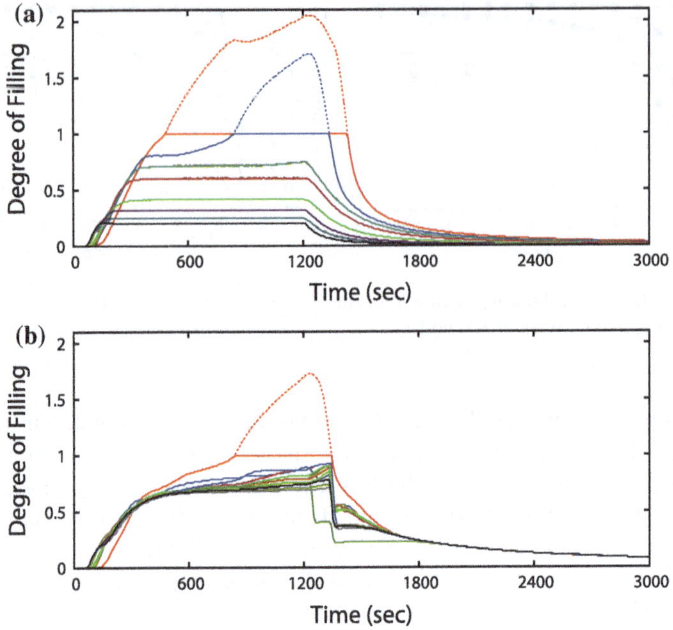

Fig. 4 Comparison of water loads (degree of filling) in different conduits, represented with different colors: base-line **a** versus controlled **b** scenarios. The areas above 1 and under the dashed lines represents the flooding volume of each conduit

exceeding stormwater volume is accumulated and hold within the pipeline itself thus avoiding (or limiting) a water overflow on the side-walks and street paving. Agents coordinate each other by running a gossip-based algorithm [53] devoted to compute the average of the water load. Such average value is then used by each gate-agent for tuning its gate so as to bring water loads closer to that average. This approach permits the systems to dynamically adapt also to unforeseen events occurring in the drainage network like damage, occlusions and so on.

The behavior of the system was assessed trough simulation. In particular, the EPA SWMM software [54], which is widely used in the hydraulic scientific community, was exploited in order to simulate the behavior of realistic urban drainage networks which include gates during severe rain events.

Simulation results proved that our algorithm, used as a real time controller for an urban drainage system equipped with a series of movable gates, is really able to manage flooding by exploiting the storage capacity of the less overwhelmed conduits of the system. Results are shown in Fig. 4. They demonstrate that the approach significantly reduces the flooding volume (the area between 1 and the dashed lines) in an urban drainage network during a heavy rainfall event. Other results can be found in [39].

4.4 An Open-Air Smart Museum

The *iSapiens* platform and related concepts were also exploited for the the realization of a smart museum in the city of Cosenza (Italy) [45]. The realized smart museum, namely the e-MAB smart museum, was developed by instrumenting and enhancing the MAB (Museo all'Aperto Bilotti) open-air museum.

The MAB is a permanent open-air museum which is located in the main commercial thoroughfare of Cosenza. The museum is located on the pedestrian area of Corso Mazzini, which is the main commercial street in the town comprehending shops, historical buildings, and cafes. The museum collects some prestigious sculptures of artists like Salvador Dalí, Giacomo Manzú, Sasha Sosno, Giorgio De Chirico, and by some other Calabrian artists. It was born thanks to the donation of Carlo Bilotti, a wealthy collector native of Cosenza but immigrant to America. Figure 5 reports the location of the MAB along with the displacement of the artworks in it.

The primary objective of the e-MAB is that of strengthening both safety and security of the artworks. The basic idea behind the smart museum is equipping the statues with "virtual senses" and turning them into *cognitive entities* which are *aware* of what is happening in their neighborhood so as to discover/recognize/react to potentially dangerous situations. Each statue has been purposely enhanced with a set of sensor/elaboration/actuator devices enabling the perception of the surrounding environment and the execution of deterrent actions (e.g., asking for help). Physical devices have been deployed in safe places close to the statues in order to avoid artwork damages. More in particular, the *virtual sight* has been implemented through cameras, the *virtual hearing* through microphones, the *virtual touch* through IR proximity sensors, and the *virtual speech* through speakers. Virtual senses were used to infer when a situation become dangerous. For instance, a situation is considered as dangerous when a statue is hit or when too many visitors are close to the artwork. Deterrent actions comprehends light flashing at the statue and the start of recorded alert messages on the speakers.

The second objective of the e-MAB is to augment its attractiveness and enjoyability with respect to visitors. Some services were developed for (i) enabling visitors to automatically take a "selfy" with an artwork, (ii) obtaining detailed info of an artwork, (iii) gathering and making it available information about the "comfort" level of

Fig. 5 The MAB location and a map of its artworks

the museum (in terms of crowding and climatic wellness), (iv) promoting information exchange among the town administrators and visitors (alert about strikes, public service interruption and so forth), commercial activities and visitors (for advertisements and promotions), and visitors and visitors (virtual post-it service). For this second purpose, a web-based application has been developed.

4.5 A Smart Office Implementation

A smart office prototype [44], with the aim of monitoring environmental conditions and applying some simple actuation rules in order to avoid waste in energy consumption, was implemented at ICAR-CNR.[2] The instrumented office room contains (i) three working desks having some computers and lamps on them (ii) one chair for each desk and another set of chairs for meetings (iii) a big window on the east side of the room. The fundamental features of the *isapiens* platform were exploited for the realization of the prototype.

From a design point of view, a first step in developing the system was devoted to defining the functionalities of the smart office. Subsequently, such functionalities were related to the physical devices needed to support them. From Fig. 6 it follows that three main functionalities were considered: (i) Monitoring RoomCondition, (ii) Monitoring DeskCondition, and (iii) Controlling DeskCondition. The latter is in charge of recognizing energy waste at the desk (e.g., if no one is at the desk and the lamp is on) and turn off the lamp if required. In the figure, the basic functionalities used to offer the main ones are listed. Moreover, for each basic functionality, the required physical devices are shown. The DeviceAbstractor is an associative class, which mirrors the use of virtual objects required for managing heterogeneity issues.

For implementation purposes, a set of three physical sensor nodes consisting in Waspmote nodes[3] (see Fig. 7) were deployed in the office: one for the desk, one for the lamp, and one for the room environment. The sensor node at desk hosts a Presence Infrared (PIR) sensor coupled with a sound sensor (for monitoring the activity at the desk), and a luminosity sensor. The lamp is instrumented with a smart plug sensor node, having an electricity sensor and a relay actuator, for monitoring and controlling the status of the lamp. The room sensor node hosts a PIR, a sound, a luminosity, a temperature, and a humidity sensor. Each sensor node sends every 30 s the average of sensed data with respect to a 30 s time window.

Two *iSapiens* servers were deployed in the office on two dedicated Raspberry Pi[4] single-board computer. One server was for the desk (linked to the desk and the lamp sensor nodes) and one was for the room (linked to the room sensor nodes). Communication among sensor nodes and Virtual Objects relies on the zigbee protocol stack.

---

[2]ICAR-CNR, via P.Bucci, cubo 7-11C, 87036 Rende, Italy.

[3]http://www.libelium.com/products/waspmote/.

[4]https://www.raspberrypi.org/.

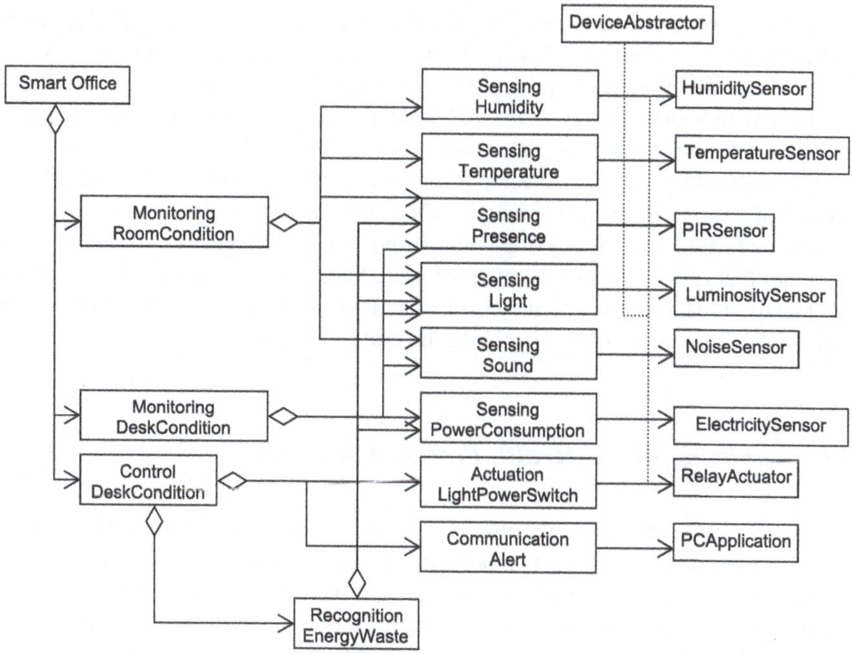

Fig. 6 A UML class diagram modelling the smart office

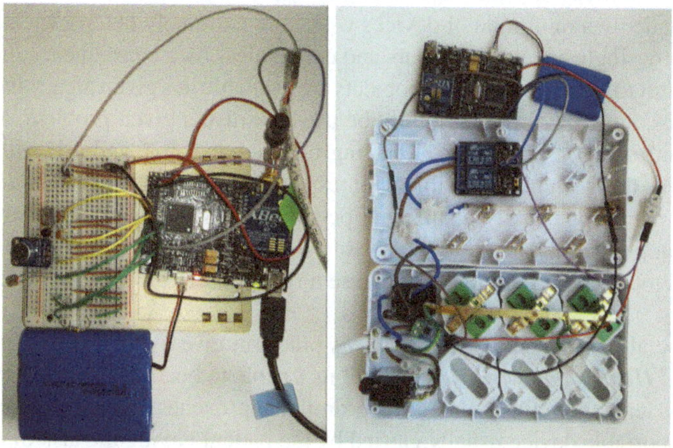

Fig. 7 The room sensor node (left) and the smart plug sensor node (right)

Accordingly to the *iSapiens* features, for each functionality a dedicated agent was developed for implementing it. The agents are deployed on the *iSapiens* server according to the closeness to the physical devices an agent has to manage. Some further Off-network services were also implemented to realize the persistence of gathered data and to signal alerts on energy waste to user clients.

The realized system is extensible as the agent servers can be further added and populated so as to furnish new functionalities which can also be obtained by composing the exiting ones. Scalability is supported, as new nodes can be deployed in the same or other offices. Fault tolerance is fostered because the computation is carried out locally in each node, and the crash of a single node does not compromise the other ones. The system is running since the middle of 2016.

4.6 Implementing e-Health Personal Services

The population of the most developed countries is constantly aging. As a consequence, it becomes of utmost importance the development of e-Health systems devoted to improve the Quality of Life of suffering/elder people. In this context, the Internet of Things (IoT) paradigm [55], proved to be effective in determining a sort of evolutionary leap in e-Health systems [56].

Despite the huge potentialities arising from the exploitation of IoT-based technologies in the domain of e-Health systems, the full integration between IoT and E-Health systems is still far from an actual implementation. Iot-based systems are usually made by a multitude of devices which are extremely heterogeneous and are deployed by single and specific users and companies. As a consequence, it becomes very difficult to estimate how many objects are available in a given area, what are their capabilities, and how they can jointly be used to solve a given problem. Moreover, as an additional constraint, the managing of integrated IoT and e-Health systems should be as transparent as possible with respect to the patients that should use them. In fact, a patient is usually a suffering person who is not to be troubled with any further complications related to the use of e-Health systems. The failing of this latter requirement could paradoxically cause a reduction in the Quality of Life of the patients.

The social components of the *iSapiens* platform give us the abilities to develop a *Personal Health Gateway* (*PHG*) (see Fig. 8) able to become the link between the IoT and e-Health domains. According to similar approaches in the literature [57], the *PHG* is a portable device usually carried by the patient having the task of collecting Health related measures from the patient and to make that information available for the Health related applications, which, in our case are implemented as *iSapiens* agents. A *PHD* is equipped with two interfaces: (i) an *E-Health* interface devoted to collect measures coming from the biomedical sensors worn by a patient, (ii) a *IoT* interface toward the *iSapiens* platform. While the latter interface is specific of the *iSapiens* platform, the E-Health interface is based on the specification of the Continua Health alliance [58] and hence there are no compatibility issues with the

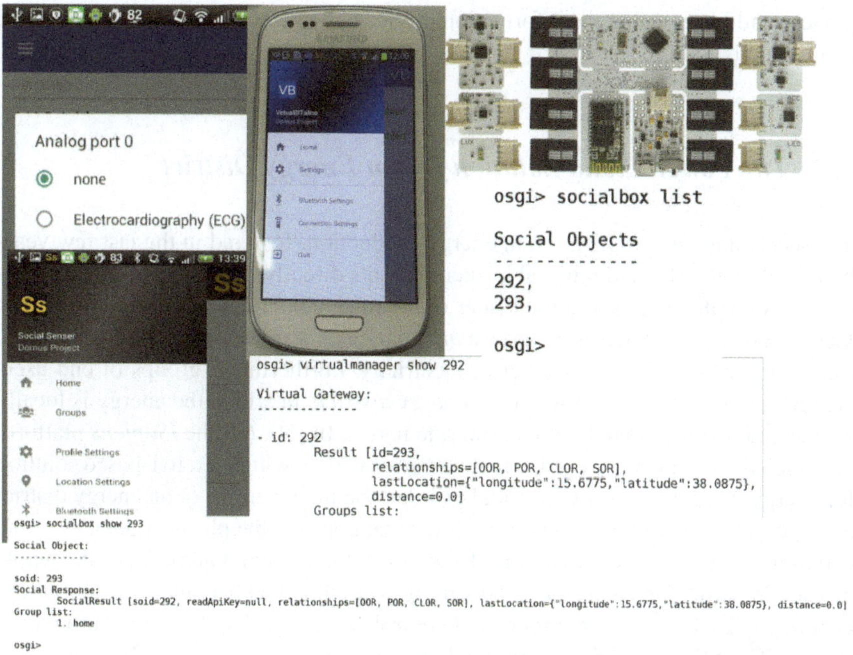

Fig. 8 The Personal Health Gateway (PHG), some snapshots of the user interface and a sample log of the correspondent Virtual Object on the *iSapiens* Platform

legacy E-Health devices. The *PHG*, as an iSapiens object, is able to discover other similar objects surrounding the patient and to create with them social ties according the rules of SIoT [10]. Hence, the medical applications running on the iSapiens platform might therefore access and process all the information made available by objects in the social network of the *PHG*.

As a case study, we implemented a prototype *PHG* on the basis of a commercial Android smartphone (see Fig. 8). We connected the smartphone to a BITalino board[5] through its Bluetooth interface in order to collect and acquire biometric data from a patient. Moreover, the *PHG* has been retrofitted with two customized software modules in order for it to become connected to the *iSapiens* platform. The first module encloses the acquired data from the Bitalino on-board sensors into properly formatted JSON messages, and forwards them to the *iSapiens* platform. The second module, the Social Senser, periodically overhears the transmission over the Wi-Fi and Bluetooth interfaces to search for other nearby social-enabled devices. This information, once forwarded to the *iSapiens* platform, is used to determine mutual contacts between social objects and, eventually, to create a social tie between those objects. Simulations confirmed the effectiveness of the social approach in dynamically discovering new

[5]http://www.bitalino.com/.

devices and in acquiring the information produced by those devices. More extensive results can be find in [59].

4.7 The Power Cloud Solution for an Energy District

The increasing decentralization in energy production, fostered in the last few years by the adoption of small renewable energy plants directly connected to the grid, has been placing the end user at the center of the management of electrical systems. In fact, the end user has becoming to be a *prosumer*, i.e., it is becoming to be at the same time both producer and consumer of electricity. Furthermore, groups of end users are encouraged to form the so-called *energy districts* in which the energy is locally exchanged and negotiated without intermediaries. In [50, 60] the *iSapiens* platform and related concepts were used to realize *Power Cloud*, which is a IoT-based solution for energy districts. Power Cloud makes smart the management of an energy district by supplying a distributed computing layer that controls the physical devices of the end users, and interacts with a centralized cloud component that manages the whole district. In particular, each end user dwelling is equipped with a Smart Energy Aware Gateway (SEAG), which monitors and controls a nanogrid and a home automation system. The SEAG enforces an optimal energy strategy for the local dwelling. This is achieved by means of complex prediction algorithms that are purposely executed on the cloud.

5 Conclusions

This chapter has been devoted to the review of iSapiens, an agent-based platform for the development of Smart Environments and Smart Urban Ecosystems. The platform favors functional and geographical extendibility, support online/offline processing, fault tolerance and software/hardware maintenance, and allows interoperability among different systems and device heterogeneity management. These features are provided by a proper combination of edge/cloud computing paradigm, agent metaphor and the Social Internet of Things. Moreover, the chapter has reviewed a significant set of case studies and applications, related to Smart City, Smart Water Infrastructure, Smart Healthcare, Smart Grid, and Smart Office. Such examples involve the exploitation of the concepts and the features of the platform, thus proving the effectiveness of iSapiens in developing Smart Environment and Smart Urban Ecosystem applications.

Acknowledgements This work has been partially supported by the "Smart platform for monitoring and management of in-home security and safety of people and structures" project that is part of the DOMUS District, funded by the Italian Government (PON03PE_00050_1).

References

1. Y. Oh, J. Han, W. Woo, A context management architecture for large-scale smart environments. IEEE Commun. Mag. **48**(3), 118–126 (2010)
2. F. Cicirelli, A. Guerrieri, G. Spezzano, A. Vinci, O. Briante, A. Iera, G. Ruggeri, An edge-based approach to develop large-scale smart environments by leveraging SIoT, in *Proceedings of the 14th IEEE Internationall Conference on Networking, Sensing and Control (ICNSC2017)*, May 2017
3. F. Cicirelli, G. Fortino, A. Guerrieri, G. Spezzano, A. Vinci, Metamodeling of Smart Environments: from design to implementation. Adv. Eng. Inform. (ADVEI) **33**, 274–284 (2017)
4. F. Cicirelli, A. Guerrieri, G. Spezzano, A. Vinci, O. Briante, A. Iera, G. Ruggeri, Edge computing and social internet of things for large-scale smart environments development. IEEE Internet Things J. **PP**(99), 1–1 (2017)
5. F. Cicirelli, A. Guerrieri, G. Spezzano, A. Vinci, An edge-based platform for dynamic smart city applications. Future Gener. Comput. Syst. **76**, 106–118 (2017)
6. M. Daz, C. Martn, B. Rubio, State-of-the-art, challenges, and open issues in the integration of internet of things and cloud computing. J. Netw. Comput. Appl. **67**, 99–117 (2016)
7. P. Garcia Lopez, A. Montresor, D. Epema, A. Datta, T. Higashino, A. Iamnitchi, M. Barcellos, P. Felber, E. Riviere, Edge-centric computing: vision and challenges. SIGCOMM Comput. Commun. Rev. **45**(5), 37–42 (2015)
8. F. Bonomi, R. Milito, J. Zhu, S. Addepalli, Fog computing and its role in the internet of things, in Proceedings of the First Edition of the MCC Workshop on Mobile Cloud Computing, ser. MCC '12. ACM, New York, NY, USA, pp. 13–16 (2012). https://doi.org/10.1145/2342509.2342513
9. N.R. Jennings, On agent-based software engineering. Artif. Intell. **117**(2), 277–296 (2000)
10. L. Atzori, A. Iera, G. Morabito, M. Nitti, The Social Internet of Things (SIoT)—when social networks meet the internet of things: Concept. Architecture and network characterization. Comput. Netw. **56**(16), 3594–3608 (2012)
11. G. Fortino, A. Guerrieri, W. Russo, C. Savaglio, Middlewares for smart objects and smart environments: overview and comparison, in *Internet of Things Based on Smart Objects* (Springer, 2014), pp. 1–27
12. O. Vermesan, P. Friess, P. Guillemin, H. Sundmaeker, M. Eisenhauer, K. Moessner, F.L. Gall, P. Cousin, *Internet of Things Strategic Research and Innovation Agenda* (2013)
13. G. Fortino and P. Trunfio, *Internet of Things Based on Smart Objects: Technology, Middleware and Applications* (Springer, 2014)
14. A. Guerrieri, V. Loscri, A. Rovella, G. Fortino, *Management of Cyber Physical Objects in the Future Internet of Things: Methods* (Springer, Architectures and Applications, 2016)
15. L. Atzori, A. Iera, G. Morabito, Understanding the Internet of Things: definition, potentials, and societal role of a fast evolving paradigm,. Ad Hoc Netw. **56**, 122–140 (2017) (Supplement C)
16. L. Yang, W. Li, M. Ghandehari, G. Fortino, People-centric cognitive internet of things for the quantitative analysis of environmental exposure. IEEE Internet Things J. **PP**(99), (1–1) (2017)
17. D. Cook, S. Das, *Smart Environments: Technology* (Wiley-Interscience, Protocols and Applications, 2004)
18. O. Evangelatos, K. Samarasinghe, J. Rolim, Syndesi: A Framework for Creating Personalized Smart Environments Using Wireless Sensor Networks, in *Proceedings of the, IEEE International Conference on Distributed Computing in Sensor Systems*, ser. DCOSS '13. Washington, DC, USA, vol. 2013 (IEEE Computer Society, 2013), pp. 325–330
19. H. Vahdat-nejad, K. Zamanifar, N. Nematbakhsh, Context-aware middleware architecture for smart home environment. Int. J. Smart Home, **7**(1) (2013)
20. F. Cicirelli, G. Fortino, A. Giordano, A. Guerrieri, G. Spezzano, A. Vinci, On the design of smart homes: a framework for activity recognition in home environment. J. Med. Syst. **40**(9), 200 (2016)

21. M. Amadeo, O. Briante, C. Campolo, A. Molinaro, G. Ruggeri, Information-centric networking for M2M communications: design and deployment. Comput. Commun. **89**, 105–116 (2016)
22. A. Socievole, S. Marano, Exploring user sociocentric and egocentric behaviors in online and detected social networks, in *2012 2nd Baltic Congress on Future Internet Communications*, Apr 2012, pp. 140–147
23. A. Savidis, C. Stephanidis, Distributed interface bits: dynamic dialogue composition from ambient computing resources. Pers. Ubiquitous Comput. **9**(3), 142–168 (2005). https://doi.org/10.1007/s00779-004-0327-2
24. J.E. Bardram, R. E. Kjær, M. Pedersen, Context-Aware User Authentication—Supporting Proximity-Based Login in Pervasive Computing," in *UbiComp 2003: Ubiquitous Computing*, ser. Lecture Notes in Computer Science, vol. 2864, ed. by A. Dey, A. Schmidt, J. McCarthy (Springer, Berlin, Heidelberg, 2003), pp. 107–123. https://doi.org/10.1007/978-3-540-39653-6_8
25. M. Román, C. Hess, R. Cerqueira, A. Ranganathan, R.H. Campbell, K. Nahrstedt, A middleware infrastructure for active spaces. IEEE Pervasive Comput. **1**(4), 74–83 (2002). https://doi.org/10.1109/MPRV.2002.1158281
26. A.K. Dey, G.D. Abowd, D. Salber, A conceptual framework and a toolkit for supporting the rapid prototyping of context-aware applications. Hum.-Comput. Interact. **16**(2), 97–166 (2001). https://doi.org/10.1207/S15327051HCI16234_02
27. A. Zanella, N. Bui, A. Castellani, L. Vangelista, M. Zorzi, Internet of things for smart cities. IEEE Internet Things J. **1**(1), 22–32 (2014)
28. M. Bain, Sentilo - Sensor and Actuator Platform for smart Cities 2014, https://joinup.ec.europa.eu/community/eupl/document/sentilo-sensor-and-actuator-platform-smart-cities. Accessed 21 Apr 2017
29. P. Ballon, J. Glidden, P. Kranas, A. Menychtas, S. Ruston, S. Van Der Graaf, Is there a need for a cloud platform for european smart cities? in *eChallenges e-2011 Conference on Proceedings of the IIMC International Information Management Corporation*, pp. 1–7 (2011)
30. Z. Khan, A. Anjum, K. Soomro, M.A. Tahir, Towards cloud based big data analytics for smart future cities. J. Cloud Comput. **4**(1), 2 (2015)
31. Y. Sun, H. Song, A.J. Jara, R. Bie, Internet of things and big data analytics for smart and connected communities. IEEE Access **4**, 766–773 (2016)
32. H. Asl, A. Iera, L. Atzori, G. Morabito, How often social objects meet each other? Analysis of the properties of a social network of IoT devices based on real data, in *2013 IEEE Global Communications Conference (GLOBECOM)*, Dec 2013, pp. 2804–2809
33. P.D. Meo, K. Musial-Gabrys, D. Rosaci, G.M.L. Sarnè, L. Aroyo, Using centrality measures to predict helpfulness-based reputation in trust networks. ACM Trans. Internet Technol. **17**(1), 8:1–8:20 (2017). https://doi.org/10.1145/2981545
34. J.A. Golbeck, Computing and applying trust in web-based social networks, Ph.D. Dissertation, College Park, MD, USA, 2005, aAI3178583
35. J. Golbeck, J. Hendler, *Accuracy of Metrics for Inferring Trust and Reputation in Semantic Web-Based Social Networks* (Springer, Berlin, Heidelberg, 2004), pp. 116–131. https://doi.org/10.1007/978-3-540-30202-5_8
36. S. Valenzuela, N. Park, K.F. Kee, Is there social capital in a social network site? Facebook use and college students' life satisfaction, trust, and participation. *J. Comput.-Med.Commun.* **14**(4), 875–901 (2009). https://doi.org/10.1111/j.1083-6101.2009.01474.x
37. M. Nitti, R. Girau, L. Atzori, Trustworthiness management in the social internet of things. IEEE Trans. Knowl. Data Eng. **26**(5) (2014)
38. K.-D. Kim, P.R. Kumar, Cyber–physical systems: a perspective at the centennial. Proc. IEEE **100**, 1287–1308 (2012) (Special Centennial Issue)
39. A. Giordano, G. Spezzano, A. Vinci, G. Garofalo, P. Piro, A cyber-physical system for distributed real-time control of urban drainage networks in smart cities, in *2014 Proceedings of the Internet and Distributed Computing Systems—7th International Conference IDCS 2014, Calabria, Italy, 22–24 Sept 2014*, pp. 87–98, https://doi.org/10.1007/978-3-319-11692-1_8

40. A. Giordano, G. Spezzano, A. Vinci, A smart platform for large-scale networked cyber-physical systems, in *Management of Cyber Physical Objects in the Future Internet of Things Methods* (Springer, Architectures and Applications, 2016)
41. O. Vermesan, P. Friess, *Internet of Things: Converging Technologies for Smart Environments and Integrated Ecosystems* (River Publishers, 2013)
42. G. Fortino, A. Giordano, A. Guerrieri, G. Spezzano, A. Vinci, A data analytics schema for activity recognition in smart home environments, in *Proceedings of the Ubiquitous Computing and Ambient Intelligence. Sensing, Processing, and Using Environmental Information—9th International Conference, UCAm I*, Puerto Varas, Chile, 1–4 Dec 2015, vol. 2015 (2015) pp. 91–102
43. F. Cicirelli, G. Fortino, A. Guerrieri, G. Spezzano, A. Vinci, A meta-model framework for the design and analysis of smart cyber-physical environments, in *Proceedings of the 2016 IEEE 20th International Conference on Computer Supported Cooperative Work in Design*, May 2016
44. F. Cicirelli, Edge enabled development of smart cyber-physical environments, in *Proceedings of the 2016 IEEE Internationall Conference on Systems, Man, and Cybernetics (SMC 2016)*, Sept 2016
45. F. Cicirelli, G. Spezzano, Concept Hierarchies For Sensor Data Fusion In The Cognitive IoT, in *Proceedings of the 30th ECMS 2016*
46. M. Wooldridge, *An Introduction to Multiagent Systems* (Wiley, 2009)
47. Social Internet of Things (SIoT), www.social-iot.org
48. L. Atzori, A. Iera, G. Morabito, SIoT: giving a social structure to the internet of things. IEEE Commun. Lett. **15**(11), 1193–1195 (2011)
49. G. Garofalo, A. Giordano, P. Piro, G. Spezzano, A. Vinci, A distributed real-time approach for mitigating cso and flooding in urban drainage systems. J. Netw. Comput. Appl. **78**, 30 – 42 (2017). http://www.sciencedirect.com/science/article/pii/S1084804516302752
50. A. Burgio, A. Giordano, A.A. Manno, C. Mastroianni, D. Menniti, A. Pinnarelli, L. Scarcello, N. Sorrentino, M. Stillo, An iot approach for smart energy districts, in *2017 IEEE 14th International Conference on Networking, Sensing and Control (ICNSC)*, May 2017, pp. 146–151
51. S. Kosta, A. Mei, J. Stefa, Small world in motion (SWIM): modeling communities in ad-hoc mobile networking, in *2010 7th Annual IEEE Communications Society Conf. on Sensor Mesh and Ad-Hoc Communications and Networks (SECON)*, June 2010
52. P. Piro, M. Carbone, G. Garofalo, Distributed vs. concentrated storage options for controlling cso volumes and pollutant loads. Water Pract. Technol. **5**(3), 071 (2010)
53. M. Jelasity, A. Montresor, O. Babaoglu, Gossip-based aggregation in large dynamic networks. ACM Trans. Comput. Syst. (TOCS) **23**(3), 219–252 (2005)
54. L. Rossman, N. R. M. R. L. (U.S.), *Storm Water Management Model, Quality Assurance Report: Dynamic Wave Flow Routing*. (U.S. Environmental Protection Agency, Office of Research and Development, National Research Management Research Laboratory, 2006), https://books.google.it/books?id=3mSRAQAACAAJ
55. D. Miorandi, S. Sicari, F. De Pellegrini, I. Chlamtac, Internet of Things: Vision, applications and research challenges. Ad Hoc Netw. **10**(7) (2012)
56. M.C. Domingo, An overview of the Internet of Things for people with disabilities. J. Netw. Comput. Appl. **35**(2) (2012)
57. D.F.S. Santos, H.O. Almeida, A. Perkusich, A personal connected health system for the internet of things based on the constrained application protocol. Comput. Electr. Eng. **44** (2015)
58. R. Carroll, R. Cnossen, M. Schnell, D. Simons, Continua: an interoperable personal healthcare ecosystem. IEEE Pervasive Comput. **6**(4) (2007)
59. G. Ruggeri, O. Briante, A framework for iot and e-health systems integration based on the social internet of things paradigm, in *2017 International Symposium on Wireless Communication Systems (ISWCS)*, Aug 2017, pp. 426–431
60. G. Belli, A. Giordano, C. Mastroianni, D. Menniti, A. Pinnarelli, L. Scarcello, N. Sorrentino, M. Stillo, A unified model for the optimal management of electrical and thermal equipment of a prosumer in a dr environment. IEEE Trans. Grid (2017). https://doi.org/10.1109/TSG.2017.2778021

Smart City Platform Specification: A Modular Approach to Achieve Interoperability in Smart Cities

Arianna Brutti, Piero De Sabbata, Angelo Frascella, Nicola Gessa, Raffaele Ianniello, Cristiano Novelli, Stefano Pizzuti and Giovanni Ponti

Abstract The development of our cities towards the Smart City paradigm is one of the challenges facing today's society. This means, among other things, continuously developing and adopting ICT technologies in order to create platforms on which governments, businesses and citizens can communicate and work together and providing the necessary connections between the networks (of people, businesses, technologies, infrastructures, energy and spaces) that are the base for the services of the city. The incredible vastness and diversity of applications that are emerging in this context generates an enormous amount of data of different types and from heterogeneous sources to be shared and exchanged. In this article we propose an approach and describe a methodology and a modular and scalable multi-layered ICT platform to address the problem of cross-domain interoperability in the context of Smart City applications.

Keywords Smart city · Interoperability · Specifications

1 Introduction

The world's urban population is expected to double by 2050, by 2030, six out of every ten people will live in a city and by 2050 this figure will run to seven out of ten. In real terms, the number of urban residents is growing by nearly 60 million people every year. As the planet becomes more urban, cities need to become smarter. Major urbanisation requires new and innovative ways to manage the complexity of urban living; it demands new ways to target problems of overcrowding, energy consumption, resource management and environmental protection; thus there is an increased

A. Brutti · P. De Sabbata · A. Frascella · N. Gessa · C. Novelli · S. Pizzuti (✉) · G. Ponti
Italian National Agency for New Technologies (ENEA), 00196 Rome, Italy
e-mail: stefano.pizzuti@enea.it

R. Ianniello
Department of Computer Science and Engineering (DISI), University of Bologna, 40126 Bologna, Italy

F. Cicirelli et al. (eds.), *The Internet of Things for Smart Urban Ecosystems*, Internet of Things, https://doi.org/10.1007/978-3-319-96550-5_2

demand for intelligent, sustainable environments that offer citizens a high quality of life. This is typically characterized as the evolution to Smart Cities as a key strategy to tackle poverty and inequality, unemployment and energy management. At its core, the idea of Smart Cities is rooted in the creation and connection of human capital, social capital and information and communication technology (ICT) infrastructure in order to generate greater and more sustainable economic development and a better quality of life; in this scenario, the Internet of Things (IoT) is a vital enabler of Smart Cities. As nodes of such a vast network get more and more intelligent, IoT becomes the backbone of smartification and the grounds of innovation. However, managing a plethora of heterogeneous connected devices is a laborious task that poses relevant challenges, demanding appropriate attention from industry, practitioners, and the scientific community alike.

Smart Cities have been further defined along six axes or dimensions [1, 2]: smart economy, smart energy, smart mobility, smart environment, smart living and smart governance. The linkages between economic, societal and environmental development are not scalable as cities expand and are difficult to predict precisely. Their beneficial evolution must therefore be facilitated by a combination of framework conditions and information and communications infrastructures. In this way a platform is provided on which governments, businesses and citizens can communicate and work together, and track the evolution of the city. We have seen that what makes a city a Smart City is the use of ICTs, which are used to optimise the efficiency and effectiveness of useful and necessary city processes, activities and services. This optimisation is typically achieved by joining up different elements and actors into a more or less seamlessly interactive intelligent system. In this sense, the concept of a Smart City can be viewed as recognising the growing and indeed critical importance of technologies (especially ICT) for improving a city's competitiveness, as well as ensuring a more sustainable future, across networks of people, businesses, technologies, infrastructures, consumption, energy and spaces. In a Smart City, these networks are linked together, supporting and positively feeding off each other. The technology and data gathering used in Smart Cities, should be able:

- constantly to collect, analyse and distribute data about the city to optimise efficiency and effectiveness in the pursuit of competitiveness and sustainability
- to communicate and share such data and information around the city using common definitions and standards so they can be easily re-used
- to act multi-functionally, providing solutions to multiple problems from a holistic city perspective.

Overall, ICT enables a Smart City to:

- make data, information, people and organisations smart
- redesign the relationships between government, private sector, non-profits, communities and citizens
- ensure there are synergies and interoperability within and across-city policy domains and systems (e.g. transportation, energy, education, health and care, utilities, etc.)

- drive innovation, for example through so-called open data, living labs and tech hubs.

ICT initiatives based on these characteristics aim to connect existing and improved infrastructure to enhance the services available to stakeholders (citizens, businesses, communities) within a city, thus IoT naturally becomes the nerve centre giving life to Smart Cities and opens up a vast road of promising potentials for innovation.

Nowadays, each city mainly carries out a multitude of heterogeneous solutions related to the different vertical application domains (e.g. Mobility, Buildings, Energy Grids) and the most common approach is that where each solution is a closed proprietary implementation not able to communicate neither with other solutions nor with the other city stakeholders (municipality, citizens).

Therefore, in order to exploit the Smart City vision potentials, we need to drive the solutions towards two fundamental concepts: open data and interoperability.

Open data is definitely an important enabler of urban smartification contributing to innovation with citizen and business value-added applications and services. Recent developments towards [3] opening up data in the process of urban "smartification" have demonstrated that making machine-readable information freely available can foster citizen empowerment, enhance public services through participation, leverage new business models, and ultimately change the paradigm on which governments operate. However, many issues still remain to be appropriately addressed so that open data can be explored to their full potential. Most infrastructure data in a city is still locked away and incompatible data formats and access methods, and various semantic interpretations of data consequently prevent open-data stakeholders to offer citizens and business value-added applications and services.

Interoperability is still at a very early stage. Most technology waves go through a similar innovation cycle—often referred to as the innovation S curve. There is a rapid explosion of innovation, many new systems and solutions appear on the market, and companies scramble to promote their approach. During this phase, standardization is hard and often gets overtaken by events. As the rate of innovation levels off (top of the S curve) standardization efforts are possible—they are usually led by companies with strong market positions as they try to impose their own proprietary solution. At the moment, the IoT space is still somewhat chaotic but there is a possibility for high level frameworks that provide some degree of standardization.

In this scenario the work described in this article is going to tackle the issues related to the two concepts mentioned above by providing a reference framework of modular specifications for stakeholders willing to implement ICT platforms with the aim of exploiting the Smart City vision potentials and therefore provide new services for citizens.

The most innovative aspects of the proposed approach are related to the Information and the Semantic interoperability levels. The usual issue, speaking of interoperability by use of shared data formats, is how to find the correct balance between too prescriptive specifications (which guarantee interoperability, but risk to inhibit innovation) and more elastic specifications (which have a lot of potential deficit with respect to real interoperability). This problem becomes more urgent in a context, like

Smart City, with a lot of interacting heterogeneous systems. In order to overcome it, the approach proposed by the chapter is to have a very light and elastic format (at Information Level) able to represent a very large set of data, moving at Semantic Level the strict definition of the data. The underlying idea is that this light approach can be easily applied also on existing systems with just small intervention on them.

Thus, the work is structured as follows. Section 2 describes the state of the art as well as the general background of interoperability and the replicability view for Smart Cities, Sect. 3 briefly outlines the reference model, the methodology and some aspects about use case definition, the following sections go in details about the different levels of the interoperability stack (Functional, Collaboration, Communication, Information, Semantics) od the reference model, highlighting the main original contributions of the proposed approach; the conclusion section sums up the main pillars and concepts of the proposed approach and outline future directions.

2 Problem Definition and Research Context

2.1 The Interoperability Problem in the (Smart) Cities

The Smart City paradigm is gaining momentum in the recent years as a holistic approach to the digitalization and convergence of the complexity of services and infrastructures resting on the same territory; for this reason, the Smart City is often thought as a system of systems.

Furthermore there are evidences [4, 5] that in recent years many investments have been implemented, for example by multi-utilities operating in the cities, for creating digital infrastructures while many Smart City projects have been developed with the aim of the vertical integration within existing services with the result of improving digitalisation, collecting more data through new IoT technologies and offering new services [6] to the citizens. The result is a lack of horizontal data flows, between vertical applications and between service suppliers and the city administration and citizens: in short interoperability is lacking between applications that have been developed like self-consistent silos, able to exchange data from the field up to the decision support systems and the dedicated control dashboards but unable to interoperate with other systems.

Two kinds of barriers should be broken in order to favour interoperability: barriers between the higher levels of the silos that hamper the monitoring and the exchange of the data between the top of different systems (e.g. global indicators about traffic, parking, public transportation and urban planning related to an area); barriers between the field level of the silos, that hamper the possibility of fully reuse existing data collection infrastructures for different services (e.g. the same sensors for parking, traffic, security, environmental monitoring). This chapter has its main focus on the first group of barriers.

2.2 *Interoperability, Definition and Researches*

Even though there is a general agreement on the need for enabling horizontal data flows it is less clear how this objective can be realistically pursued.

In literature the discussion about how the integration between independent systems can be achieved has been largely addressed since 1990 in the domain of Enterprise Interoperability (see [7] for an overview), and, the theme of interoperability was dealt with by a large number of successive European projects, starting from the IDEAS project (2002), then continuing in the INTEROP (2002), ATHENA (2004), Abilities (2006), COIN (2008) projects, just to name a few of the most important ones; [8] summarizes some of the main points that have been identified during these research activities. These efforts have also led to the drafting of a European document for interoperability between digital public services [9]. Coming from the world of enterprise systems, the interoperability concept has been adopted in the domain of public infrastructures and Smart Cities with the same definition: "the capability of two or more networks, systems, devices, applications, or components to exchange and readily use information, securely, effectively, and with little or no inconvenience to the user" [10].

In the field of Enterprise Interoperability, characterized by the continuous evolution of the business processes and products, the research focus is addressing the means for enabling interoperability and the concept of sustainable interoperability: the crucial point, in fact, is assuring the capability of the systems to interact while they and the external ecosystem evolve over time, with new technologies, new services, new requirements, new categories of data and applications [11].

Meantime the Smart Cities seem growing thanks to both digitization of existing services and composition of new services upon the existing ones; the number of potential new applications and services (and data flows to deploy) is rapidly increasing [12, 13]. Thus, in this case, the interoperability hampering factors seem to be the number of already existing solutions with different institution and organisations in charge, architectures in concurrence with the lack of convergence in the field of standardization initiatives.

2.3 *Approaches and Means for Achieving Interoperability*

A look at the landscape of existing initiatives can give the sense of the strength of the feeling that exists on this subject in the Smart City community. The wide landscape of the initiatives can be split into the following categories: Models, Tools (technologies and platforms), and Standards.

With the word "**models**" we mean analysis frameworks defining high level requirements for Smart City Platforms. Such kind of frameworks aims at providing shared languages for describing system basics and interoperability levels, for categorizing platform and comparing different architectures.

Some examples of such frameworks are:

1. The SGAM [14] (Smart City Architecture Model): an interoperability oriented analysis framework in the smart grid context, built by the Smart Grid Coordination Group (which join CEN, CENELEC and ETSI). It enables identification of data exchange interfaces, standard classification and mapping of different architectures on the same reference model. Even if it is thought for a different field, smart grids—similarly to Smart Cities- are made of complex ecosystems of inhomogeneous and independently born applications and services, as a consequence they present similar interoperability issues.
2. The SCIAM [15] (Smart City Infrastructure Model): it uses the same SGAM interoperability levels, changing the set of application domains. At the moment it is a proposal and needs to be consolidated.
3. GSCAM [16] is another proposal for extending the SGAM to the Smart City context (Generic Smart City Architecture Model), which adds to the SGAM new dimensions applicable to the domain.
4. Moreover, the main European SDOs are trying to replicating the success of the SGAM with a similar initiative: the SF-SSCC (Sector Forum on Smart and Sustainable Cities and Communities) [17], built on the previous Smart and Sustainable Coordination Group.

Other interesting models are presented within initiatives with more operative objectives, like those at the base of:

- SMArc [18] (Smart, Semantic Middleware Architecture Focused on Smart City Energy Management), a middleware proposal for Smart Grids
- U-City (Ubiquitous Eco-City Planning, in Korea), a project aiming to create a ubiquitous city model [19]
- ITU-T, a proposal about the Smart Sustainable Cities [20].

If the previous models give us the language, more technological **tools** are available. Let us give a look just at some of them, choosing only among existing open tools and splitting them into three main categories:

1. General purpose development framework. One of them, very active in the Smart City application context is FIWARE [21], managed by an open community, it provides public and royalty-free set of APIs and the related open source reference implementations.
2. Smart City Platforms. For examples Open City Platform [22] is a cloud platform aiming to sustain cloud service offer by public administration; KM4City [23] is an open source system providing customizable dashboards for managing the city; E015 [24] has a different approach and presents a set of guidelines for enabling the definition of interoperable Smart City services.
3. IoT Middleware platforms. Some examples: Kaa IoT Development Platform [25] is a middleware platform which allows building complete end-to-end IoT solutions; OM2M [26] is an autonomic ETSI-compliant M2M service platform; OpenIoT [27] is a middleware open source platform for getting data from a cloud of sensors.

But, in order to have the sense of the incredible complexity of the situation, a look at the landscape of Internet of Things technologies (so just the third point of the previous set) made by Matt Turk is really instructive. The conclusion of the author, looking at this landscape, is that "there is no dominant horizontal platform and not enough mature, cheap and fully reliable components just yet" [28].

The last important point about Smart City interoperability is the set of standards. A really interesting work about the categorization of Smart City related standards was made by the British Standard Institution (BSI) [29]. In its report BSI identified 100 Smart City standards (only considering inter-domain standards) split into three levels (technical, process and strategic standards).

Because of the previous complexity, NIST set up the IES-City (Internet of Things-Enabled Smart City Framework) initiative, involving ENEA and other international organisations. The aim is to distil a common set of architectural principles (Pivotal Points of Interoperability—PPI) from the comparison of existing architectures. PPI are very basic (e.g. to convert XML to JSON, to use REST APIs, etc.), but they show a way for facing the problem: finding common principles upon which new architectures should be designed.

2.4 *The Still Open Issues*

The work conducted by IES-City initiative [30] is a first important step to create environments facilitating the reuse and the automated and interoperable interaction between different systems and applications; nevertheless it is not enough. If we state that the public administrations can play a key role in breaking the barriers between the Smart City subsystems, they should be provided with a common global set of standards and guidelines to be used in their call for tender.

This chapter presents an approach to this issue: concretely it consists in the definition of a set of modular, general specification (the Smart City Platform Specification) for implementing horizontal ICT platforms, in order to enable interoperability among the vertical silos.

The idea of specifications comes from the need of the city administrations for avoiding the "vendor lock-in" and to clearly state the data that the public service providers are required to supply.

In parallel, an ICT platform would allow to get and monitor data from different utilities and urban service providers (Fig. 1).

The specifications consider also the following requirements:

- split in modules, according to interoperability levels (Functional, Communication, Information, Semantic, Collaboration levels) [31];
- focus on the interfaces;
- the city viewed as system of heterogeneous systems (building, lighting, …) with different aggregation levels (sources, local platforms, Smart City Platform);

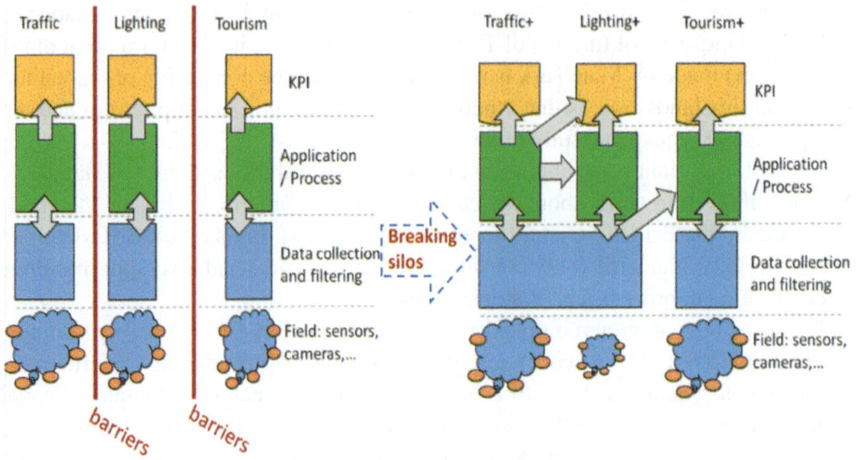

Fig. 1 Breaking down the silos barriers for creating smart cities, from [13]

- scalability and composability: the specification modules have been thought to be used also separately and to be adaptable to contexts and domains not known at their design time.

3 The Reference Model for the Smart City Platform Specification

The starting point for solving the "silo" problem, explained in the previous section, is a close look to real world applications and the consequent identification of a clear methodology. For this purpose, a reference model was defined based on a "customization" of the SGAM, considering also the SCIAM proposal and the other examined models.

All these models present some common elements:

1. At their base there is a data acquisition layer, made of sensors plus the physical infrastructure for connecting them.
2. Over the acquisition layer there is a layer where data are aggregated, inserted in (often distributed) databases and some elaborations (e.g. data fusion, statistical analyses, decision support) are performed.
3. On the top there are the Applications that exploit the collected and elaborated data for offering services to the users.

The previous scheme represents the usual data cycle in a single domain application: the application gets the data from the field, collects and elaborates them and uses the results for providing applications to the users. But, in order to get

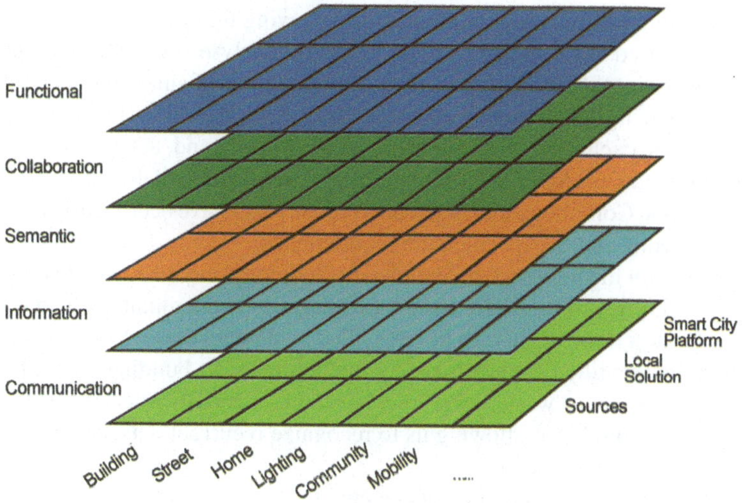

Fig. 2 Reference model for the smart city platform specification

inter-application interoperability, a higher layer has to be added, enabling data exchange between domain centric applications (i.e. the aforementioned silos). Considering a SGAM-like vision, the ICT layers of our model (the z-axis) are the following (see Fig. 2): Sources/Field (sensors and related infrastructure, layer 1), Local Platform/Solution (collection, elaboration and user services, it joins layer 2 and 3 of the previous list) and Smart City Platform (the inter-application interoperability layer we add).

The y-axis of the SGAM-like model represents the interoperability levels. Re-elaborating these levels on the base of the Enterprise Interoperability Framework [8], the levels that have to be addressed for answering to interoperability issues are:

- **Functional**: key concepts, actors, architectural model, component and functionalities;
- **Collaboration**: information expressing the collaborations among the different actors and the configuration of the interoperable communications;
- **Semantic**: semantic of the common language, for sharing the data meaning, avoiding ambiguity;
- **Information**: the common language data format, involving both data models and syntaxes;
- **Communication**: definition of data exchange interfaces, including transport protocols.

We adopted these levels for the definition and organization of the Smart City Platform Specification (SCPS).

The last axis (the x-axis) represents the application domains and it has to be intended as an "elastic" axis, since the domains can change depending on the city context.

The resulting model is represented in the following figure.

We firstly used this reference model to deeply analyse (by UML diagrams and textual description) different use cases, intended as a set of smart applications from city (e.g. smart lighting, smart building, waste water treatment plant) in order to identify: actors, exchanged data and the related processes and, at the end of the analysis, requirements for SCPS definitions. Then Functional, Collaboration, Semantic, Information and Communication aspects of the use cases have been clearly described, helping in identifying the specification requirements.

Together with the reference model, the methodology and the template defined by the IEC 62559 [32] were used for use case analysis. The template, which contains the fields for describing the use cases, was filled by domain experts with the information about the identified applications (e.g. including smart building, smart lighting, smart home, and smart waste water treatment). On this basis inter-application data exchanges were modelled allowing us to recognize recurrent data structures and to fix specification requirements.

In particular, some common points emerged:

1. The different spatial data can be aggregated, in one of the following levels:
   - Items: punctual data (e.g. at single sensor level);
   - Facility: data related to the whole monitored facility (an entire building, street, ...);
   - Aggregation of facilities: data from an application that collects them from a group of facilities;
   - City: monitoring data from the whole city;
   - Region: data related to infrastructures extended beyond a city (e.g. energy infrastructures).
2. Similarly, the different data can be aggregated, at spatial level, in one of the following way:
   - Static: anagraphic data (e.g. geographical localization of a facility);
   - Instantaneous: instantaneous measurements from sensors;
   - Average: average data, elaborated starting from instantaneous ones, in a fixed time range;
   - Total: sum of instantaneous data in a fixed time range
   - Forecast: forecast (future) values (e.g. weather forecast).
3. The exchanged data present a set of common properties that can be shared at data model level (e.g. the timestamp, the spatial coordinates, the reference time period, the updating frequency, etc.).

In the following section we detail the five levels of the SCP Specification.

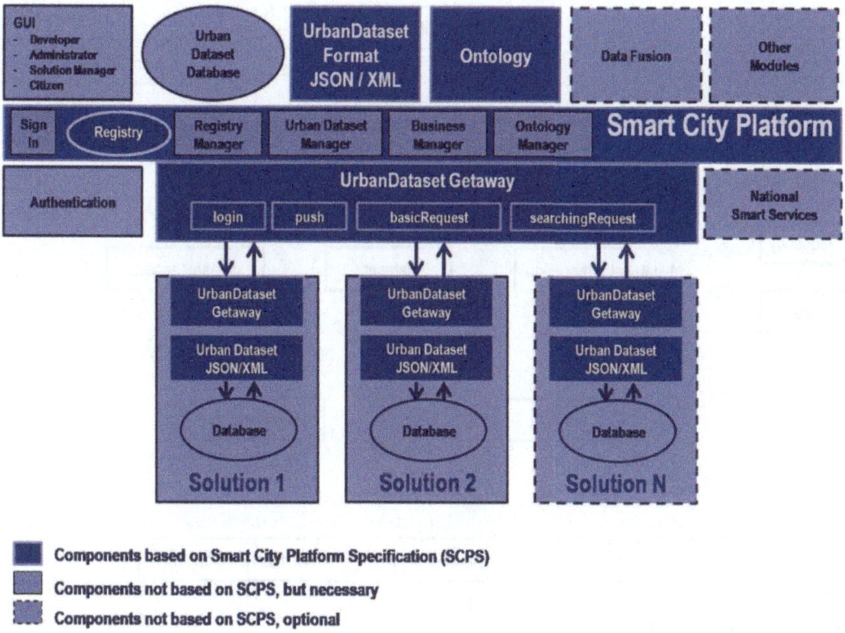

Fig. 3 SCP reference architecture

3.1 The Functional Level and the Architecture

The core of the Functional Level is the description of the Smart City Platform Architecture, its components and functionalities, and the interactions with the connected solutions: Fig. 3 depicts its schematic representation that includes:

- a horizontal Smart City Platform;
- a series of vertical Solutions (i.e. local platforms for the management of a single application context);

where, ideally, the data collected by the city will "rise" vertically, from the solution management in its own application context, up to the integrated ICT platform. In the figure, we represent in blue the basic components that must adhere to the proposed Smart City Platform Specification: these components will be described later in this chapter. We defined the components with a modular approach: each of them is independent from the others, and the adoption of the specifications can be carried out gradually, choosing step by step what to implement. For example it is possible to start adopting only the data format, and only afterwards adding the other components according to own priorities.

The Architecture shown in Fig. 3 depicts how each vertical Solution, managing specific application contexts, exports, from its local database, the information in the "Urban Dataset" JSON/XML format and sends them to the horizontal Smart City

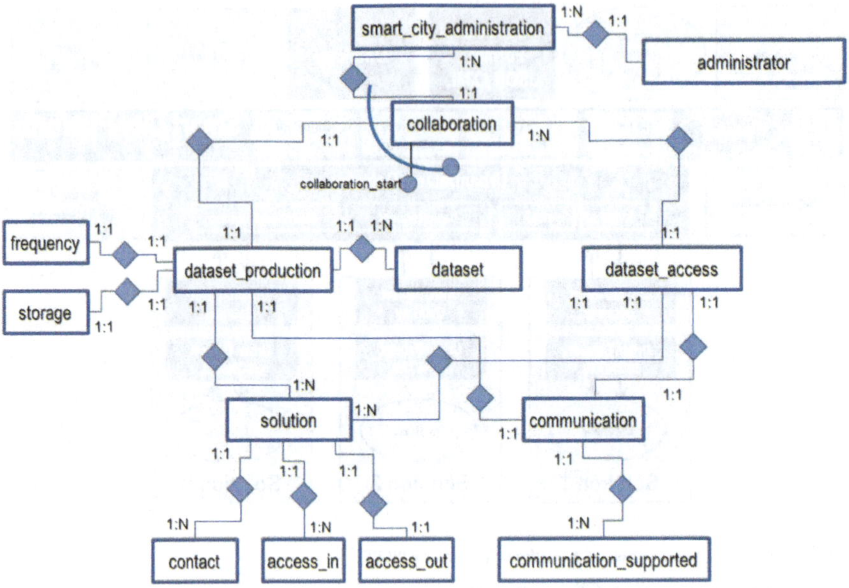

Fig. 4 Schema E-R registry

Platform through the transport service "Urban Dataset Gateway". The Smart City Platform has to handle the dataset production and access (Fig. 4).

Moreover, the Functional Level provides a description of:

- key concepts and the Components (there is a 1–1 correspondence between the key concepts and the main Components of the Architecture);
- users;
- macro-functionalities (through Use Case Diagrams).

The following are some Key Concepts:

An **Urban Dataset** (UD) is a set of data exchanged between the city's vertical solutions and the Smart City Platform, according to the SCPS, characterized by a univocal and centralized semantic description (the Ontology component), and by common formats defined by the Abstract Data Model and Syntactical Implementation (JSON and XML).

The **Ontology** is the component defining the semantic structures of the Urban Datasets, as well as classifying them into categories and sub-categories. The Ontology is an independent component, external to the Smart City Platform, to allow a shared use with other SCPS-based platforms, in the way that a process of convergence is at the base of every communication.

The data **Transport** Service allows to send and receive Urban Datasets, and it is defined in the SCPS by defining a single Web Service: Urban Dataset Gateway. The Web Service is provided with three patterns (push, request/response, publish/subscribe), a common definition of service interface and two implementations (REST and SOAP) based on the same interface.

The **Registry** is a database that manages the information related to the collaborations between Smart City Platform and Solutions, from the point of view of who produces and accesses the Urban Datasets.

There are four kinds of **Users** defined for the Smart City Platform:

1. *Developer*: it deals with administering the platform from the technical point of view and the software components that implement the platform;
2. *Administrator*: it manages the horizontal platform (typically it is a representative of the municipality or multi-utilities that manages the city/district) handling the collaborations between Smart City Platform and the different vertical Solutions;
3. *Solution Manager*: responsible for the vertical Solution platform connecting to the Smart City Platform; it accesses to the data describing the situation of its Solution (personal data, web services access credentials and parameters, Urban Datasets to be produced or accessed and data sent view);
4. *Citizen*: unregistered generic citizen user who accesses the services offered by the platform.

These four kinds of users can access to the functions of the SCP. The SCPS Functional Level provides also four use case diagrams to describe the main steps for the set-up, instantiation and use of the SCP:

1. SCP configuration for City/District;
2. Solution configuration;
3. Interoperable Communication;
4. Solution deleting.

3.2 The Collaboration Level

The dialogue between the horizontal Smart City Platform and vertical Solutions, through the datasets exchange, implies the management of a remarkable set of information (e.g. the produced datasets, who produce them, who will access them, in which format the information will be represented, which data transport protocol will be adopted, etc.).

The **SCPS Collaboration Level** provides an approach to manage this set of information, through the description of:

- the definition of proper roles for each user of the Smart City Platform (developer, administrator, solution manager, citizen) and the way each of them interacts with the Platform through the Graphical User Interface (GUI);
- the Registry database, which allows recording and storing information about the Solution and the managed Urban Datasets, as well as the information regarding their production.

The collaborations are handled, in the Registry, defining four groups of information about:

1. Smart City Administration: administration of the Smart City Platform;
2. Vertical Solutions: registered Solution, reference contacts and credentials for the access to the transport data services (for the Smart City Platform that has to access to the vertical Solutions, and for the vertical Solutions that has to access to the Smart City Platform too);
3. Urban Datasets: declarations of which dataset (registered in the Ontology) will be produced by the vertical Solutions in a given collaboration, as well as the ownership of that datasets;
4. Complementary aspects: complementary information to the collaborations, such as the production frequency of an Urban Dataset, the related ownership and the access by other Solutions (if the UD is declared as OPENDATA, it will be accessed by everyone), the transport protocols, the formats used in the transmission, etc.

Each group of information has been organized in the following E-R scheme and their detailed description is provided with the related tables. It should be noted that the collaboration table is the core of the whole E-R schema since a **collaboration** describes the production of a specific Urban Dataset, by a specific vertical Solution, towards a specific Smart City Platform, in a given time period.

Collaboration between a vertical solution and the SCP is defined, in fact, as an agreement with the aim to produce a particular Urban Dataset. Between a solution and the SCP, there might be multiple collaborations (production of different Urban Datasets and/or production of the same Urban Dataset in different periods) and different accesses to the produced Urban Dataset. This collaboration provides a connection between the Smart City and the Dataset Production of a Solution, at a certain timestamp. Since this aspect is crucial and it is employed in several parts of the Smart City Platform, it is useful to have an ad hoc identifier to easily achieve this information without involving Registry accesses. In this respect, the identifier *collaboration_id* is created from the sequence of the identifiers of Smart City Platform, solution, urban dataset and timestamp of the reference period.

The *collaboration_id* is the link among the different SCPS specifications that can be used in a modular way but, in any case, they offer their maximum potential when used jointly. There are, in fact, two main moments in which the *collaboration_id* is treated:

1. In GUI navigation to search for available/accessible Urban Datasets, from which it is possible to obtain a *collaboration_id*;
2. Using the web service to produce or access a particular Urban Dataset (this will be evident in the web service interface, in the communication part).

3.3 The Information and Communication Levels

The data exchange is managed starting from the definition of two main aspects:

- **Information**: representation of the exchange data by defining an abstract model of Urban Datasets and related common formats (JSON and XML);
- **Communication**: description of the data transport through the architectural patterns and the web service interface.

These two aspects correspond to the homonymous sub-specifications which, we recall, can be adopted in a modular and gradual way.

The Information level defines a format able to make interoperable the exchange of Urban Dataset between heterogeneous systems or applications.

It is defined by:

- an Abstract Data Model;
- the syntactic implementation of the Model; currently, two reference implementations are available: XML (eXtensible Markup Language) and JSON (JavaScript Object Notation).

The Abstract Data Model gives a syntax-independent representation of the content that is mandatory for a document used to exchange Urban Dataset. Because the Smart City scenario is continuously evolving, it has been designed to be scalable during the time and across different contexts; in this perspective, it satisfies the following requirements:

- independence from syntaxes and communication protocols;
- ability to represent any kind of Urban Datasets coming from any vertical system or application;
- ability to represent new, unexpected Urban Datasets, which were not foreseen at platform design time and do not come from the initial use cases.

To meet these requirements the Abstract Data Model has a flexible structure (it is unaware respect to the Urban Dataset properties) and we also define a procedure to customize it without loss of interoperability. The result is an Abstract Data Model composed of the following three parts:

- **Specification**: it contains meta-information describing the Urban Dataset (e.g. the Urban Dataset specification reference, its properties,...);
- **Context**: it provides information needed for contextualizing the exchanged values (e.g. the time zone related to the timestamps);
- **Values**: it contains the measured data on the Urban Dataset properties, organized in key-value pairs.

The customization procedure consists in binding the Abstract Data Model to a specific Urban Dataset and its properties, following its semantic definition (in the SCPS it is given by an Ontology). The aim of this approach is to facilitate the mapping between different syntactical implementations used by different systems or applications and to enable interoperable data exchanges.

Actually multiple syntactic implementations are possible; currently, two **reference syntaxes** are available:

- XML, formally expressed by an XML Schema conformant with the XML Schema 1.0 Specification;
- JSON, formally expressed by a JSON Schema conformant with the JSON Schema Draft 06 Specification.

The following figure shows an XML fragment related to data measured on an Urban Dataset (this fragment implements the "Values" section of the Abstract Data Model). It is important to emphasize that the same structure is able to contain very different kinds of Urban Dataset properties and related values.

The Communication Level (or data transport) deals with two main actions that a vertical Solution can undertake to communicate with the Smart City Platform:

1. Production of an Urban Dataset: export, from a vertical Solution, of the urban datasets that will be received from the Smart City Platform;
2. Access to an Urban Dataset: vertical Solution access to the Urban Datasets that the Smart City Platform has previously published.

Note that, if the first action enables the retrieving of data from a Solution to the Smart City Platform and the second action enables transfer of data from Smart City Platform to a Solution, the coordinated set of the two actions allows the exchange of data from a vertical Solution to another vertical Solution through the Smart City Platform (Fig. 5).

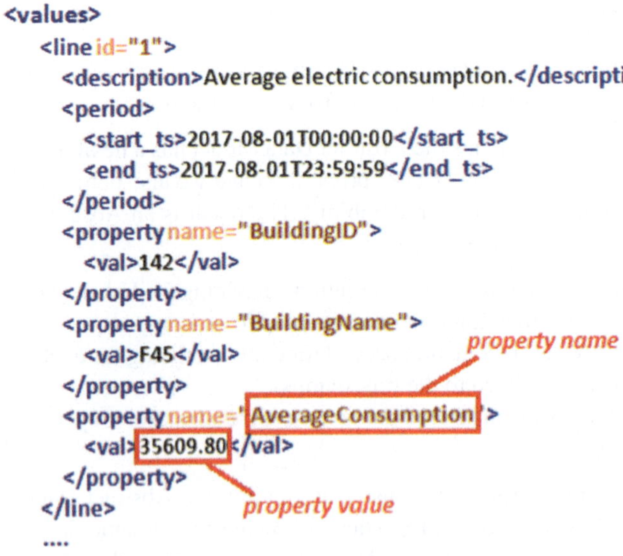

Fig. 5 Urban dataset XML fragment

The Communication, to perform the two actions above, must be defined through the configuration of three aspects:

1. The Architectural Patterns,
2. The Web Service Interface,
3. The Transport Protocol.

The SCPS Communication Level describes the set of possible configurations that the Smart City Platform and the Solution can agree on. SCPS Communication Level provides three architectural patterns [33] defined as Client-Server interactions:

- Request/Response: it allows a client to request an Urban Dataset to a listening server (e.g. a Vertical Solution, as a client, requires an Urban Dataset to the Smart City Platform's service, as a server, which responds with the Urban Dataset required);
- Push: it allows a client to directly send an Urban Dataset to a listening recipient (e.g. a vertical Solution, as a client, sends an Urban Dataset to the Smart City Platform's service that, as a server, receives the Urban Dataset);
- Publish/Subscribe: it allows the transmission of data between a vertical Solution "publisher" and N vertical "subscriber" Solutions, through the Smart City Platform that acts as a "broker".

These patterns can be used to implement the communication among the vertical Solutions and the Smart City Platform, depending on the needs inherent in the case to be managed.

We report, as example, the schema of the "Request/Response" pattern.

Note that, depending on the situation, the Client and Server roles can be interpreted by both a vertical Solution and the Smart City Platform.

The patterns can be realized through the implementation of the "Urban Dataset Gateway" web service, whose interface is defined with various methods (detailed descriptions are provided for each method and each parameter specified). For example, the "basicRequest" and "searchingRequest" methods implement the "Request/Response" pattern (see Fig. 6). Note that the description of the web service interface is independent from the implementations, based on the REST protocol in combination with the JSON format or SOAP (described through WSDL, descriptor of the service) in combination with the XML format (Fig. 7).

3.4 The Semantic Level

The core functionalities of the Smart City Platform aim to make efficient the exchange of information between the different interfaced application contexts. Clearly this leads to interoperability problems, due to the use of different languages, paradigms, software, data formats. A way to tackle this problem is the reduction or elimination of terminological confusion and with shared knowledge and terminology through the definition of a unifying framework that could serve as a basis for interoperability.

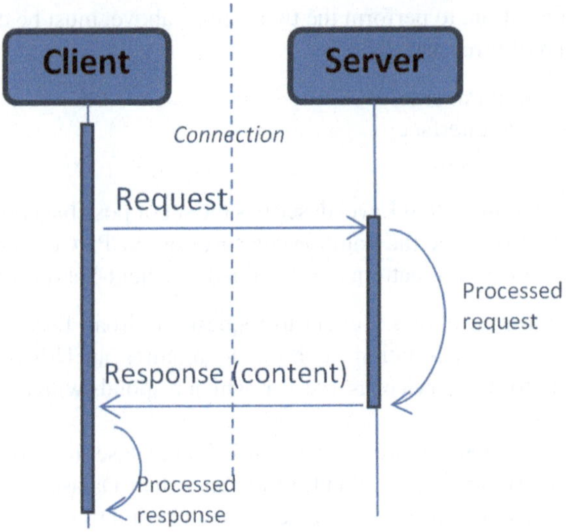

Fig. 6 Pattern request/response

Service	UrbanDatasetGateway
Methods	login (username, password)
	push (token, collaboration_id, dataset)
	basicRequest (token, collaboration_id)
	searchingRequest (token, collaboration_id, from_utc_datetime, to_utc_datetime, coordinate_center, radius)

Fig. 7 Methods of the urban dataset gateway

Ontology definition can help in this task, and could represent a common ground to face up some of the aforementioned issues, playing a relevant role in the communication management and solving semantic ambiguity. Modularity and reusability are two key characteristics that make ontologies a proper tool to be used inside the platform.

The use of standard technologies for the definition of Ontology, such as those defined by the World Wide Web Consortium (W3C), makes available a whole series of protocols and languages for communication. These tools favour the automatic discovery of the data structures and their meaning through the organization of the concepts in a hierarchy and the link with related information, making possible also the automatic generation of the relative documentation. An additional advantage using Ontology is the possibility to define, in a simple way, rules for generating new information autonomously with the appropriate software tools. In fact, there are different mechanisms and languages that allow to define rules to obtain new information starting from the information stored in the knowledge base.

In the task of defining and creating the Ontology, the initial phase was focused on the analysis and exploration of existing ontologies that deal with the Smart Cities in various aspects. We started analysing the initiatives listed in the http://smartcity.linkeddata.es/ portal of the READY4SmartCities FP7 CSA [34] project, whose aim is to increase awareness and interoperability for the adoption of semantic technologies in the energy field to reduce consumption and CO2 emissions at the Smart City community level. The portal contains several ontologies for the most diverse services related to a Smart City. During the analysis phase we explored different projects that tried to achieve the objective of cataloguing and data integration through ontologies. The aim was to observe the state of the art in order to understand how to move in this area and what are the pros and cons of different approaches. Below are some of the identified solutions.

- The DIMMER project (http://www.dimmerproject.eu) integrates information on the topology of buildings with real-time data from sensors and users to analyse and correlate the use of buildings and provide information on their behaviour on energy consumption in real time. The defined Ontology is very simple and defines a relationship between services provided by a sensor, building and physical location.
- The City Protocol project (http://cityprotocol.org/), conducted by the city of Barcelona, whose goal is the creation of a common system for cities and the development of protocols that allow innovation through the interaction between different sectors and areas of city life. The project provides the definition of Ontology [35] that defines a fundamental structure of the city through the definition of basic and general concepts for the different types of activities of the city from both an infrastructural and social, economic and environmental point of view.
- Km4City Ontology: Ontology developed by the DISIT of the University of Florence [36] with the aim of integrating information on transport and mobility with the dataset provided by the municipality of Florence and the Tuscany region.
- The CITYkeys project (http://www.citykeys-project.eu/) aims to improve the exchange of information between the various subjects providing services within a Smart City. The project identified a set of KPIs (Key Performance Indicators) to help the city in implementing strategies by linking various services. CITYkeys does not define any Ontology, but helps collecting and detailing a wide set of indicators that can be used in the evaluation of impacts on Smart Cities.

The analysis led also to the identification of two existing ontologies that have been integrated in our platform for greater compatibility with the outside world:

- Ontology of units of Measure (OM) [37], it has a strong focus on units, quantities, measurements, and dimensions.
- PROV Ontology (**PROV-O**), provides a set of classes, properties and restrictions that can be used to represent and exchange information of sources, generated in different systems and different contexts

In its definition, the Ontology revolves around the concept of Urban Dataset (defined as a subclass of Entity of the PROV Ontology). This concept refers to any data, aggregated or not, that an application context is able to process based on the

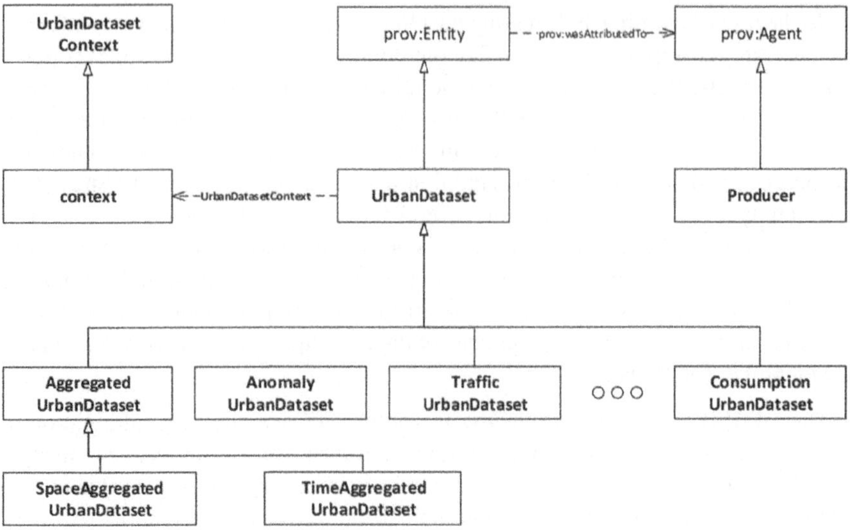

Fig. 8 Main concepts related to urban dataset

data collected by the sensors located in the Smart District. This is the crucial node of all communication and the main information that must be exchanged within the infrastructure.

Figure 8 shows the main concepts of the Ontology:

- **UrbanDatasetContext**: each Urban Dataset has its own context that serves to better describe and characterize the information (e.g. the context includes the geographic position to which data refer, language used for descriptions, time and time zone). Information from a context is modelled as a property.
- **Property**: when an Urban Dataset is generated, it is associated with other information (both specific and contextual) describing it, such as the creation time, the position of the sensor from which it is produced, its description or its identifier and the version (that is the specificationRef). This information is defined as a property of the Urban Dataset or a context.
- **Producer**: an important aspect of a data set is its origin, that is, who is the data provider. PROV Ontology, introduced earlier, in this case performs this function. In the specific case, Producer, subclass of "Agent", (see Fig. 8) describes the entity that creates the data. The property "wasAttributedTo" of PROV-O can be used to report the producer.
- **DataType**: this class was created to manage and list the canonical data types. As said, the OM Ontology is used to link Urban Dataset to the units of measurement of the international system.
- **ApplicationContext**: in addition to the specific properties, the Urban Dataset is also associated with a field that indicates the application context to which it refers (e.g. SmartBuilding for the case of anomalies in a building).

The individual properties of the different Urban Dataset are in turn the subclasses of Property:

- **ContextProperty**: groups all the property instances used to describe the context (e.g. coordinates, language, time in which the data was collected, etc.);
- **UrbanDatasetProperty**: groups all the property instances used to describe specific properties of the Urban Datasets (e.g. the number of anomalies, the number of occupied parking spaces, the average energy absorption, etc.).

Finally, an Urban Dataset often contains data that can be attributed to a particular application context (e.g. smart building, smart mobility, etc.). For this reason, the existence of a property that relates an instance of Urban Dataset with an applicative context has been foreseen in the Ontology. The Ontology has been formalized in the OWL language and made available online and downloadable at the link http://smartcityplatform.enea.it/specification/semantic/1.0/ontology/scps-ontology-1.0.owl.

3.5 The Original Contribution of the Approach

The SCPS with respect to existing development framework and platforms:

- does not impose a reference technology or the use of an existing implementation;
- looks only at the interfaces at the higher ICT level (z-axis in Fig. 2), leaving completely free the implementation of each vertical application;
- is highly customizable, since:
  - the developers can choose the subset of the specification they want to implement (favouring incremental approach)
  - the flexibility of the format enables to exchange a huge kind of data. So the set of data can be defined on the base of the vertical applications that need to be connected.

Moreover, the SCPS approach to the data format definitions is based on two principles, aiming to join the flexibility with the capability of checking conformance to the specification:

1. *Need for minimum set of constraints on the Information Level*: this has been got by the definition of a general, huge and elastic data model able to contain a wide set of measured data, coming from different contexts and managed by different applications. This data model can be imagined as a table, which names of columns are defined case by case. So at this level only the conformance to this general light format is checked.
2. *Moving of conformance checking on the semantic level*: this has been got by defining the semantic structure of the data model within the Ontology, so assuring not only the semantic interoperability thanks to shared meaning, but also a part of the formal validation of the data format.

Moving on a more political and organizational level, another interesting point is the recognition of the key role played by the Public Administrations in the effort to activate the city's change. It resulted in the identification of the public calls for tenders as a powerful leverage for applying common interoperability principles between Smart City Applications.

4 Conclusions and Future Directions

Nowadays, each city carries out a multitude of heterogeneous solutions related to the different vertical application domains (e.g. Mobility, Buildings, Energy Grids) and the most common approach is the one where each solution is a closed proprietary implementation which is not able to communicate neither with other solutions nor with the other city stakeholders (municipality, citizens). Meanwhile the Smart City holistic vision is rapidly spreading out thanks to the digitization of existing services and the creation of new services upon the existing ones. In fact, what makes a city a Smart City is the use of ICT in order to optimise the efficiency and effectiveness of useful and necessary city processes, activities and services. This optimisation is typically achieved by joining up different elements and actors into an interactive intelligent system. For that reason, in order to break the silos and break the barriers in the ecosystem of urban solutions, we need to drive them towards interoperability concepts.

In this scenario we started from the analysis of real contexts and the processed and exchanged data in the Smart Cities; this analysis made it possible to identify multiple aspects related to interoperability among systems and to group them into 5 groups/levels; then a concrete solution was given for each of the 5 levels, providing a precise specification; the 5 resulting specifications are modular, connected but independent of each other: this makes it possible to establish a process of gradual convergence towards full interoperability among the city's vertical solutions with the aim of exploiting the Smart City vision potentials and therefore provide citizens with new services, fostering competitiveness and avoiding the 'vendor lock-in' given by proprietary implementations.

As a result, we proposed an approach inspired by the SGAM Model, an interoperability oriented analysis framework in the smart grid context, which enables the identification of data exchange interfaces, standard classification and mapping of different architectures on the same reference model.

Re-elaborating the SGAM in a Smart City scenario five levels have to be addressed in order to face the interoperability issues: Functional, Collaboration, Semantic, Information, Communication.

The Functional Level describes the Reference Architecture, the Key Concepts, the Users and the macro-functionalities through Use Case Diagrams. The Reference Architecture assumes a schematic representation which includes a horizontal Smart City Platform and a series of vertical Solutions where the data move from the Solution management in its own application context up to the Smart City Platform.

The Collaboration Level manages the set of information that characterizes the cooperation between the SCP and any solution, through the description of the basic functionalities needed for the interaction with any user and the registry database which is the repository of the solution and of the managed Urban Datasets, as well as their relationships. The collaborations are handled, in the registry, as a set of information organized in the following four groups of information: Smart City Administration, Vertical Solutions, Urban Datasets, and Complementary. Each group of information is explained in a data model through the definition of the appropriate E-R scheme and detailed descriptions of the related tables.

The Semantic level deals with the definition of an Ontology which provides reduction or elimination of terminological confusion by means of shared knowledge and terminology. Ontology definition plays a relevant role in the communication management and solving semantic ambiguity. Modularity and reusability are two key characteristics that make ontologies a proper tool to be used inside the platform. In particular, the Ontology is essential for the right interpretation of the information of the Urban Dataset structure. The Ontology has been formalized in the OWL language and made available online.

The Information Level defines the format able to make interoperable the exchange of significant information about the city, namely the Urban Dataset, among heterogeneous systems or applications. The format has been thought to be self-consistent (it is mandatory to be compliant with the SCP specifications but it can be adopted also outside the framework) and it is defined by an Abstract Data Model and the syntactic implementation of the model in XML (eXtensible Markup Language) and JSON (JavaScript Object Notation). The Abstract Data Model gives a syntax-independent representation of the content that a document, used to exchange Urban Dataset, must have. Because the Smart City scenario is under continuous evolution, it has been designed to be scalable during the time and across different contexts.

The Communication Level (or data transport) defines the actions that a vertical Solution can undertake in order to communicate with the Smart City Platform, which are: the production of Urban Datasets (information which a solution exports towards the Smart City Platform) and the access to Urban Datasets (a solution retrieves information that the Smart City Platform has previously published). Therefore, if the first action enables the retrieving of data from a Solution to the Smart City Platform and the second action enables transfer of data from Smart City Platform to a Solution, the coordinated set of the two actions allows the exchange of data from a vertical Solution to another vertical Solution through the Smart City Platform. The communication, to perform the two actions above, must be defined through the configuration of three aspects: the Architectural Patterns (Request/Response, Push, Publish/Subscribe), the Web Service Interface and the Transport Protocol.

All these features of the SCP specifications are modular; it means that each stakeholder can implement not necessarily all of them but also a subset of them.

Thus, as already stated, since the absence of a univocal and 'holistic' standard, what we have proposed here is a high level reference framework aimed at enabling interoperability in Smart Cities, i.e. able to include all the multiple aspects of the Smart City with a coherent vision. For the future this means the need to face the issue of standardization, confronting emerging initiatives or even proposing the SCP Specification as a contribution to the construction of standards and it requires the engagement of the main national/international standardization initiatives (as the IES-City initiative led by NIST) in order to achieve a convergence on all the technological issues related to the proposed model.

This convergence process through open specifications is the opposite of an imposition of a single centralized and proprietary solution; in fact it allows existing solutions to join without having to be replaced by another technology and allows interested parties to contribute on specific aspects of the overall design.

Nevertheless it is not enough. If we state that the public administrations can play a key role in breaking the barriers among the Smart City subsystems, they should be provided with a common global set of standards and guidelines to be used in the call for tender.

At the moment the specification are being experimented in a laboratory context, implementing the communication among existing testbed applications (for example smart buildings within ENEA site and another existing WebGIS application, and so on). The first results are very encouraging. Moreover the approach has been shared with an important group of stakeholders (including city, industry and research representatives) in an ENEA initiative called Italian Convergence Table and it has received a very positive welcome. The next step will be the experimentation in a real urban context. In this phase it is expected to get a credible evaluation of the approach and to be able to identify possible criticalities in it and to modify the specifications in order to solve them.

Lastly, SCP Specifications do not aim to solve all the issues related to communication in the Smart City field, but the potential impact, at the national or European level through a first convergence path based on interoperability, can be noticeable in socio-economic terms and in the efficient energy management, pursuing an improvement in the quality of life of the citizen.

References

1. A. Caragliu, C. Del Bo, P. Nijkamp, Smart cities in Europe. J. Urb. Technol. **18**(2), 65–82. https://doi.org/10.1080/10630732.2011.601117
2. Smart Cities, Ranking of European Medium-Sized Cities (2007), http://www.smart-cities.eu/
3. X. Masip-Bruin, G.-J. Ren, R. Serral-Gracia, M. Yannuzzi, Unlocking the value of open data with a process-based information platform, in *IEEE 15th Conference on Business Informatics (CBI), Vienna*, 15–18 July 2013, pp. 331–337
4. Global City Team's Challenge Super Cluster Activity, https://www.nist.gov/el/cyber-physical-systems/smart-americaglobal-cities. Accessed 15 Jan 2018
5. A. D'Acunto, M. Mena, G. Di Pasquale, A. Bua, M. Cavalli, S. Contini, S. Dafano, C. Riso, S. Ital, Rapporto Smart City Index 2016, EY
6. IEC, Orchestrating infrastructure for sustainable Smart Cities. White paper (2014), http://www.iec.ch/whitepaper/smartcities. Accessed 13 May 2017
7. H. Panetto, J. Cecil, Information systems for enterprise integration, interoperability and networking: theory and applications. Enterp. Inf. Syst. **7**(1), 1–6 (2013), https://doi.org/10.1080/17517575.2012.684802

8. D. Chen, Enterprise Interoperability Framework, in *Proceedings of Enterprise Modelling and Ontologies for Interoperability*, EMOI-Interop (2006)
9. European Interoperability Framework—Implementation Strategy, European Commission, Brussels, 23–3-2017
10. G. Locke, P.D. Gallagher, NIST Special Publication 1108—NIST Framework and Roadmap for Smart Grid Interoperability Standards, Release 1.0. NIST (2010)
11. C. Agostinho, Y. Ducq, G. Zacharewicz, J. Sarraipa, F. Lampathaki, R. Poler, R. Jardim-Goncalves, Towards a sustainable interoperability in networked enterprise information systems: trends of knowledge and model-driven technology, in *Computers in industry*, vol. 79, June 2016, pp. 64–76
12. R.G. Hollands, Will the real Smart City please stand up? in *City: analysis of urban trends, culture, theory, policy, action*, vol.12, no. 3 (2008) pp. 303–320
13. V. Bhatt, A. Brutti, M. Burns, A. Frascella, An Approach to Provide Shared Architectural Principles for Interoperable Smart Cities
14. Reference, CEN-CENELEC-ETSI Smart Grid Coordination Group, "Smart Grid Reference Architecture, http://gridscientific.com/images/Smart_Grid_Reference_Artichtecture.pdf. Accessed Nov 2012
15. M. Gottschalk, M. Uslar, C. Delfs, Smart city infrastructure architecture model (SCIAM), in The Use Case and Smart Grid Architecture Model Approach (Springer, 2017), pp. 75–76
16. C. Neureiter, S. Rohjans, D.E. Dominik, C. Danejas, M. Uslar, Addressing the complexity of distributed smart city systems by utilization of model driven engineering concepts, in *Conference VDE Kongress 2014 At Frankfurt*
17. CEN CENELEC Web Site, CEN-CENELEC-ETSI Sector Forum on Smart and Sustainable Cities and Communities, https://www.cencenelec.eu/standards/Sectors/SmartLiving/smartcities/Pages/SSCC-CG.aspx. Accessed 22 Jan 2018
18. J. Rodríguez-Molina, J.-F. Martínez, P. Castillejo, R. de Diego, SMArc: a proposal for a smart, semantic middleware architecture focused on smart city energy management. Int J. Distrib. Sens. Netw. (2013)
19. M. Jang, S.-T. Suh, U-City: New Trends of Urban Planning in Korea based on Pervasive and Ubiquitous Geotechnology and Geoinformation, in *ICCSA 2010*
20. ITU-T, Smart sustainable cities — a road map. ITU News, https://itunews.itu.int/En/5217-Smart-sustainable-cities-a-road-map.note.aspx (2014)
21. G. Aiello, M. Alessi, L. Marasso, City Enabler: a FIWARE based tool for crawling, collecting and rendering on a map valuable data at urban scale. ngineering Ingegneria Informatica S.p.A, Research and Development Laboratory, Palermo, Italy (2017)
22. E. Leone, Open City Platform—Il Cloud Computing per Smart Government, Firenze, 4 Aprile 2013
23. C. Badii, E. Bellini, P. Bellini, D. Cenni, A. Difino, P. Nesi, M. Paolucci, Km4City: una soluzione aperta per erogare servizi smart city, in *GARR-Conf16-proceedings*
24. C. di Milano, Milano Smart City—Progetti e principali interventi (2013)
25. R. Liviu, DUMITRU, IoT platforms: analysis for building projects, Inform. Econ. **21**(2) (2017)
26. M. Ben, Y. Alayaab, T. Banouarab, C. Monteilab, K.Drira Chassotab, OM2 M: extensible ETSI-compliant M2 M service platform with self-configuration capability. Proc. Comput. Sci. **32**, 1079–1086 (2014)
27. C. Perera, P. Jayaraman, A. Zaslavsky, P. Christen, D. Georgakopoulos, Dynamic configuration of sensors using mobile sensor hub in internet of things paradigm, in *Proceedings of the IEEE 8th International Conference on Intelligent Sensors, Sensor Networks and Information Processing (ISSNIP)*, Melbourne, Australia, Apr 2013, pp. 473–478
28. M. Turck, Internet of Things: Are We There Yet? (The 2016 IoT Landscape) (2016)
29. BSI, Mapping City Standards (2016)
30. The Internet of Things-Enabled Smart City Framework, or IES-City Framework (pronounced "YES-City"), NIST, 2017, https://s3.amazonaws.com/nist-sgcps/smartcityframework/files/IoT-EnabledSmartCityFrameworkWP.pdf

31. NIST, NIST Framework and Roadmap for Smart Grid Interoperability Standards, Release 1.0 (2010)
32. M. Gottschalk, M. Uslar, C. Delfs: Use Cases—The IEC 62559 methodology, in *The Use Case and Smart Grid Architecture Model Approach* (Springer, 2017) pp. 11–39
33. F. Buschmann, R. Meunier, H. Rohnert, P. Sommerlad, M. Stal, Pattern-oriented software architecture, in *A system of patterns,* Vol. 1
34. READY4SmartCities FP7 CSA, Project Reference: 608711, Subprogramme Areae: FP7-SMARTCITIES-2013. Available at the address, http://www.ready4smartcities.eu/
35. V. Guallart, F. Meneses, D. Frogheri, D. Ibañez, R. Rubio, F. Giralt, "City Anatomy: A Framework to support City Governance, Evaluation and Transformation"
36. Km4City: Smart City Model and Tools Main Documentation Web Page, http://www.disit.org/drupal/?q=node/6056
37. H. Rijgersberg, M. van Assem, D. Willems, M. Wigham, J. Broekstra, Top, "Ontology of units of Measure (OM)", https://github.com/HajoRijgersberg/OM#om

Integrated Cyber Physical Assessment and Response for Improved Resiliency

P. Sivils, C. Rieger, K. Amarasinghe and M. Manic

Cyber-physical systems (CPS) are control systems that facilitate the integration of physical systems and computer-based algorithms. These systems are commonly used in control system and critical infrastructure for control and monitoring applications. The internet-of-things (IoT) is a subset of CPS in which multiple physical embedded devices and sensors are connected via a distributed network to communicate and transfer data while being driven by computational algorithms for data delivery and decision-making tasks. The various devices in IoT results in a large amount of heterogenous data that is handled by the system. The system needs to ensure that the provided data is trustworthy in order to ensure that the system is working efficiently and producing accurate, actionable outputs. Additionally, human-in-the-loop elements are crucial to any CPS infrastructure, especially IoT and must be considered. This chapter provides a summary and analysis of crucial concepts in under-standing cyber-physical degradation assessment, heterogeneous data-fusion, and visualization under a smart city IoT architecture. These concepts will provide a basis for enhancing the effectiveness of human response to physical and cyber-events within the scope of smart city infrastructure.

1 Introduction

Cyber-physical systems (CPS) are commonly used architectures for critical infrastructure applications such as smart urban environments. The Internet-of-Things (IoT) is a network based architecture in which physical devices, sensors, embedded elec-

P. Sivils (✉) · K. Amarasinghe · M. Manic
Virginia Commonwealth University, Richmond, USA
e-mail: sivilspg@vcu.edu

C. Rieger
Idaho National Laboratory, Idaho Falls, USA

F. Cicirelli et al. (eds.), *The Internet of Things for Smart Urban Ecosystems*, Internet of Things, https://doi.org/10.1007/978-3-319-96550-5_3

tronics, software, vehicles, etc. can communicate and transfer data with each other over an internet connection. IoT technology is a subset of CPS in which physical procedures are controlled and affected by computational processes while the computations are simultaneously altered based on the physical system state. In expansive CPS designs, such as those required for IoT enabled urban ecosystem applications, expansive networks are implemented to collect, transfer, monitor, and analyze data. This data is processed for automated control applications, generate predictive models, and provide enhanced understanding for operators and consumers [1].

For cyber-physical systems, as in control systems and critical infrastructure, ensuring the resilience of a system is critical to its long-term efficiency and security. CPS uses many automated systems to perform tasks and handle anomalies. These systems can be very complex, have different methods of integration, and involve varying levels of human interaction. Resilience is a multi-disciplinary effort that ensures changes and anomalies experienced by a system are tolerated by the system design. Resilient control systems maintain system monitoring, awareness, cyber-security, and decision making at an acceptable level of operation to facilitate normal and necessary functions.

Ensuring the trustworthiness of data coming from the various devices, sensors and network traffic should be one of the top priorities in designing any cyber-physical system (CPS). Trustworthy sources ensures that the available data is being used efficiently and the outputs are accurate. This is especially important in smart city infrastructure where outputs and decisions affect people's daily lives. Untrustworthy data can lead to poor automated system control, difficulty in decision making, and frustrating end-user experiences. This data is very often very large in scale and comes from heterogeneous sources. Moreover, information technology (IT) and operation technology (OT) operators who now find themselves responsible for cybersecurity come from a variety of backgrounds, differing decision support requirements, and knowledge capabilities. To effectively abstract the complexity of cybersecurity and simultaneously address the variety of roles, knowledge, and need, a design is needed that performs much of the required analysis for the user and presents only relevant information in a consistent way. IoT infrastructure on a city-wide scale requires many different data nodes in the form of devices and sensors to perform specific, individualized tasks. This means that the datasets created from a city-wide IoT system will be high-dimensional and heterogeneous.

This chapter provides a summary and analysis of crucial concepts in understanding cyber-physical degradation assessment, heterogeneous data-fusion, and visualization under a smart city IoT architecture. These concepts will provide a basis for enhancing the effectiveness of human response to physical and cyber-events within the scope of smart city infrastructure. It is important to understand that CPS degradation analytics provide the source information that a data-fusion engine will use to tailor context awareness to the human. Visualization presents this information to ensure a reproducible response for each operational role regardless of background (e.g., cyber, operational, scientific, etc.) or level of performance or the humans involved at any particular time.

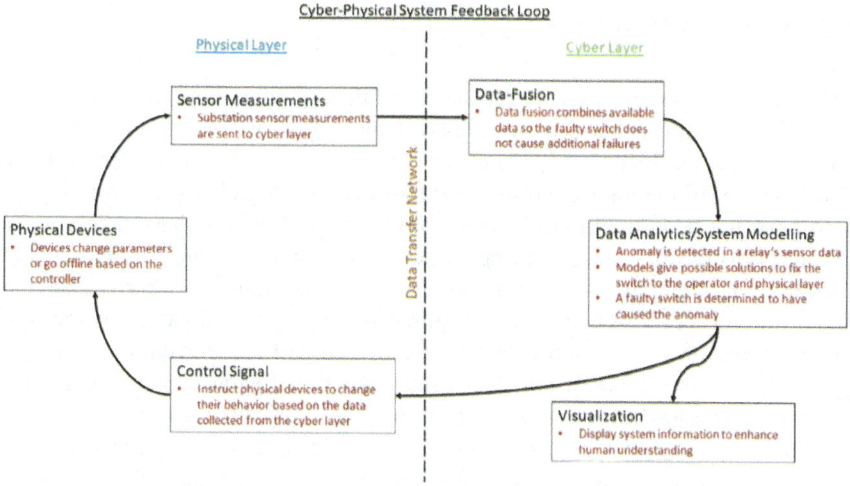

Fig. 1 CPS data feedback loop between the cyber and physical layers

1.1 *What Are Cyber-Physical Systems?*

Cyber-physical systems are usually used to monitor and control critical infrastructures. Examples include smart grids, autonomous vehicles, and smart buildings. CPS combines physical components and computer-controlled algorithms for monitoring and process control. CPS integration is achieved through the implementation of feedback loops in which both the cyber and physical aspects of a system affect each other (Fig. 1). The physical processes are monitored and controlled by the computational algorithms embedded in the cyber aspects of a CPS, and computations are simultaneously altered by the physical state of the CPS. This is accomplished through data analysis, system modelling, both data-driven and physical, and data fusion to enhance knowledge discovery from heterogeneous sources of cyber-physical data. In general, OT systems are one type of CPS. Processing heterogeneous, high-dimensional data from acquisition through data-fusion is a critical task that must ensure capability and scalability in CPS [2, 3].

The IoT is a subset of CPS. IoT architectures establish communication networks between multiple decentralized, heterogeneous CPS. In this sense, the IoT is the second layer of CPS that enables digital integration and communication. IoT applications are often provided by embedding sensors, software, devices for computational or networking applications, etc. into existing objects within CPS architectures. These are usually physical objects that are not usually designed for computational tasks (appliances, toys, vehicles, etc.). The communication and data-transfer between the various IoT devices takes advantage of pre-existing network infrastructure such as the internet, reducing overhead for large-scale distributed system applications [1].

1.2 Challenges in CPS

Concern over cyber-attacks has led to the thoughtless proliferation of tools focused on addressing pressing cybersecurity needs without long-term considerations. Much of what has been developed originates in the IT sector, and has been inherited with little customization by the control systems world. As a result, control systems professionals who have not been traditionally responsible for security, now have a role in the cybersecurity of OT systems, and they lack tools customized to the OT environment. Therefore, it is safe to assume that various roles exist within the security equation, and a different level of expertise is required for the many individuals working in these complex systems. Even within IT, the diversity of manufacturers, number of cyber security appliances installed, and sheer number of parameters monitored can create data overload in the most adept user. This overload situation can grow as an increasing number of security solutions are fielded to protect the system, each with its own stream of monitoring data. The resulting data deluge likely produces many false negatives because humans cannot examine all the data that has become available. Additionally, the common use of overly conservative alarm thresholds produces numerous false positives that human controllers learn to ignore since the majority are frivolous.

Based upon cyber, physical, and interacting cyber-physical characterizations of both host and network patterns, CPS models can be used to distinguish conditions and behaviors indicative of a cyber-attack from benign, unintended actions or physical failures. As a feedback loop to recognize performance, latency and integrity will provide the fundamental attributes that will be correlated by measurable cyber-physical parameters [2, 4, 5]. Through the use of leveraged and data-driven models, cyber-physical parameters specific to use cases can be evaluated to correlate performance impact. That is, given a change (normal or abnormal) in the cyber-physical environment, unacceptable variations will be evaluated and codified. The resulting process will establish a network performance baseline, a direct measurement of resilience, and a diverse approach to recognizing distributed threats across the interwoven layers of the OT/IT architecture.

Computer networks remain the primary vector for cyber-attacks, and yet detecting cyber-attacks over computer networks remains limited—"misuse-based intrusion detection" [6], which relies on static "signatures" of "known bad" activity, has significant value, but also remains blind to unknown attacks, and frequently even to slight variations on attacks. "Anomaly-based intrusion detection" [6], which analyzes statistical variations from "normal," helps to address some of these issues [7], but is limited by the massive imbalance in "good" data to "bad" data in-training sets, the high cost of false positives to human operators, the so-called "semantic gap" between flagging anomalous events and understanding the cause of those events, and the raw diversity of network attacks [7]. Moreover, the extremely high and diverse types of network traffic and computing environments present in IoT infrastructure makes these problems even worse. Indeed, many commercial solutions are ineffective since these are often either not capable of performing on networks with bandwidths as high as those on IoT networks, or not tuned for such environments. A basic misapplication of the anomaly-detection approach is to assume that all anomalous behaviors are necessarily suspicious. The terms, "anomalous," "errant," and "malicious" all have different meanings, but "malicious" carries a value judgment with it. Normal activities may vary significantly with software upgrades and network changes.

Combining multiple heterogeneous data sources can introduce multiple challenges for data-fusion, system-modeling, and visualization applications. The collected datasets may contain information with varying resolutions, is incomplete or uncertain, or is unsynchronized due to various offsets in the measurement devices. Every node in a CPS, whether it is recording data or providing an end-user interface, will have its own method of recording and saving data. So within a large-scale IoT system, the vast amounts of available data sources can provide their data in many different formats that must be combined so that an analytics model can make sense of it. Additionally, the available data is a mix of cyber-data, e.g. network traffic, and physical data, e.g. temperature readings. Both types of data are part of the same system and should be looked at as a whole, rather than separate entities with their own architecture. Data-fusion techniques combine heterogeneous sources of data into a new representation by exploiting existing interdependencies in the dataset. These techniques are used to help improve the performance, scalability, and reliability of the control and monitoring analytics systems that implement these datasets.

2 Cyber-Physical Analytics for Resilience and Assessment

This section outlines the anomaly detection analytics to be used in combination with high-fidelity models to recognize and mitigate cyber-attacks and enhance system resilience. Providing complete, reliable, and actionable information is essential for system resiliency and decision-making within an IoT system. In order to ensure the trustworthiness of available sensors in an IoT system design, analytics that ascertain accurate health measurements of sensors is essential. Failures within highly interdependent and complex environments can lead to cascading adverse consequences. As the digital footprint of these environments has continued to evolve, the potential ramifications of conjoined CPS failures has not been considered and has instead become more obscure. If recognized and characterized quickly and consistently at the source, however, the adverse effects can be localized and cascading failures prevented. With this focus in mind, methodologies should be implemented to characterize a diverse range of behaviors found on OT/IT networks and classify them according to their degree of normality. Application of these methodologies will lead directly to measurable improvements in system resilience.

A cyber-physical approach towards IoT infrastructure will notably ascertain degradation—both cyber and physical—to distinguish cyber-attack from physical failure. Information on blended security attacks (both cyber and physical) should also be characterized. Analytics systems should remain robust under various degradation scenarios resulting in partial or unreliable information. IoT enabled sensors and devices, as well as the networks between them are susceptible to malicious tampering, unforeseen failures, and accidents; as well as degradation that arises naturally from normal operations. Robust cyber-physical analytics designs should enhance the response of decision-makers by identifying and providing actionable information about how and where degradation is occurring.

Computer networks within a system are often the primary vector for cyber-attacks on IoT infrastructure. However, established intrusion detection system (IDS) methodologies are limited when it comes to distinguishing between anomalous and malicious network behavior. "Misuse-based intrusion detection" requires a priori knowledge of network traffic patterns that indicate malicious activities. This type of network monitoring, while useful, is unable to detect unknown, or sometimes slightly altered, cyber-attacks. Another IDS scheme, "anomaly-based intrusion detection," implements various statistical metrics to determine how far off network traffic is from baseline activity. Statistical analysis is useful for mitigating the problems inherent in misuse-based IDS, but it introduces another problem with differentiating between non-malicious anomalous traffic patterns from cyber-attacks identified by the IDS. Due to vast diversity of network traffic patterns, which is further exacerbated by the vast diversity of cyber-attacks and the large amounts of traffic inherent in a city-wide IoT system, many anomalous patterns can emerge that are not malicious in nature. Since the simple approach of assuming all anomalous network traffic is malicious is impractical at scale, false-positives flagged by an IDS requires some human analysis to understand the causes of the events and to provide the necessary response. This presents a further problem as backgrounds, knowledge, and physical capabilities vary among individuals. As such, wide-scale IDS systems for IoT applications should aim to provide as few, if any, false positives as possible. These systems should be able distinguish degradation arising from cyber-attacks and physical failures. Additionally, a design focus on quickly identifying and localizing failures enables fast and accurate response times for human operators. A design focused on localization can also help mitigate cascading failures. As IoT systems grow larger and more interconnected, the potentiality of one failure leading to other, possibly unforeseen, failures in multiple other areas increases.

2.1 State Awareness and Anomaly Detection

Developing an accurate state awareness system with robust anomaly detection techniques is crucial to providing users with the relevant information and models needed to recognize threats and coordinate effective responses. These systems are typically implemented either through data-driven or physics-based models. Data driven models encompass computational intelligence (CI) and statistical models. These designs use collected data to generate models that represent the system as it should be when fully-functional. These state estimations can be compared to current system states which can help identify when and where system anomalies are occurring and predict how a system will behave in the future given the current state. Additionally, there are data-driven models that use 'online' learning algorithms. These algorithms continue to refine their initial models using new data collected after deployment. These algorithms allow the models to be further refined over time to better handle new or unforeseen challenges/attacks. While data-driven system models are powerful tools for IoT analysis applications, there are some potential issues that should be kept in mind during the design process. Data-driven approaches use some form of pattern matching to build their models and therefore require large datasets describing the systems activity, with data on multiple safe/unsafe scenarios often being required. Not having enough training data can result in a model with poor performance leading to unsatisfactory results. This means that designing useful models can require extensive, potentially cost-prohibitive, data collection. The pattern matching nature of data-driven models means they are also susceptible to overfitting, where the trained model too closely resembles the training data. Overfitting can cause the model to have very little room for error, meaning normal data patterns that were missing from the training data could be flagged as abnormal resulting in many false positives. Additionally, pattern matching schemes are susceptible to spoofing. Spoofing is when an attacker reverse engineers the model allowing them to know what inputs are needed to create a desired output. In CI applications, most CI models are 'black-box' algorithms. This means that observers can only know the inputs and outputs of the model but cannot get information about the model's inner workings. This makes understanding and explaining the behavior of a model extremely difficult, if not impossible.

Computational intelligence (CI) and machine-learning methods have been used to provide anomaly detection systems for a variety of applications. Artificial neural networks (ANNs), such as multilayer perceptron (MLP), are a popular machine-learning technique that have been shown to provide excellent results in anomaly detection applications. In cybersecurity, ANNs have been used in anomaly-based network intrusion detection systems that can handle a large variety of cyber-attacks [8]. Designs were presented in literature that were able to achieve very high detection rates (~99%) [9–12] on various datasets. Additionally, [10, 13] presented ANN designs that can not only achieve high-detection rates, but also report very few false positives and false negatives (~0.1%). Recently, [12] presented an improvement to standard MLP anomaly detection in medical CPS by reducing the detection process to multiple 2-class classifications to improve accuracy and reduce false reports. ANNs used for anomaly detection systems in control system architectures, such as smart grids, have been presented with relatively high detection rates [14–16], though the reported false alarm rates are often high as well. Integrity attacks can compromise a system by spoofing system data so that it appears to an operator that the system is in a safe-state, while an attacker gains access to critical resources. An ensemble modeling design, where multiple models are aggregated to increase accuracy, was presented in [17] for these types of anomalies that performed better than other prior MLP methods with a small percentage of false reports (~2%).

Another area of CI that has shown promise in CPS anomaly detection is fuzzy logic. Fuzzy logic systems (FLSs) embed human understanding and knowledge into a system in the form of 'if-then' rules. These rules are then converted into crisp values, which can then be used for decision-making tasks. Traditionally, the rule base for a FLS is derived from expert knowledge of the system, though since some CPSs can be extremely complex, various publications have explored adaptive fuzzy architectures that can generate their own rules [18] or allow them to be dynamically altered [19]. FLSs are grey-box algorithms, where some inner workings of the system can be known but other aspects remain unknowable. This is an improvement over other CI algorithms in terms of understandability and system awareness.

Recently, deep-learning neural network algorithms, such as convolutional and recurrent neural networks, have emerged as the state-of-the-art in machine-learning due to their ability to produce more robust models than standard ANN architectures. Literature has presented deep learning algorithms for cyber-attack anomaly detection, showing robust detection against a variety of attacks on CPSs [20, 21]. Additionally, a simple convolutional neural network design was presented in [22] to monitor motor conditions in real-time with a high accuracy when tested on real-world motor datasets.

Classical CI algorithms are still in use, though in recent research they are mostly used for comparison and validation of newer CI methods. These include K-Nearest Neighbor (KNN) [23], Support Vector Machine (SVM) [23, 24], and Self-Organizing Map (SOM) [25, 26].

Various other statistical modeling algorithms have been explored for enhancing anomaly detection applications. These algorithms do not necessarily fall under the CI category, though they are similarly data-driven, and are often used alongside CI classifiers to enhance performance. Brahma et al. and Biswal et al. [24, 27] present a detailed comparison of multiple statistical modeling methods. Multiple experiments were run using Discrete Fourier Transform (DFT), Discrete Wavelet Transform (DWT), Principal Component Analysis (PCA), S-Transform, and Shapelet algorithms on Phasor Measurement Unit (PMU) fault and generation-loss datasets. Each method was verified using various KNN and SVM classifiers. The authors show that the more recent Shapelet methods outperformed all other compared methods. Interestingly, the authors also note that feeding raw data into the classifiers often outperformed classic statistical methods. PMU data was also used in [23] to test a kernel principle component analysis (kPCA) method for anomaly detection in high-resolution micro-PMUs. The kPCA was combined with a novel 'partially hidden structured' SVM to classify the type of anomaly detected. The authors showed the combined algorithms outperformed standard decision algorithms such as Ada Boost and Decision Trees.

These methods have also been explored for the detection of cyber-anomalies. A Reduction of Quality (RoQ) attack detection method is presented in [28]. Since a quality drop in network traffic does not always imply that a system is under attack, a more robust algorithm is needed to detect anomalies, as well as identify that the anomalies are malicious. The two-stage design uses wavelet analysis to detect abrupt changes in quality in the first step. Autocorrelation analysis is then used to identify attack characteristics in the local network traffic. The proposed design achieved a 3% false negative rate with 0% false positives. A Chi-square statistics algorithm was proposed by [29] for tiered intrusion detection applications with multiple alarms. The

Chi-square algorithm achieved detection rates of 71–100%, depending on various conditions, with no false positives.

Physics-based models implement prior expert knowledge to exploit known relationships in the available data to detect when an anomaly has impacted the system. Their advantage over data-driven models is that they are white-box by nature, allowing users to have access to all of the information of the inputs, outputs, and inner-workings of the model. Another advantage of physics-based models is that they do not rely on data for training meaning no data collection is necessary. However, the required expert knowledge needed to design an adequate model may not always be readily available. Physics-based models can also miss useful intricacies in a system that data-driven models can exploit. Hybrid models that incorporate aspects of both data-driven and physics-based models can help alleviate the weaknesses found in both modeling schemes.

Physics-based modelling techniques can be seen in [24, 30] where the authors presented a physics-based method that uses energy information from PMUs to reconstruct a model of the system as a whole that can then be used to detect anomalies present in the system. A physics-of-failure mechanism is presented in [31] using particle-filtering, a probability density algorithm, to detect anomalies in brushless DC motors. A different model-based anomaly detection method is proposed in [32]. This method uses 'gaps' between data points to define and detect anomalies in the data. This approach does not require any prior knowledge of the system and uses only a local subset of data points. The authors detail its potential for anomaly detection in big data and data stream applications due to its computational efficiency.

2.2 Distributed Systems

The distributed nature of IoT systems for city infrastructure applications adds its own challenges in the development of data analytics systems. The many devices and sensors in a city-wide IoT system can be spread out across large distances, and all of this data needs to be collected, analyzed, and acted upon quickly. Classic distributed system architectures implemented a central data-fusion center where all data is funneled to a single location, analyzed, and output analytics are sent back out to the system and users. While this approach is 'efficient' in that only one location has to be built, maintained, and protected, there are some downsides that become more problematic on large-scale distributed architectures. First, data transfers take time. In a city-wide IoT, emergency decision making requires all relevant information be provided as quickly as possible. This means that data transfers need to be as fast as possible to enable real-time analysis and system monitoring. Second, central data-fusion centers inherently create single-points-of-failure across the system. This increases the chance of crippling security risks and makes localizing degradation difficult. Lastly, not all data may be necessary for global degradation analysis. Certain data-streams from devices and sensors in the network may only be relevant to the health of their respective device or subsystem meaning including them in a global degradation analysis design may be unnecessarily increasing the computational overhead of the system.

A common design method used to alleviate these issues is to decentralize the system such that multiple localized subsystems perform their degradation analysis as independently as possible. Using decentralized subsystems allows for shorter data transfers as each subsystem data center should be local to the data sources fed into it. Subsystems reduce single points of failure as each subsystem can continue to run independently if another subsystem is shut down. Subsystem analytics could also help identify the source and cause of the degradation. The computational overhead inherent in high-dimensional heterogeneous data sets can be reduced since each subsystem only uses the data it needs. Additionally, if a central data-fusion center is needed for global degradation analytics, subsystems can report device and sensor information as needed, or only provide a subsystem overview reducing the amount of network traffic and necessary computational overhead and the central location. However, it should be kept in mind that in any complex system like smart city IoT infrastructure, crucial interdependencies may exist across all data sources. In a system where degradation in one area can have huge impacts on other aspects of the system and/or the wellbeing of city residents, care must be taken to ensure the communication of the interdependencies between subsystems is not lost in the architecture design or within the implementation of knowledge- and data-fusion techniques.

A fully decentralized CPS design is proposed in [33] that uses relay-assisted sensor networks. Their design makes accurate estimations by only exchanging information between neighboring sensors, using relay nodes to transmit information to the rest of the network. The authors show their design scheme can handle sensor failures, fading channels, and noisy data without making assumptions about the communication topology, which further enhances resilience. Another design for a decentralized resilient monitoring system is presented in [34]. The system quantifies the trustworthiness, or data quality, of the sensors by comparing readings to a known trustworthy source. The system is divided into subsystems through process-variable probabilistic-mass-function adaptations to alleviate the high-dimensionality of data is CPS, and knowledge fusion techniques are incorporated to ensure important interconnected information between subsystems is not lost. Distributed CPS designs also have a layer of communication that must be taken into account. When parts of a CPS are separated, potentially by large distances, reliable communication is just as important as the control algorithms. Cao et al. [35] introduced a joint optimization framework for control and communication. It proposes a simulated annealing-based optimization approach to minimize the number of control tasks sent, with the sub-objective of minimizing energy consumption across devices during communications, and was successfully implemented on a heating, ventilating, and air conditioning (HVAC) dataset.

The increasing interconnectedness of information and communication technologies in CPS make in-depth design analysis crucial to ensure that both physical safety and cybersecurity requirements are met. Friedberg et al. [36] presented a formal methodology to integrate safety and security analysis into a single framework. Their method, System-Theoretic Accident Model for Safety and Security (STPA-SafeSec), offers a top-down approach to analyze and identify constraints in both areas of cyber-physical systems simultaneously. STPA-SafeSec focuses on the desired outcomes of a system, rather than using existing threats as the basis for security requirements. The framework was applied to various scenarios in micro-grid systems and was shown to help system analysts identify critical system components along with safety and security impacts arising from specific vulnerabilities.

2.3 Resilient Monitoring

In CPS, particularly critical infrastructure, safeguards must be taken to keep the system stable despite device degradation from malicious attacks or natural damage/degradation. Resilient system-state awareness, estimation, and anomaly detection methods are crucial for system monitoring and decision-making under highly dynamic conditions. Additionally, CPS can encompass large numbers of distributed networks. These networks can contain different data types and may span large distances, increasing both the computational and temporal overhead needed to perform analysis. Many CPS use a central data-fusion center to perform tasks, though this can create a single-point-of-failure for the entire system.

As with anomaly detection, CI techniques have been a promising area for resilience in state-of-the-art CPS architectures. Wijayasekara et al. [18] presented a fuzzy-neural data-fusion engine to model systems to create resilient state awareness. The design implements an ANN that models current and future system states based on historical data to enhance state awareness when system data is unreliable or unavailable. This was implemented alongside standard threshold-based alarms and a fuzzy logic-based anomaly detection system to further enhance the state awareness architecture. Similarly, an adaptive neuro-fuzzy controller is presented in [19] for use in a nuclear power plant and was shown to be highly tolerant to faults in a variety of control tests.

A model-based state estimation method using Satisfiability Modulo Theory (SMT) is presented in [37]. SMT uses first-order logic to create and verify models. The proposed design models a CPS using Boolean and convex constraints; then, the authors' Imhotep-SMT uses the constraints to estimate the state and identify which sensors are under threat. This method was verified using simulated data, as well as successfully controlling an unmanned ground vehicle under adversarial attacks and noisy sensor data.

CPS resilience can also be improved through the design choices made. Fawzi et al. [38] described a theoretical framework for secure state estimation in CPS. This work discusses two ideas for system design towards state estimation: (1) system-states cannot be accurately reconstructed if more than half of the sensors are under attack, and (2) if a system can be stabilized in the presence of sensor attacks, then its state can be accurately estimated as well. The authors use this framework to show the importance of designing resilient controllers that can be used for secure state estimation, rather than separately implementing secure controllers and secure estimators.

3 Data-Fusion and Data Alignment

This section provides an overview of challenges and possible solutions for data-fusion in the domains of IoT and CPS. Data-fusion is the process of combining the available data streams into new, less unwieldy representations by exploiting existing interdependencies in the dataset [18]. These representations are then used to increase the reliability and consistency of state awareness, control, modeling, and predictions for a CPS. Datasets generated by large scale CPS are usually high-dimensional (data is collected from multiple sources) and heterogeneous (data of different types and modalities is collected and reported). Furthermore, the collected data can have low or multiple resolutions, report events out of sync with other data sources, or just be missing.

Human effectiveness is a challenge in all contexts of human interface with information technology (IT) and operation technology (OT), including cybersecurity. Even within common role and responsibility area, the backgrounds, best methods of learning/comprehension, and physical performance vary among individuals. In moving towards a repeatable and low latency response to cyber-attacks, both human and automated response actions to cyber-events must be considered. This includes consideration of the most resilient effectiveness that can be achieved by adding human influence within an OT system. The human aspect of OT systems requires an extension of research and evaluation of the fusion and presentation of cyber-physical analytics that characterize the cyber-posture. Considering the importance of the cognitive and social environment, multidisciplinary exercises are planned to characterize the performance of research solutions. To this end, data-fusion designs should be created with a focus on delivering appropriate representations of information to the operator/consumer (Fig. 2).

An extensive survey on recent data-fusion techniques and applications can be found in [1].

3.1 Multi-modal Data

Multi-modal data describes data that is collected from multiple sources under varying conditions, such as data acquisition techniques in CPS. System modeling and decision-making tasks often rely on high-frequency data to achieve sufficiently accurate results. Data sources can have different resolutions, or sampling rates, meaning that one sensor may record data at 100 times per second, while another sensor records data only once a second. This is detrimental to the performance of the system as the analytics system will be forced to compare 99 fresh data points from the first sen-

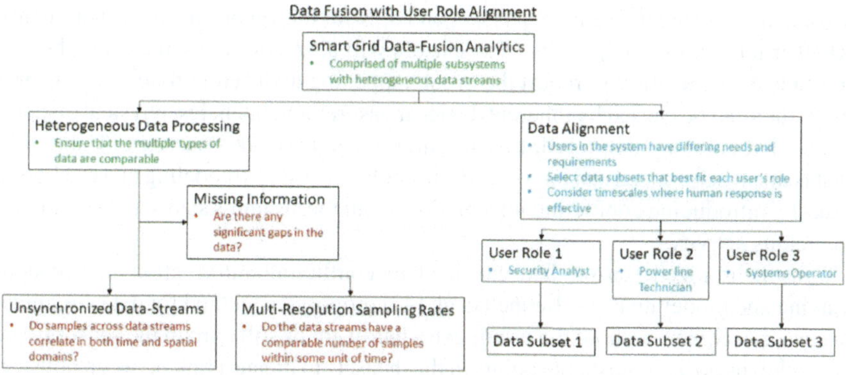

Fig. 2 Data Fusion with a Focus on User Role Requirements

sor to the same static data point collected from the second sensor. Any important concurrent changes or interdependencies between the two sensors is completely lost within that second until the second sensor makes its next measurement. It is often too expensive and/or time consuming to ensure that all devices meet the same standards, even more so if part of the infrastructure is already implemented. As technology improves, newer devices may be introduced to the system that outperform the old ones and not every device can, or even needs to, be consistently upgraded to keep up. Therefore it is necessary to have a software solution to handle low/multiple sample rates to ensure that there exists enough data points from each sensor to generate accurate analytics as often as needed.

A method using ANNs is presented in [39] to increase state awareness by increasing the spatial resolution of data. The authors' method implemented data downscaling to gain increased spatial resolution, and they validated their algorithms on real-world CO_2-concentration datasets. Similarly, [40] presented an online ANN method that can predict high-resolution temporal data using lower resolution sensors. The authors validated their algorithm on a real-world building energy management dataset, showing their method was more accurate than classical predictive models and that it could adapt to changes in building behavior. Another solution to the problem of data-fusion for multi-resolution data is presented in [41]. The proposed surface modeling algorithm combined features with varying resolutions through surface reconstruction and registration using Gaussian Process (GP)-modeling. The algorithm was validated on simulated and measured multi-surface data, and was significantly faster than the commonly used weighted least squares data-fusion (WLSDF).

Due to the heterogeneous nature of IoT data, it is important to keep the various data streams synchronized. Since system input is often collected from heterogeneous sources, synchronization of the data streams is important for state awareness and control. The potentially large number of heterogeneous data sources allows for singular events to be observed by multiple devices at once. These data sources can have varying latencies due to sample rates, transmission times, and missing data/noise. Multiple sensors monitoring the same object can end up with unsynchronized measurements, whether temporally or spatially. The sensors will all record the same event, but the multiple data streams will report the event happening at different times. Unsynchronized data can be caused by inherent device measurement latencies, physical distance from what is being monitored, transmission times, network issues, etc. Using data that is not synchronized can increase the difficulty of system-modeling and prediction tasks by introducing conflicting information about the system's status and potentially malicious activity.

Unsynchronized sensor data can lead to conflicting information in decision-making and modeling tasks. Bennett et al. [42] proposed a method for data alignment in sensor networks without the use of extra hardware or software in the sensors. The algorithm uses distinguishable points in the datasets to create a link between physical events and time. The authors showed their method successfully reduces the temporal offset between sensors, though testing was done with only two measurement devices. Another method for synchronizing the constant spatiotemporal offsets in measurement devices is shown in [43]. The authors presented an offline method to calculate the offsets using continuous state representation using a single estimator rather than the two-stage designs of previously established methods. This was shown to generate highly accurate offset estimations on several combinations of heterogeneous sensors.

Data-streams can also be incomplete or uncertain. Data sources can be compromised, temporarily disabled, or physically broken causing uncertainty in the IoT control systems. Trust metrics should be implemented so when conflicting data is presented to the system, determining which data sources can be trusted and which sources should be discarded can be done quickly to avoid unwanted consequences on the control side. Incomplete data should notify operators so that measures can be taken to restore the data source to its operational state, and safeguards should be in place to adjust the control and state awareness models in the presence of the missing data, or at the least notify users of a potential decrease in operational accuracy so any necessary risk assessments can be properly made before any actions are taken.

To help analysis on incomplete data, [44] presented a framework for rapid knowledge discovery from potentially incomplete datasets called structured data-fusion (SDF). SDF implements multidimensional arrays (tensors) so that users can quickly create and change libraries of processing methods, or factorizations. This allows users to work towards finding solutions from incomplete data faster and with reduced overhead.

When multiple sources provide conflicting data about an event, a method for choosing which sources to trust and which to disregard needs to be in place. Li et al. [45] presented an algorithm for conflict resolution in heterogeneous datasets through "source-reliability estimation." The authors implemented an optimization model where source trust weights are continuously updated for each source based on their distance from known truths derived from confirmed reliable sources in the system. The optimization problem was tested on multiple real-world and simulated multi-source datasets and was shown to outperform other popular conflict resolution methods, such as the Gaussian Truth Model. A multi-sensor data-fusion approach to fall classification is presented in [46]. The authors presented a novel approach for daily activity and fall detection for individuals using accelerometer and gyroscope data collected from a smartphone, along with user-specific measurements (i.e., height, weight, etc.). The authors combined the available data using Receiver Operating Characteristic (ROC) theory and then input the combined data into a threshold classification algorithm.

A fast-growing field that relies on cyber-physical data-fusion is autonomous vehicles. These systems require fast and accurate data-fusion to make life-and-death decisions in real-time. The unique challenges presented by autonomous vehicles may produce solutions and provide insight into problems faced by other control systems. Shan et al. [47] proposed a data-fusion method to increase performance from incomplete multi-sensor systems, specifically for auto traffic in urban areas. The authors used Multiple Linear Regression (MLR) models with historical taxi global positioning system (GPS) data to extract spatiotemporal traffic-state correlations. This information is used to fill in data gaps to increase the accuracy of other sensors. Historical data correlations could be implemented to make predictions in case of sensor failure, enabling other devices to make accurate decisions without ignoring the information of the failed device. Authors also noted the use of data-centric parallelization for handling large multisensory datasets, but no specifics were noted. Hu and Vasilakos [48] proposed a new model for describing moving objects using meta-information from multiple sources and global relationships to other objects to reduce measurement and sampling errors present in existing models. The authors used this model to develop a new map-matching algorithm, IF-Matching (Information Fusion Matching). This algorithm outperformed Spatio-Temporal Matching and performed as well as the ACM Geographic Information Systems Cup 2012 winner on city-wide trajectory data.

3.2 *Data Alignment/Tailoring*

Because of the potentially large scale of CPS data, displaying all of the information that is available can be overwhelming and unnecessary for the consumer. Finding and reporting only the information that is relevant to the consumer and the current/predicted system state may help improve decision-making response-time, accuracy, and reproducibility.

Since CPSs often contain large amounts of data from multiple sources, feature selection is a crucial component in anti-system architecture. Selecting the most relevant data and sources can improve both the accuracy and speed of decision-making tasks, as well as enhance understanding and system awareness. Wang et al. [49] showed a factor analysis method for feature selection using probabilistic kernels that outperformed common factor analysis methods for physical degradation monitoring. The selected feature sets were verified using support vector regression. A unique feature selection paradigm is model predictive control (MPC), which is a commonly used and effective technique for process control in industry. This technique models the dynamics of a plant to optimize input data based on predictions of future behavior. A MPC algorithm for modeling non-linear systems using neural networks is proposed by [50]. The authors use feed-forward neural networks to create a parameter-varying MPC for use in tracking a system with partially unknown dynamics over a set of operating points. The algorithm was validated on simulated tank and tubular reactor systems.

In critical infrastructure systems, the final decision is left to a human operator. The operator must look at the data and possible countermeasures supplied by the system and make a choice on how best to deal with the situation. Often, all countermeasures are given equal weight regardless of the situation. Bernieri et al. [51] proposed a method using the analytic hierarchy process to suggest alternate recovery options to the operator based on specific criteria (e.g., economic, security, social, environmental). This method was then applied to an intrusion detection system that could suggest alternative actions based on the input criteria.

In large-scale CPSs, end-users require efficient access to the available data. A data management system for healthcare applications that focuses on data accessibility is presented in [52]. The proposed cyber-physical patient-centric healthcare system incorporates cloud and big-data analytics for data collection, management, and application service layers in the system. The data-collection layer combines heterogeneous data nodes with adapters to give nodes access to the system and vice versa. The data-management layer incorporates a distributed file-storage system with a distributed parallel computing framework to ensure efficient data storage and retrieval, along with real-time and offline data analysis of the data. Finally, the application service layer supplies operational resources for developers and end-users, such as data access and visualizations, testbeds, security management, etc.

4 Data Visualization

This section provides a brief review of recent research conducted in visualization, with a focus on data-fusion for cyber-physical systems. Accurate visual representations of information is essential for any large scale system. Visualizations inform operators and consumers of how the system is running, health metrics, and can identify problem areas. Intuitive visualizations aid and simplify human-machine interaction by making relationships in the data more apparent and facilitating efficient decision making. The large size of the datasets available in a city-wide IoT application makes visualization a difficult task. It is not realistic to present all of the data in a single interface because the interface will be cluttered and unintuitive. While data-fusion reduces and exploits correlations among huge data streams, visualization plays a critical role in human understanding of this data. Visualization simplifies human-machine interaction. Accurate visualization of information can make previously unknown relationships apparent and facilitate faster, better decision-making.

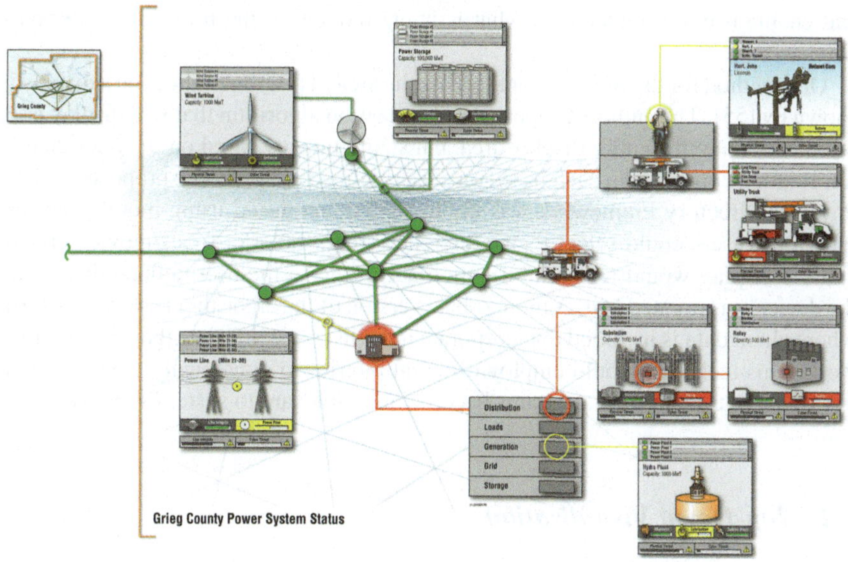

Fig. 3 Example of a Cyber-Physical Visualization Scheme for Power Grids

4.1 Dynamic Visualization

Dynamic visualizations change how the data is represented in real time which allows users to quickly understand where changes in the system are happening and if they have cause cascading failures. Dynamic data visualization allows for visual representations of data that can be altered as the represented data changes. This can allow for quick and intuitive understanding of changes in the system as size, color, length, etc., is altered. in an IoT design, devices can be connected to and communicate with each other over a network. This can allow for an efficient peer-to-peer resource sharing system to be implemented across the system. Resource sharing adds an additional layer to visualization by allowing local visual implementations to change dynamically based on global data as well. It is also important that different visualization nodes are consistent in their presentation to users. This helps users learn how the display layout implemented by the system works, and ensures users won't have to relearn the basic navigation when introduced to a new display (Fig. 3).

Of course, any visualization system that emphasizes some data over others is deciding on behalf of the user. Although it may do so according to user-stated preferences, this decision can also form a filter bubble [53], where important information may be suppressed because the user has not deemed it important. Filter bubbles become a self-perpetuating confirmation bias that may blind rather than reveal. Visualization developers implicitly trust their creations, but for critical tasks such as cybersecurity, operators tend to distrust decisions and simplifications made by visualization [54]. Thus, all visualization research should take user-trust into account and enable users to understand what is being hidden by the things that are being emphasized.

One method for dynamically optimizing the area of relevant images on a display is shown by [55]. The authors propose an optimization algorithm that will display a set of objects using the entire display area, but sizing images based on the relevance of information contained in each based on a user-defined context. The proposed Cyber-Physical Directory Framework was implemented and tested using mobile devices. Individual users could select their preferences (e.g., favorite foods/movie genres), and the display would resize information according to the user-defined data when the device was pointed at it. A more robust version of the algorithm is shown by [56], where cloud data is included to incorporate more user-specific data than the previous work. This technique could employ different metrics to dynamically resize system information as certain data streams become more relevant to system stability.

4.2 *Top-Down Visualization*

In a smart urban environment application, the system infrastructure will be spread out widely with many small components. This is a difficult system to visualize as operators and consumers may need access to very low level system information, but not all of the data can be shown at one. A potential visualization scheme to alleviate this problem is a high level, top-down design with an information 'drill-down' to access lower level data representations. An initial visualization may only show a map of the city with the major areas or landmarks marked. Users could then select an area to open a new visualization with more detailed information about the selected domain. This tiered design can continue for as many layers as are needed, though too many layers can be counter-productive.

A visualization scheme that could incorporate the top-down design previously discussed as well as incorporate dynamic data representations is presented in [57]. The authors presented a study of several techniques that can be used to visualize entity interoperability in cyber-physical systems, specifically the Node-Link Diagram (NLD) and corresponding balloon layout. An NLD is a tree-like graph structure with nodes being entities in the tool chain and links being relationships between the tools. Each node and link can be resized and color coded to represent different qualitative and quantitative information. However, the authors note that an NLD could become unwieldy and ambiguous as the size of the represented system increases. To alleviate this problem, the balloon layout is suggested. This layout clusters tools into smaller NLDs contained in parent nodes, and the parent nodes create a higher level NLD.

4.3 Visualization Techniques

A visualization design that is directly applicable to CPS is shown in [58]. The authors proposed a visual analytics concept to better handle the problems of critical infrastructure monitoring, cascading effects in infrastructure, and crisis response management. Their method proposes various visualization techniques to combine data from multiple infrastructures into a unified overview to assist operators in decision-making. Specifically, the authors' method aims to highlight important events, portray crisis events towards understanding in interdisciplinary teams, and present system details and controls when needed. The presented design was tested on a real power grid with an interconnected digital communication grid, and was shown to consistently highlight current and future events in the tested infrastructures.

Certain applications may require end-users to visualize data directly rather than relying on a high-level system representation. A method for visualizing and exploring patterns in temporal multivariate data is presented by [59]. The authors create temporal multidimensional scaling (MDS) plots that consider temporal-event information from the multivariate data to create sequences of one-dimensional similarity mappings. In the MDS plots, the x-axis is time and the y-axis is a similarity metric which visually groups similar events together over time. Furthermore, the multivariate nature of the data is visualized through a sequenced diversity matrix shown underneath the temporal MDS plots as a heat map. The colors of the matrix elements represent the diversity metrics, with black representing low diversity and white high diversity, to show the correlations between features. This visualization, coupled with a clustering algorithm to assist in event detection, allows users to find, analyze, and define reoccurring patterns in the data.

A newer area in visualization technology are 3D visualizations. Using 3D technology, more information can be kept within the reduced dataset while maintaining an accessible form for human visualization and intuition. 3D technology can also allow users to 'step into' and immerse themselves in the data. However, 3D visualizations also induce occlusion, hiding artifacts behind one another.

Visualizing self-organizing maps (SOMs) in a 3D environment was presented in [60] by using an immersive visualization technology called the Cave Automatic Virtual Environment (CAVE). SOMs are typically used for dimensionality reduction and feature selection to transform high-dimensional data into lower spaces that can be easier for humans to understand. The CAVE technology uses motion tracking to allow users to fully immerse themselves in the data. The virtual environment is updated as users move around and interact with the data using a wand tool. CAVEs excel at collaborative visualization where multiple users simultaneously explore a dataset, but they have practical limitations. Often, direct manipulation of CAVE objects is difficult or impossible, and they are not yet considered practical for use in other than exploratory and academic environments where there is no time sensitivity. It remains to be seen whether the technology can make the leap to operational usability. The authors tested the CAVE-SOM method on several benchmark datasets and a wind-power dataset.

Miyachi et al. [61] presented a project aiming to provide end-users with effective use of quantitative and qualitative 3D visualization and 2D analysis for big data applications. This work implements and improves upon a visualization method called Particle Base Volume Rendering (PBVR) for parallel volume rendering. PBVR and its improvements can run on a Graphics Processing Unit architecture, allowing for interactive use without having to pause to reload information. Other visualizations include contour maps and multidimensional transfer functions. A 4D cross-correlation, volume-data environment is in development.

5 Conclusions

This chapter outlined several challenges identified in cyber-physical system design, with potential solutions from recent publications. Discussions for establishing resilient CPS architectures that leverage IoT devices were outlined. Providing complete, reliable, and useful information is essential for the systems and operator's understanding and decision-making. Developing an accurate state awareness system, along with robust anomaly detection techniques, is crucial to providing users with the relevant information and models needed to recognize threats and coordinate effective responses. Data-driven CI algorithms have shown promising solutions for state awareness and prediction, attack detection, and decision-making with compromised data. Many of these algorithms are 'online,' meaning they can continue to learn after deployment to potentially adapt to new anomalies in the system, and provide information that is most relevant to the system and/or the decision-makers' needs. However, the pattern-matching nature of data-driven algorithms are subject to problems, such as spoofing, limited data-sets, and overfitting. Additionally, the black-box nature of CI designs do not allow for in-explanations or in-depth analysis of why the system behaves a certain way. Model-based and physics-based designs are other commonly used methodologies. These systems do not rely on data for generation and are white-box, which allows end-users and analysts to understand the inner workings of the system. However, creating these models require expert knowledge of how the system operates, and they won't be able to capture potential intricacies in the data that data-driven pattern matching methods can. These two approaches to modelling could be used together in a hybrid-modelling approach so that each methods strengths can cover the others weaknesses, though to the best of the authors' knowledge, no work has been presented on this topic.

Managing large-scale CPS provides its own challenges. Many architectures rely on a central data-fusion center that collects data from distant locations, makes decisions, and sends data back out. These data transfers take time and create single-points-of-failure across the network. A common solution in many large-scale systems is decentralizing the system into multiple, localized subsystems that collect and analyze data independently. Allowing the subsystems to make calculations and decisions from local data can reduce the computational overhead inherent in high-dimensional heterogeneous datasets and would require only essential information be sent to a system controller/operator on an as-needed basis. It also allows for multiple, subsystem specific algorithms to be implemented, though steps should be taken to ensure important interdependencies between subsystems are not lost by implementing knowledge-fusion or other techniques.

CPS datasets are usually high-dimensional, heterogeneous datasets that may have low/multi-resolutions, temporally unsynchronized features, etc. Because of the large amounts of data needed for system-modeling and prediction, it is crucial that the information be consistent and synced temporally. Machine-learning approaches have promising applications for datasets containing low or varying sample rates. Artificial neural networks in particular were shown to be able to increase both temporal and spatial resolutions on building sensor data. Another statistical machine-learning method, Gaussian Process Modeling, was used to combine multiple features with different resolutions. Due to varying transmission times and device latencies, CPS devices may report measurements as being recorded simultaneously, though in reality are not matched temporally.

Though CPS can provide a lot of data, not all of it may be relevant to a consumer. Recognizing and reporting the most relevant system information can increase the speed and reproducibility of the decision-making and analysis process. An algorithm to improve prediction quality for human-in-the-loop control systems was shown for emergency response applications. Using the analytic hierarchy process, the algorithm can suggest a hierarchy of recovery options derived from user-defined metrics indicating how important different criteria are (e.g., economic, security, environmental) for that system. These metrics can be defined for the needs of an individual, plant, or company to help ensure design-specific requirements are met in critical decision-making. Displaying user-relevant information is also an important aspect in visualization. Too much data can clutter a screen and make it difficult to locate what the consumer needs, but too little information will not provide robust system representation. However, the information needs to be sourced from the system globally and not be too focused on local activity. Visualization designs throughout a system should be consistent in their presentation to help users navigate all displays without having to relearn basic navigation of a display.

References

1. F. Alam, R. Mehmood, I. Katib, N.N. Albogami, A. Albeshri, Data-fusion and IoT for smart ubiquitous environments: a survey. IEEE Access **5**, 9533–9554 (2017)
2. T. Vollmer, M. Manic, Cyber-physical system security with deceptive virtual hosts for industrial control networks. IEEE Trans. Ind. Inform. **10**(2), 1337–1347 (2014)
3. O. Linda, D. Wijayasekara, M. Manic, C. Rieger, Computational intelligence-based anomaly detection for building energy management systems, in *2012 5th International Symposium on Resilient Control Systems* (2012), pp. 77–82
4. D. Wijayasekara, M. Manic, C. Rieger, Fuzzy linguistic knowledge-based behavior extraction for building energy management systems, in *2013 6th International Symposium on Resilient Control Systems (ISRCS)* (2013), pp. 80–85

5. T. Vollmer, M. Manic, O. Linda, Autonomic intelligent cyber-sensor to support industrial control network awareness. IEEE Trans. Ind. Inform. **10**(2), 1647–1658 (2014)
6. D.E. Denning, An intrusion-detection model. IEEE Trans. Softw. Eng. **13**(2), 222–232 (1987)
7. R. Sommer, V. Paxson, Outside the closed world: on using machine learning for network intrusion detection, in *2010 IEEE Symposium on Security and Privacy* (2010), pp. 305–316
8. N. Ádám, B. Madoš, A. Baláž, T. Pavlik, Artificial neural network-based IDS, in *2017 IEEE 15th International Symposium on Applied Machine Intelligence and Informatics (SAMI)* (2017), pp. 000159–000164
9. N. Sen, R. Sen, M. Chattopadhyay, An effective back propagation neural network architecture for the development of an efficient anomaly-based intrusion detection system, in *2014 International Conference on Computational Intelligence and Communication Networks* (2014), pp. 1052–1056
10. J. Esmaily, R. Moradinezhad, J. Ghasemi, Intrusion detection system based on multi-layer perceptron neural networks and decision tree, in *2015 7th Conference on Information and Knowledge Technology (IKT)* (2015), pp. 1–5
11. Z. Jadidi, V. Muthukkumarasamy, E. Sithirasenan, M. Sheikhan, Flow-based anomaly detection using neural network optimized with GSA algorithm, in *2013 IEEE 33rd International Conference on Distributed Computing Systems Workshops* (2013), pp. 76–81
12. N. Mowla, I. Doh, K. Chae, Evolving neural network intrusion detection system for MCPS, in *2017 19th International Conference on Advanced Communication Technology (ICACT)* (2017), pp. 183–187
13. C. Callegari, S. Giordano, M. Pagano, Neural network-based anomaly detection, in *2014 IEEE 19th International Workshop on Computer Aided Modeling and Design of Communication Links and Networks (CAMAD)* (2014), pp. 310–314
14. A.M. Kosek, Contextual anomaly detection for cyber-physical security in smart grids based on an artificial neural network model, in *2016 Joint Workshop on Cyber-Physical Security and Resilience in Smart Grids (CPSR-SG)* (2016), pp. 1–6
15. M. Ghanbari, W. Kinsner, K. Ferens, Anomaly detection in a smart grid using wavelet transform, variance fractal dimension and an artificial neural network, in *2016 IEEE Electrical Power and Energy Conference (EPEC)* (2016), pp. 1–6
16. V. Ford, A. Siraj, W. Eberle, Smart-grid energy fraud detection using artificial neural networks, in *2014 IEEE Symposium on Computational Intelligence Applications in Smart Grid (CIASG)* (2014), pp. 1–6
17. S. Ntalampiras, Detection of integrity attacks in cyber-physical critical infrastructures using ensemble modeling. IEEE Trans. Ind. Inform. **11**(1), 104–111 (2015)
18. D. Wijayasekara, O. Linda, M. Manic, C. Rieger, FN-DFE: fuzzy-neural data-fusion engine for enhanced resilient state-awareness of hybrid energy systems. IEEE Trans. Cybern. **44**(11), 2065–2075 (2014)
19. E. Hatami, N. Vosoughi, H. Salarieh, Design of a fault tolerated intelligent control system for load following operation in a nuclear power plant. Int. J. Electr. Power Energy Syst. **78**, 864–872 (2016)
20. J. Goh, S. Adepu, M. Tan, Z.S. Lee, Anomaly detection in cyber-physical systems using recurrent neural networks, in *2017 IEEE 18th International Symposium on High Assurance Systems Engineering (HASE)* (2017), pp. 140–145
21. C.G. Cordero, S. Hauke, M. Mühlhäuser, M. Fischer, Analyzing flow-based anomaly intrusion detection using replicator neural networks, in *2016 14th Annual Conference on Privacy, Security and Trust (PST)* (2016), pp. 317–324
22. T. Ince, S. Kiranyaz, L. Eren, M. Askar, M. Gabbouj, Real-time motor fault detection by 1-D convolutional neural networks. IEEE Trans. Ind. Electron. **63**(11), 7067–7075 (2016)
23. Y. Zhou, R. Arghandeh, I. Konstantakopoulos, S. Abdullah, A. von Meier, C.J. Spanos, Abnormal event detection with high resolution micro-PMU data, in *2016 Power Systems Computation Conference (PSCC)* (2016), pp. 1–7
24. S. Brahma, R. Kavasseri, H. Cao, N.R. Chaudhuri, T. Alexopoulos, Y. Cui, Real-time identification of dynamic events in power systems using PMU data, and potential applications #8212: models, promises, and challenges. IEEE Trans. Power Deliv. **32**(1), 294–301 (2017)

25. S.Y. Huang, Y.N. Huang, Network traffic anomaly detection based on growing hierarchical SOM, in *2013 43rd Annual IEEE/IFIP International Conference on Dependable Systems and Networks (DSN)* (2013), pp. 1–2
26. M. Du, S. Ma, Q. He, A SCADA data-based anomaly detection method for wind turbines, in *2016 China International Conference on Electricity Distribution (CICED)* (2016), pp. 1–6
27. M. Biswal, Y. Hao, P. Chen, S. Brahma, H. Cao, P.D. Leon, Signal features for classification of power system disturbances using PMU data, in *2016 Power Systems Computation Conference (PSCC)* (2016), pp. 1–7
28. K. Wen, J. Yang, F. Cheng, C. Li, Z. Wang, H. Yin, Two-stage detection algorithm for RoQ attack based on localized periodicity analysis of traffic anomaly, in *2014 23rd International Conference on Computer Communication and Networks (ICCCN)* (2014), pp. 1–6
29. M. Gu, The algorithm of information system anomaly detection, in *2013 3rd International Conference on Consumer Electronics, Communications and Networks* (2013), pp. 653–657
30. R.G. Kavasseri, Y. Cui, S.M. Brahma, A new approach for event detection based on energy functions, in *2014 IEEE PES General Meeting | Conference Exposition* (2014), pp. 1–5
31. M. Balchanos, D. Mavris, D.W. Brown, G. Georgoulas, G. Vachtsevanos, Incipient failure detection: a particle filtering approach with application to actuator systems, in *2017 13th IEEE International Conference on Control Automation (ICCA)* (2017), pp. 64–69
32. P. Angelov, Anomaly detection based on eccentricity analysis, in *2014 IEEE Symposium on Evolving and Autonomous Learning Systems (EALS)* (2014), pp. 1–8
33. S. Zhu, Y.C. Soh, L. Xie, Distributed inference for relay-assisted sensor networks with intermittent measurements over fading channels. IEEE Trans. Signal Process. **64**(3), 742–756 (2016)
34. H.E. Garcia, S.M. Meerkov, M.T. Ravichandran, Resilient plant monitoring systems: techniques, analysis, design, and performance evaluation. J. Process Control **32**, 51–63 (2015)
35. X. Cao, P. Cheng, J. Chen, Y. Sun, An online optimization approach for control and communication co-design in networked cyber-physical systems. IEEE Trans. Ind. Inform. **9**(1), 439–450 (2013)
36. I. Friedberg, K. McLaughlin, P. Smith, D. Laverty, and S. Sezer, "STPA-SafeSec: Safety and security analysis for cyber-physical systems," *J. Inf. Secur. Appl*
37. Y. Shoukry, P. Nuzzo, A. Puggelli, A.L. Sangiovanni-Vincentelli, S.A. Seshia, P. Tabuada, Secure state estimation for cyber-physical systems under sensor attacks: a satisfiability modulo theory approach. IEEE Trans. Autom. Control **PP**(99), 1–1 (2017)
38. H. Fawzi, P. Tabuada, S. Diggavi, Secure estimation and control for cyber-physical systems under adversarial attacks. IEEE Trans. Autom. Control **59**(6), 1454–1467 (2014)
39. K. Amarasinghe, D. Wijayasekara, M. Manic, Neural network-based downscaling of building energy management system data, in *2014 IEEE 23rd International Symposium on Industrial Electronics (ISIE)* (2014), pp. 2670–2675
40. D. Wijayasekara, M. Manic, Data-fusion for increasing temporal resolution of building energy management system data, in *IECON 2015—41st Annual Conference of the IEEE Industrial Electronics Society* (2015), pp. 004550–004555
41. M.J. Ren, L.J. Sun, M.Y. Liu, C.F. Cheung, Y.H. Yin, A reconstruction-registration integrated data-fusion method for measurement of multi-scaled complex surfaces. IEEE Trans. Instrum. Meas. **66**(3), 414–423 (2017)
42. T.R. Bennett, N. Gans, R. Jafari, A data-driven synchronization technique for cyber-physical systems, in *Proceedings of the Second International Workshop on the Swarm at the Edge of the Cloud*, New York, NY, USA (2015), pp. 49–54
43. J. Rehder, R. Siegwart, P. Furgale, A general approach to spatiotemporal calibration in multisensor systems. IEEE Trans. Robot. **32**(2), 383–398 (2016)
44. L. Sorber, M.V. Barel, L.D. Lathauwer, Structured data-fusion. IEEE J. Sel. Top. Signal Process. **9**(4), 586–600 (2015)
45. Y. Li et al., Conflicts to Harmony: a framework for resolving conflicts in heterogeneous data by truth discovery. IEEE Trans. Knowl. Data Eng. **28**(8), 1986–1999 (2016)
46. B. Andò, S. Baglio, C.O. Lombardo, V. Marletta, A multi-sensor data-fusion approach for ADL and fall classification. IEEE Trans. Instrum. Meas. **65**(9), 1960–1967 (2016)

47. Z. Shan, Y. Xia, P. Hou, J. He, Fusing incomplete multi-sensor heterogeneous data to estimate urban traffic. IEEE Multimed. **23**(3), 56–63 (2016)
48. J. Hu, A.V. Vasilakos, Energy big data analytics and security: challenges and opportunities. IEEE Trans. Smart Grid **7**(5), 2423–2436 (2016)
49. J. Wang, J. Xie, R. Zhao, K. Mao, L. Zhang, A new probabilistic kernel factor analysis for multisensory data-fusion: application to tool condition monitoring. IEEE Trans. Instrum. Meas. **65**(11), 2527–2537 (2016)
50. A. Gautam, Y.C. Soh, Stabilizing model predictive control using parameter-dependent dynamic policy for nonlinear systems modeled with neural networks. J. Process Control **36**, 11–21 (2015)
51. G. Bernieri, S. Damiani, F.D. Moro, L. Faramondi, F. Pascucci, F. Tambone, A multiple-criteria decision-making method as support for critical infrastructure protection and intrusion detection system, in *IECON 2016—42nd Annual Conference of the IEEE Industrial Electronics Society* (2016), pp. 4871–4876
52. Y. Zhang, M. Qiu, C.W. Tsai, M.M. Hassan, A. Alamri, Health-CPS: healthcare cyber-physical system assisted by cloud and big data. IEEE Syst. J. **11**(1), 88–95 (2017)
53. E. Pariser, *The filter Bubble: What the Internet is Hiding from you* (Penguin, UK, 2011)
54. G.A. Fink, C.L. North, A. Endert, S. Rose, Visualizing cybersecurity: usable workspaces, in *2009 6th International Workshop on Visualization for Cyber Security* (2009), pp. 45–56
55. J.L. Lamothe, J. She, M. Cheung, Cyber-physical directory: a dynamic visualization of social media data, in *2013 IEEE International Conference on Green Computing and Communications and IEEE Internet of Things and IEEE Cyber, Physical and Social Computing* (2013), pp. 2007–2012
56. M. Cheung, J. She, S. Park, Analytics-driven visualization on digital directory via screen-smart device interactions. IEEE Trans. Multimed. **18**(11), 2303–2314 (2016)
57. D. Gürdür, J. El-Khoury, T. Seceleanu, L. Lednicki, Making interoperability visible: Data visualization of cyber-physical systems development tool chains. J. Ind. Inf. Integr. **4**, 26–34 (2016)
58. S. Mittelstaedt, D. Spretke, D. Sacha, D.A. Keim, B. Heyder, J. Kopp, Visual analytics for critical infrastructures, in *International ETG-Congress 2013; Symposium 1: Security in Critical Infrastructures Today* (2013), pp. 1–8
59. D. Jäckle, F. Fischer, T. Schreck, D.A. Keim, Temporal MDS plots for analysis of multivariate data. IEEE Trans. Vis. Comput. Graph. **22**(1), 141–150 (2016)
60. D. Wijayasekara, O. Linda, M. Manic, CAVE-SOM: immersive visual data mining using 3D self-organizing maps, in *The 2011 International Joint Conference on Neural Networks* (2011), pp. 2471–2478
61. H. Miyachi, K. Koyamada, D. Matsuoka, I. Kuroki, Fusion visualization system as an open science foundation, in *2016 19th International Conference on Network-Based Information Systems (NBiS)* (2016), pp. 401–404

On the Integration of Information Centric Networking and Fog Computing for Smart Home Services

Marica Amadeo, Andrea Giordano, Carlo Mastroianni and Antonella Molinaro

Abstract Research on monitoring and control services for smart home and building management is expanding, stimulated by the growing interest in Cloud computing and Internet of Things. In addition to proprietary platforms, a common trend is to connect the smart home network to the Internet and leverage Cloud resources to run the application logic and store historical information. Recently, in many designs, intelligence is introduced at the edge of the home network to support low complexity operations. Interoperability between the different network domains is offered by the TCP/IP protocol suite and its extension for low-power nodes, i.e., 6LoWPAN. In parallel, the revolutionary Information Centric Networking (ICN) paradigm has been recently proposed to support future Internet communications and also data delivery in smart urban ecosystems, including smart home/building services. By leveraging name-based communication, in-network caching and per-packet security, ICN can largely simplify data delivery and service provisioning in instrumented environments. Moreover, by integrating ICN with Cloud technologies, a comprehensive home management system can be built. In this chapter, solutions are presented that rely on ICN for monitoring and controlling the smart environment. The integration of ICN with Cloud/Fog resources is also discussed and a reference architecture is presented as proof-of-concept, together with a preliminary testbed.

M. Amadeo · A. Molinaro
DIIES Department, University Mediterranea of Reggio Calabria,
Via Graziella Feo di Vito I, 89100 Reggio Calabria (RC), Italy
e-mail: marica.amadeo@unirc.it

A. Molinaro
e-mail: antonella.molinaro@unirc.it

A. Giordano (✉) · C. Mastroianni
CNR - National Research Council of Italy, Institute for High Performance
Computing and Networking (ICAR), Via P. Bucci 41C, 87036 Rende (CS), Italy
e-mail: giordano@icar.cnr.it

C. Mastroianni
e-mail: mastroianni@icar.cnr.it

F. Cicirelli et al. (eds.), *The Internet of Things for Smart Urban Ecosystems*,
Internet of Things, https://doi.org/10.1007/978-3-319-96550-5_4

1 Introduction

By promoting the global interconnection of billions of smart objects that produce and consume information, the Internet of Things (IoT) opens new opportunities for users, manufactures and companies to deploy smart urban ecosystems [1]. IoT objects range from quite complex systems, like smartphones and tablets, to very resource-constrained devices with slow CPUs and small memory footprints, like sensors and RFIDs. Of course, the management and processing of the huge amount of information generated by such heterogeneous devices, while also providing a common interface to different players and always-on service availability with security and privacy guarantees, is a challenging task. The dominant strategy, which largely simplifies the overall design, leverages Cloud computing technologies to host the application logic. To complement the Cloud technology, the Fog computing paradigm has been promoted [2], which foresees a set of servers at the network edge and offers low-latency services with context awareness. The IoT can largely benefit from real-time interactions with the Fog, while letting big data analytics and long term storage to the Cloud [3].

In this chapter, we show how Cloud and Fog technologies can play a key role in a representative IoT scenario, the smart home, where a variety of monitoring, automation and control functions can be deployed [4]. Smart home applications and, more generally, building management systems (BMSs) have been mainly based on proprietary protocols and specialized hardware, thus missing interoperability, flexibility and extendibility requirements. Today, however, stimulated by the research in IoT and Cloud technologies, there is a growing interest into novel solutions that allow global reachability to services and contents in such environments; IP-based solutions are under-way to facilitate the access to smart home resources and also reduce the installation and management costs, by introducing open standard hardware and software. On the other hand, very recently, smart home solutions based on the revolutionary paradigm called Information Centric Networking (ICN) are appearing. ICN has been proposed as a novel mechanism to improve the communication in the future Internet [5], including smart urban ecosystems based on IoT and Machine-to-Machine (M2M) environments with resource-constrained nodes [6, 7].

By design, ICN can naturally overcome some of the open issues of IP-based solutions in presence of challenging networks [8]. ICN implements name-based routing and leverages application-meaningful names, which can be used to address at the network layer different types of resources, including contents, services and things. In-network caching and consumer mobility are intrinsically supported, and data security is performed at the network layer. As a result, ICN can simplify the design of IoT applications and, coupled with Cloud and Fog computing, can build a comprehensive framework for effectively managing and control the IoT environment.

Therefore, the main target of this chapter is to analyse ICN-based smart homes and their integration with Cloud/Fog technologies. After shortly introducing the ICN paradigm and its exploitation in a IoT context in Sects. 2 and 3 we discuss the applicability of ICN in smart homes and building management systems, by identifying its native benefits and surveying the related literature. Then, Sect. 4 discusses the most relevant state-of-the-art solutions for the exploitation of IoT and Fog computing technologies in smart home environments. In Sect. 5, we describe the methodologies adopted to integrate the technologies discussed so far, i.e., ICN and Fog computing, in a smart home scenario. Finally, Sect. 6 shows a practical implementation of the ICN-Fog integration by presenting ICN-iSapiens, a framework deployed under an Italian academic/industrial project, and Sect. 7 concludes the chapter.

2 ICN-IoT in a Nutshell

ICN was originally conceived as a new paradigm for the future Internet where information is the first class network element and communication is based on unique, persistent, and location-independent content names, used by applications for search and retrieval. Over the years, several ICN architectures appeared that share the same common principles [5], and the model evolved towards a network where names not only identify contents, but also heterogeneous resources, like services and things [9].

In the following, without loss of generality, we refer to a very popular ICN architecture called Named Data Networking[1] (NDN). The NDN project was funded in the US by the National Science Foundation (NSF) in September 2010 and it is in constant development, with contributions from the worldwide research community. NDN uses a hierarchically structured naming scheme, with names in the form of Universal Resource Identifiers (URIs) having a potentially unrestricted number of components. For instance, the temperature values provided by a sensor located in the telecommunication laboratory of the University of Reggio Calabria could be named as */unirc/engineering/tlclab/temperature/*, while the humidity values could be named as */unirc/engineering/tlclab/humidity*, and so on. Communication is based on two named packet types, the *Interest*, used to request a resource (e.g., a content, an actuation command, a generic computation), and the *Data*, which answers the Interest by carrying the outcome of the request. In this way, ICN naturally matches the pattern of many IoT applications, which are interested in the data itself, e.g., the temperature in a specific zone, not in a well-specific data source. Moreover, thanks to the consumer-driven connectionless communication, NDN supports consumer mobility: when the consumer changes its point of attachment, it can simply re-express the unsatisfied Interests, without the need of resuming any previous data session. This is especially useful in many mobile IoT scenarios, e.g., in presence of electric vehicles.

NDN implements per-packet security: the Data includes a digital signature, created by the producer, which binds the name of Data with its content, so that the authenticity can be verified at any time. As a result, each named Data packet is a self-consistent unit, which can be cached by any in-network node, according to its policies and resources, to serve future requests without involving the original sources.

[1] https://named-data.net/.

This is a great departure from the IP model, where, due to the host-centric connectivity, caching can be implemented only in specific nodes, e.g., a Proxy that separates the IoT domain from the Internet. In-network caching is, instead, a key feature of ICN that not only limits retrieval delay and network load, but it is especially useful in presence of resource-constrained nodes, which sleep most of the time and cannot tolerate massive data access.

Caching operations in NDN also respect the freshness requirement that usually characterizes many IoT Data, e.g., a temperature value can vary over time. The original producer can set a *FreshnessPeriod* in each Data packet so that the information is considered stale after that time.

Thanks to the name-based forwarding, NDN also supports native multicast delivery. Specifically, an NDN node is provided with the following Data structures, which are used in the forwarding process: *(i)* a *Content Store* (CS), to cache incoming Data packets; *(ii)* a *Pending Interest Table* (PIT), to store incoming Interests that are not consumed by the Data packets; *(iii)* a *Forwarding Information Base* (FIB), to route Interests towards the content sources. At the Interest reception, a node *N* first looks in the CS: if a name matching is found, then the Data can be sent back immediately. Otherwise, *N* looks in the PIT: if a name matching is found, it means that the same request has been sent, and the node is waiting for the result. Therefore, *N* just updates the existing PIT entry with the Interest incoming interface. Otherwise, *N* looks in the FIB. If a matching is found, *N* creates a novel PIT entry and sends the Interest over the outgoing interface. Otherwise, the Interest is discarded and optionally a negative acknowledgement is sent back.

The Data packet follows the PIT entries back to the requester(s): *N* sends the packet to the interfaces from which the Interest was received, (optionally) caches the data in the CS, and removes the PIT entry. Data without a PIT matching are discarded. Figure 1 summarizes all the forwarding process.

Forwarding rules are defined by the *Routing Information Base* (RIB), which records routing information, used to update the FIB when needed, and by the *Strategy Choice Table* (SCT), which collects a set of named forwarding strategies, associated with the namespaces, manually chosen by the system administrators. This is another important feature of NDN: packets can be treated differently by merely considering their name prefix. Therefore, for instance, all the packets from delay-sensitive emergency applications, e.g., with main prefix*/emergency*, can be forwarded with high priority and over multiple interfaces simultaneously.

3 ICN for Smart Homes and Building Management

In the near future, homes and buildings will host a large set of small, resource-constrained devices, like sensors, actuators and controllers, that will participate in different automation and monitoring applications, like smart lighting, smart energy consumption, surveillance, and healthcare. A Gateway can act as the *Home Server* (HS) that controls the devices and bridges the home network to the Internet.

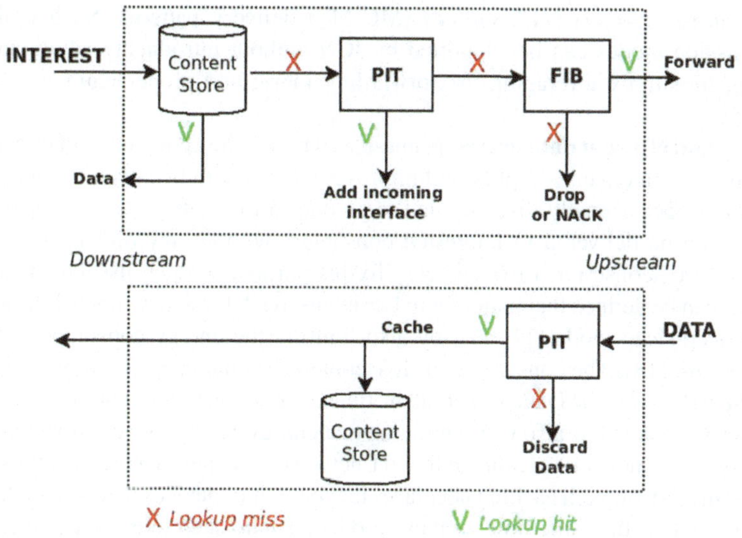

Fig. 1 NDN forwarding process

ICN holds great promise in instrumented environments by offering the following distinctive features.

Simplified configuration and management. By using application-level names, ICN simplifies network configuration and management, since there is no need of addressing each host and updating the address if the node moves, e.g., healthcare applications can be based on wearable devices that move with the patient. Names can be defined that are meaningful for the applications, but also able to support routing scalability [10]. Automatic self-configuration mechanisms, enabling practical ICN deployments on IoT networks, have been also defined [11], where each device is able to derive names locally, by satisfying the requirements of meaningfulness and uniqueness.

Improved security. Many home applications involve high sensitive information and need strong privacy, integrity and authentication support. Per-packet security implemented by ICN ensures that the data is secured, independently from the node that serves it and the channel where it travels. Data packets can also be encrypted to ensure access control. Security mechanisms can be extended to Interest packets; this is especially useful in presence of automation applications, like the lighting control, where requests need to be authenticated [12]. Security can also be implemented in presence of low-powered nodes, as in [13, 14].

Group-based communications. Home applications frequently originate group-based communications, where multiple data sources are queried at the same time and/or the data is of interest of multiple consumers. Therefore, in addition to the *single-consumer-single-source* communication model, we can identify *single-consumer-multi-sources* (SC-MS), *multi-consumers-single-source* (MC-SS) and even *multi-consumers-multi-sources* (MC-MS) delivery patterns. Such multicast transmission modes can be supported by ICN without introducing additional signalling, by simply leveraging the forwarding fabric and proper naming schemes [15, 16].

Easy and efficient data access. Name-based forwarding coupled with in-network caching can effectively simplify and improve data reachability, even when the data sources are occasionally disconnected, e.g., due to low battery level. Indeed, data in ICN can be delivered to interested consumers without any direct connectivity with resource-constrained IoT devices. To this purpose, proper distributed caching policies can be defined that maintain IoT contents available at any time while devices are in deep-sleep mode [11, 17], by also limiting the energy consumption in the IoT network [18]. The consistency of IoT contents is guaranteed by introducing the *FreshnessPeriod* in the Data packets at the time of creation, so they can be removed by the CS when stale. It is also worth noticing that energy savings are possible thanks to (even small) in-network caching in the IoT network, as experimentally demonstrated in [19]. Indeed, on-path caching decreases the number of devices involved in the data delivery and, at the same time, content producers can sleep more time, since their contents are available in the network.

In the following, by reviewing the related literature, we present the main aspects of ICN-IoT solutions for smart homes and BMSs. The discussion is organized by considering the principal implementation aspects, namely naming, security, caching, and communications patterns.

3.1 Data Naming Solutions

Although there are not a-priori restrictions in the way ICN names can be composed, it is a shared belief that some common conventions must be defined, in order to facilitate application interoperability and re-usage. The most popular, widely accepted, solution is to define user-friendly, hierarchical namespaces, which reflect the application scope and/or the data type offered by IoT devices.

Typically, ICN-IoT names include: *(i)* a root prefix, used to forward the Interest towards the IoT domain (e.g. the smart home), *(ii)* a set of middle components that identify the application type and/or the thing or the data produced by the device, and *(iii)* optionally, a final component that identifies the specific instance of the Data [9].

In [12], a namespace for the BMS at UCLA campus it defined, which reflects the physical location of the devices and the hierarchy of the building structure. A sensor data packet can be named as */ndn/ucla.edu/bms/building/melnitz/studio/1/data/-panel/J/voltage/<timestamp>*, where the main prefix *ndn/ucla.edu* indicates the IoT domain at UCLA, *bms* indicates the application type, the components *building/ melnitz/studio/1/data/panel/J/voltage/* identify the location of the sensor and its measurements (voltage). Finally, the timestamp identifies a specific voltage instance.

Similarly, in [15], sensing and control operations in the house are expressed through a hierarchical human readable namespace, having as root prefix */homeID/task*.

The component */homeID* is a unique name related to the geographic location of the house and/or to an owner identification number, and it is advertised by the HS in the ICN network thus remote consumers can reach the smart home and its services. The route prefix is followed by the components */type/subtype/location/*, which specify, respectively, the task type (sensing or action), the specific sensing or action task to be performed, and the physical position of the device in the house, e.g., the name */bobHouse/task/action/light/on/kitchen* is used to require the kitchen light fixture to turn on.

It is worth noticing, however, that human-readable names may not be required or permitted in presence of low power, low-bandwidth communications, whereby no humans are involved, and transmission technologies require small MTU (Maximum Transmission Unit). In such case, short naming schemes, e.g., compact ASCII representations, are necessary to leave more space for data payload [19]. An autoconfiguration mechanism, with no a-priori human intervention, has been deployed in [11], which derives short name prefixes from the device type and the (unique) node identifier like the vendor ID, and uses timestamps, as version numbers, e.g., */hum/DEADBEEF/1466250645*.

3.2 Security

Producer's digital signature is mandatory in each Data packet to make it verifiable from any node. A *MetaInfo* field in the Data packet header indicates the name of the public key that can verify the signature. Interest packets, instead, are not signed in the standard ICN framework.

When dealing with smart home automation services, however, Interests can be used to request different types of actions, e.g., to switch on/off lights, to open/close windows, and, of course, the packets need to be authenticated to prevent malicious intrusions in the home. The use of *signed Interest*, originally proposed in [12] for lighting control, is today a common practice: the packet is signed by the consumer with a private key, the receiver will perform the action only if it is successfully verified with the consumer's public key.

To manage and assign keys, trusted third parties are needed in the IoT domain. For instance, in [13], a framework for ICN secure sensing is proposed that introduces two entities in the IoT domain: a Configuration Manager (CM) and an Authorization Manager (AM). The CM assigns to sensors the namespace for publishing data, a unique public/private key, and the AM's public key that identifies the root of trust shared by all sensors and applications within the same domain. For each sensor, the AM distributes a signed access control list containing the identities and namespace of applications which are allowed to query it.

Although public key cryptography is widely used in standard ICN implementations, it could not be supported in presence of resource-constrained devices, like small battery-operated sensors and actuators commonly used in the smart home. This is why the research community has defined alternative less-expensive security mechanisms. NDN-ACE is an actuation framework based on symmetric cryptography that offloads key distribution and management to a powerful third party called Authorization Server (AS) [14]. Consumers and actuators first establish trust relationships with the AS. Then, each actuator delegates access control tasks to the trusted AS by periodically sharing with it a per-service secret key (seed), which is transmitted in an encrypted packet. When the consumer needs to access a service, the AS authenticates its identity and generates an access key for the client. The key is derived from the corresponding service seed and cryptographically bound to the service name and the client identity. The client sends an authenticated Interest that carries two security-related information: *(i)* the HMAC (Hash-based Message Authentication Code) signature signed with the access key, which indicates that the client has been authorized by the AS, and *(ii)* signature information, which allows the actuator—which does not maintain per-consumers key—to recompute the access key from the service seed and finally authenticate the consumer. If the Interest is successfully verified, the actuator executes the command and sends an acknowledgement Data, HMAC-signed with the same access key.

3.3 Caching

To maximize the benefits of in-network caching, proper strategies must be defined in order to limit the energy consumption in the smart home domain by also guaranteeing high content reachability, even if the data sources are in sleep mode.

A Cooperative Caching mechanism, called CoCa, for low-power devices has been deployed in [11]. CoCa leverages link-local broadcast transmissions to maximize data availability in presence of nodes with intermittent connectivity. Specifically, each Data packet generated by IoT nodes is broadcasted in the IoT domain, so that each receiving node can decide to cache it, according to its content capacity and the caching strategy, like random caching or Max Diversity Most Recent (MDMR). By doing so, when an Interest is transmitted in the IoT domain, there is (almost always) an active node with a valid copy of the content. Of course, CoCa can be successfully implemented in small network topologies like a smart home environment, but it suffers from scalability issues when dealing with high density IoT deployments, since it can lead to broadcast storm phenomena.

In this case, high-selective caching mechanisms can be implemented. pCASTING, in [18], is a probabilistic strategy that adjusts the caching probability according to the data freshness and two device features, the battery energy level and the cache occupancy of the node. At the Interest reception, the IoT node computes a caching utility function that simultaneously takes into account the above normalized parameters and computes the caching probability. As a result, pCASTING efficiently stores Data in the network by limiting the energy consumption in the nodes and, at the same time, guaranteeing low content retrieval delays.

There are cases, however, where IoT information does not need to be cached, e.g., because it is privacy-sensitive, or simply not of interest for other consumers.

Consumers and/or producers can request to not perform caching operations by including this information in the Interest or the Data. For instance, in [20], the consumer can set a special bit in the Interest, thus the corresponding content is not cached by intermediate routers.

3.4 Communications Patterns

IoT communication patterns are very heterogeneous: they differ in the number of involved actors, i.e., single/multiple consumers and producers, and in the service model, i.e., push/pull.

Multi-consumer communications (MC:SS) are inherently supported by the ICN forwarding fabric, thanks to Interest aggregation in the PIT. Vice versa, multi-source communications (SC:MS, MC:MS) need slightly modifications in the PIT behaviour to allow that multiple data can be accepted with a single Interest. The work in [16] implements multi-source retrieval by leveraging a proper-defined naming scheme, where sources of the same type publish data under the same principal prefix, e.g., the temperature sensors in the same house share the principal prefix */BobHouse/temperature* and publish Data with names */BobHouse/temperature/bedroom*, */BobHouse/temperature/bathroom*, etc. When the consumer wants to collect more Data with a single request, it sends an Interest with the principal prefix, thus multiple nodes has a partial name matching and answer the request. The pending Interest in the PIT is not deleted after the first Data reception, but it remains active to collect other packets.

So far, we considered a pull-based service model, guided by the consumer, that matches a large set of smart home applications, namely polling-based monitoring and actuation-based applications. However, many sensor-based applications require publish-subscribe communications, where devices send data to subscribers in a periodic or event-based manner.

The so-called *long-lived-Interests*, originally proposed in [21], are designed to support push-based delivery over NDN. The idea is to use Interests as subscriptions, which are not deleted after the first Data reception but remain active in the PIT to allow the forwarding of other packets. The main con of such implementation is the need of maintain soft-state information in each NDN node; however, this is not an issue in small network topologies like the smart home.

4 IoT Fog Computing for Smart Homes

In the context of Smart Homes, in the last few years many advancements and challenges have been done concerning several fields: *Home Automation and Domotics* [22], *Energy Optimization* [23], *Activity Recognition* [24], *Ambient Assisted Living* [25, 26], *Indoor Positioning Systems* [27], and *Home Security* [28]. Nowadays

Smart Home applications can be developed exploiting technologies and solutions offered by the emerging research field of the Internet of Things [29], in which many devices have been enhanced with respect to their ordinary role to be proactive and collaborative with other devices [30]. In this background, exploiting IoT for the new era of Smart Home applications [31] represents an important research field both in academia and in industry context. One of the major challenges in the development of these applications is supporting interoperability among various heterogeneous devices and their proper deployment. Thus, the need for new architectures and frameworks—supporting both smart control and actuation—has been identified by many researchers. The research community has presented several proposals regarding middlewares and frameworks for the rapid prototyping of smart environments in general, or smart homes in particular. As an example, authors in [32] presented *Voyager*, a framework conceived for the support to the development of smart environments. Voyager relies on the exploitation of small bluetooth devices, which are user programmable, used to give intelligence to developed systems. The work in [33] introduced *JCAF* (Java Context-Awareness Framework), an extensible framework that supports the creation of context-aware in-home applications. The main components of the framework are *clients*, *actuators*, *monitors*, and *services*. JCAF allows all of them to be dynamically added, updated or removed at runtime. Authors of [34] presented *Gaia*, a middleware that supports the control of heterogeneous resources in a physical space. Gaia allows the developers to see collections of individual devices as a whole by introducing programmable *active spaces*. Active spaces are programmable through a scripting language called *LuaOrb* [35]. The *Syndesi* framework was introduced in [36]. Syndesi exploits Wireless Sensor Newtorks to create personalized smart environments. The environment augmented with the Syndesi framework allows to identify people and perform specific actions based on people profiles.

Considering that urban environments usually concern large city areas, a distributed approach appears as a "natural" solution [37]. Conversely, most of the actual smart city implementations rely on centralized approaches, where a big amount of data is collected in a common data center that processes such data and provides services to citizens, or schedules actuations on physical infrastructures [38]. Moreover, many services are developed as independent monolithic blocks, and their interaction is not properly considered. In a more distributed approach, benefits can arise from the exploitation of the so-called *Fog computing* paradigm [39–41]. With Fog computing, rather than processing information in a monolithic Cloud layer, the computation is pushed close to the data sources or, said in another way, at the "edge" of the network. Other benefits of this computing paradigm include: *(i)* a faster reactivity to events, *(ii)* a better exploitation of the communication bandwidth, as data is processed locally and only aggregated information is propagated across the system, *(iii)* an increase in reliability and scalability, since the Fog computing paradigm fosters the use of distributed algorithms.

5 Integrating ICN with Fog and Edge Computing

Fog computing and ICN are key technologies for future IoT applications. On the one side, the Fog is able to extract knowledge from IoT raw data and deploys the application logic that process IoT data and monitors and controls IoT environments. It is complemented by the Cloud, when needed. On the other side, ICN can largely improve data retrieval and service access, mainly thanks to name-based forwarding and in-network caching.

It is worth observing that, when dealing with IoT environments, some research works refer to the *Edge computing*, instead of the Fog computing. Indeed, edge computing should focus more toward the things side [42], with IoT end-devices (EDs) acting both as data consumers and producers, and the intelligence and processing power installed even into simple, low-cost devices like programmable automation controllers or Raspberry-Pi boards. At the same time, some research works use the terms Fog computing and Edge computing interchangeably, while others consider the Fog a broad concept that includes also Edge computing [43].

In the following, we use the term Fog for the sake of brevity, although the discussion can consider aspects of Edge computing.

Although research on the integration between Fog/Cloud computing and ICN is still at the beginning, two main approaches can be identified.

The first one considers a full ICN-based network model and assumes that name-based computing requests are resolved by in-network nodes, possibly close to the client. In this case, ICN nodes dynamically act like Fog servers: computing capabilities are not statically hosted by well-defined edge servers, but are distributed in ICN routers. When the task cannot be resolved in the network, the request is forwarded to the Cloud.

The second approach, instead, considers ICN deployments only at the network edge, and uses Fog servers as computing nodes that also bridge the ICN domain with the IP-based Internet by leveraging proper *translation* mechanisms.

In the following, we discuss the mentioned approaches by surveying the related literature.

5.1 *Full ICN-Based Design*

The authors in [44] present Named Function Networking (NFN), an ICN extension where consumers request computations over contents (e.g., a video compression) and the requests are routed towards in-network nodes that will execute them. Functional programs are in the form of λ-calculus expressions and the ICN forwarding machine is augmented with a λ-expression resolution engine, which processes all the Interests that have the postfix name component */NFN*. Therefore, unlike the traditional Fog computing, where Fog servers are fixed hosts with dedicated functionalities, in NFN the network becomes a distributed and dynamic Fog platform, where each content router, depending on its local capabilities, can act as computing unit. However, how to distribute the processing load between NFN nodes is still an open issue.

In [45], NFN is customized for IoT networks. The proposal, called PIoT (Programmable IoT), is an application layer solution, statically installed in the more powerful nodes of the network, which consists of the following services: *(i)* Expression Resolver, which resolves the λ-expression into sub-tasks; *(ii)* the Expression Pusher, which creates new Interests to fetch the IoT data needed to execute the task; *(iii)* the Expression Evaluator, which executes the task and generates the resulting Data packet(s). In [46], the same authors extend PIoT with a computation service management (CS-Man) that enables IoT devices to offer in-network computing operations, based on their own capability. In CS-man, a network entity acts as Service Manager (SM), which maintains a Service Repository (SR) with the names of the Service Executor (SE) nodes. So, when a client needs a processing task, it can retrieve the identity of the SE(s) from the SR.

Similarly to NFN, a framework for in-network function execution called Named Function as a Service (NFaaS) has been deployed in [47]. NFaaS supports processing with lightweight virtual machines in the form of named *unikernels*. A *Kernel Store* in ICN nodes stores function codes, and decisions on which functions to execute or delete are based on a score function that takes as inputs the requests popularity and the latency/bandwidth requirements of the applications. Nodes advertise the functions they can execute according to a routing protocol, so that task requests are forwarded towards the more appropriate node. The authors consider as an example two forwarding strategies, for delay-sensitive and bandwidth-hungry services, respectively, and demonstrate that functions move to the right direction (i.e., delay-sensitive services towards the edge of the network, and bandwidth-hungry services towards the core).

In [48], ICN supports service provisioning in challenged networks with intermittent connectivity. Services are built as containers coupled with a semantic naming scheme, thus they can be easily requested and cached in the network. The ICN framework is enhanced with a Delay Tolerant Networking (DTN) interface, which communicates with an underlying DTN implementation that handles intermittence by encapsulating Interest and Data packets in DTN bundles.

5.2 Hybrid IP-ICN Design

Large-scale clean-slate ICN implementations are still far from being a reality. The Internet is currently based on the TCP/IP protocol suite, and ICN protocols will be more likely implemented in a virtual environment or as an overlay over the TCP/IP transport. Vice versa, edge domains, like the smart home, can easily deploy new implementations from scratch. Therefore, in the following we focus on a hybrid scenario where the IoT domain is ICN-based, while the Internet is IP-based. The IoT domain leverages a Fog layer to support local storage and processing capabilities. In addition, Fog servers bridge the ICN domain to the IP-based Internet and

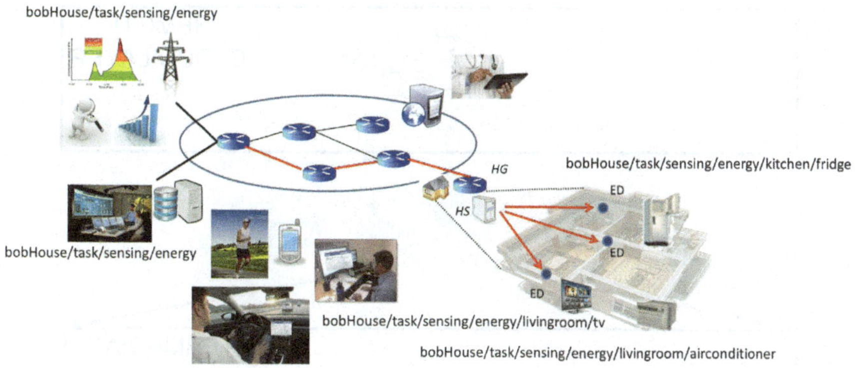

Fig. 2 Reference scenario

guarantee full reachability to IoT information, in a way that is independent from the technological and networking details of IoT devices.

A Fog computing-based gateway (FOGG) that bridges the Internet with ICN-IoT domains is proposed in [49]. The FOGG gateway is a powerful node that collects information from sensors by using the lightweight ICN implementation called *ccn-lite* (www.ccn-lite.net/), and makes it available to external users by running some translation mechanisms. It also controls actuations, with inputs from remote consumers, and manipulates raw data to extract meaningful information.

In the context of smart homes, Fog (or Edge) computing capabilities can be implemented in the smart home server (HS), which natively monitors and controls the environment, by interacting with ICN end-devices, and with remote clients and the Cloud through the Internet. Therefore the HS acts also as a Home Gateway (HG), as shown in Fig. 2.

The result is a three-layered architecture with a physical ICN layer, the Fog as intermediate layer and the Cloud as upper layer. In the next section, we present a real implementation of this framework, which has been described in a previous paper [50]. The core component of the framework is *iSapiens*, an IoT platform that allows the development of general cyber-physical with edge computing capabilities [51, 52].

6 The Case of the ICN-iSapiens Framework

6.1 Architecture Design

Figure 3 shows the three-layered ICN-iSapiens framework.

The *Physical Layer* includes all the EDs, which deploy sensing and automation tasks in the smart home and communicate with the Home Server via ICN. Each ED

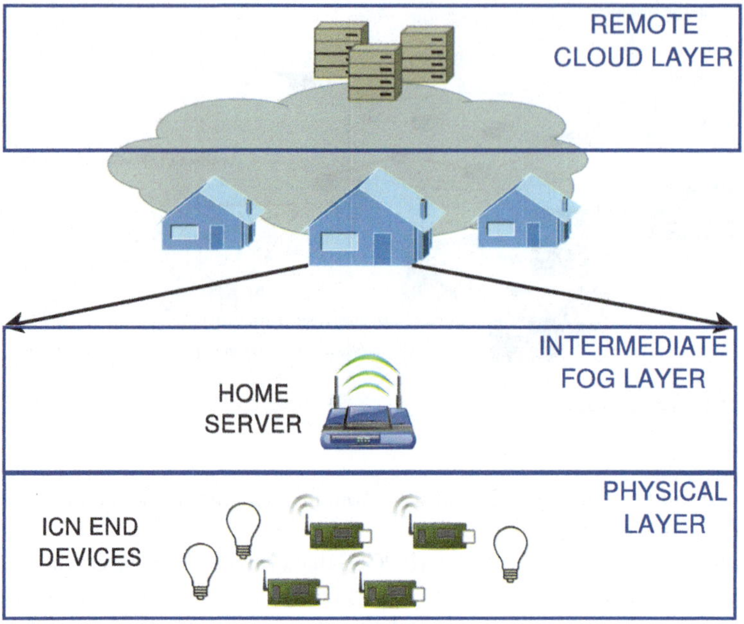

Fig. 3 ICN-iSapiens framework

is configured on a specific hierarchical namespace that describes the resource(s) it offers, i.e., sensing measurement(s) and automation services. The namespace identifies *(i)* the task type, sensing or action; *(ii)* the task subtype, e.g., energy and temperature for sensing tasks, light and heating for actuation tasks; *(iii)* one or more task attributes, e.g., the location of the EDs like bedroom or kitchen. Moreover, the name hierarchy is structured in a way that multi-source communications are enabled, where a single Interest is used to simultaneously query groups of EDs sharing some part of the same namespace, e.g., the Interest "/sensing/temp" requests the temperature information to every temperature sensor of the house.

The *Intermediate Fog Layer* includes the Home Server (HS) that hosts the ICN protocols to communicate with EDs and the TCP/IP protocol stack to communicate with remote entities through the Internet. Moreover, the HS hides the details of ICN EDs through a name-based *virtual objects abstraction* and hosts a multi-agent software application to monitor and control the EDs.

Each ICN ED is indeed represented as a virtual object (VO), that is a high-level standardized description of the device's functionalities that hides its heterogeneity in terms of technological and networking details. Each ICN VO can include a collection of sensing and actuation *named functionalities*, and a collection of standard *methods* used by Agents to monitor/control such functionalities via ICN. Agents use VOs methods to pull monitoring and action tasks, and to be asynchronously notified about some events; indeed, VOs methods call ICN primitives to send named Interest and

Data. There is a one-to-one correspondence between VO functionalities and ICN names, so the VO name is directly used by ICN to interact with the related ED.

Finally, the Cloud layer addresses all those activities that cannot be executed by the HSs, e.g., tasks requiring high computational resources or long-term historical data. Moreover, the data analysis executed by the Cloud can be used to optimize the Agents' behaviour.

6.2 Testbed

As shown in Fig. 4, an ICN-iSapiens demonstrator with low-cost off-the-shelf devices has been deployed. The HS is implemented over a Raspberry Pi device (www.raspberrypi.org/), equipped with an Ethernet interface, for Internet communications, and a IEEE 802.11g external interface for wireless communications with the EDs. *Raspbian* (http://www.raspbian.org) is the selected operating system. EDs are different kinds of sensors and actuators, like temperature and motion sensors, light actuators, attached to Intel Galileo boards, with *Yocto* (http://ark.intel.com/products/78919/) as operating system.

We use four Galileo boards in our testbed, each one symbolically located in a different room of the smart home (bedroom, kitchen, bathroom, living) to monitor and control it. Each room can be partitioned in two or more zones, each one identified

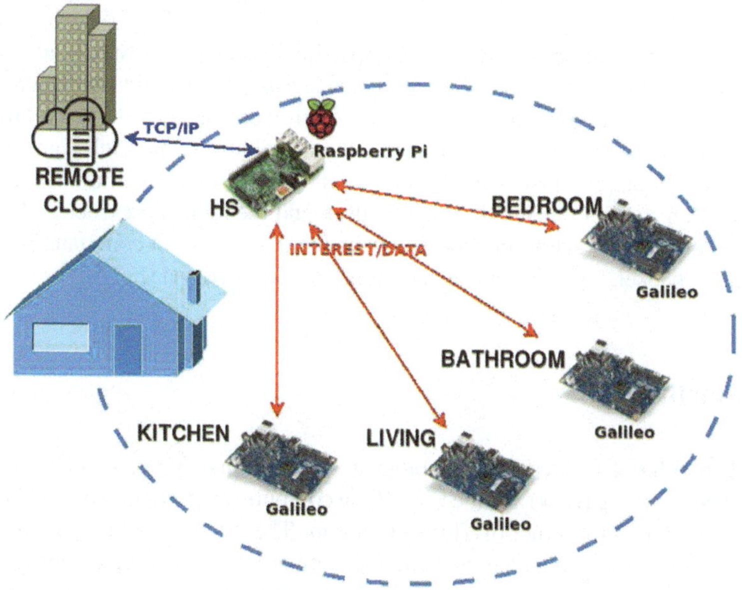

Fig. 4 ICN-iSapiens testbed

by a number, e.g., zone1 and zone2. Each Galileo is one-hop away from the HS and uses IEEE 802.11g shields for wireless communications with it. Finally, a workstation is used to host the remote Cloud applications. It is connected to the campus network and communicates with the HS through standard TCP/IP protocol.

ICN communications are deployed with the CCN-Lite software (www.ccn-lite.net/), a lightweight implementation of the CCN/NDN protocols that has been properly extended to support smart home services. The Raspberry Pi also hosts the *iSapiens* core components operating at the Fog Layer, which consist of: *(i)* the *Agent Server*, a runtime environment for Agents execution and *(ii)* the *VO Container*, an entity that manages the VOs. A set of methods are defined to allow Agents to control and monitor ICN EDs, e.g., the *check* method returns sensing information. Complex application logics, based on specific rules, can also be defined. VOs are named resources and such names can be directly used to access them at the physical layer via ICN. The Raspberry Pi hosts a set of C programs that take as input ICN names and allow to send Interests and extract information from Data packets.

As application example, we consider a lighting system, which adjusts the lights in each room on the basis of people presence/movements and the current illuminance. At this purpose, we define a *Virtual Light* object, whose structure can be replicated for each room of the house thus creating the Virtual Kitchen Light, the Virtual Bedroom Light, and so on. Each virtual light object includes the following action functionalities: *(i) light-on* and *light-off*, to switch on or off the light; *(ii) increase-light* and *decrease-light*, to increase/decrease the light brightness in the zone. In addition, two sensing functionalities are deployed: *(i) illuminance*, to identify the illuminance in the zone and *(ii) near-people*, to identify the number of people in the zone.

When the application is active, the HS periodically sends Interest packets carrying the relative names, e.g., an Interest with name *sensing/illuminance/kitchen/zone1* is issued to query the illuminance sensor in the zone1 of the kitchen. The *Virtual Kitchen Light* collects the sensed values and makes them available to the LightAgent, which controls the lights. For instance, it switches on the light when the illuminance value is lower than a target threshold set by the user, and the human presence is detected. In this case, the VO acting method is invoked and a switch-on command is sent in an Interest packet, e.g., with name action/light/on/kitchen/zone1.

7 Conclusions

This chapter has discussed the technological solutions that integrate the Information Centric Networking (ICN) and the Cloud/Fog computing paradigms so as to combine their features in an Internet of Things scenario. The Fog computing paradigm has been promoted to complement the Cloud technology so as to provide intelligence at the edge of the network and offer low-latency services with context awareness. In parallel, ICN has been proposed as a novel mechanism to improve the communication in the future Internet, including IoT and Machine-to-Machine (M2M) environments with resource-constrained nodes.

In this chapter we focused on a representative IoT scenario, i.e., the smart home, where a variety of monitoring, automation and control functions are required. A reference architecture is presented as proof-of-concept, together with a preliminary testbed. The deployed framework, ICN-iSapiens, conceived under an Italian academic/industrial project, shows a practical implementation of the ICN-Fog integration.

Acknowledgements This work was partially funded under grant PON03PE_00050_2 DOMUS "Cooperative Energy Brokerage Services", MIUR.

References

1. G. Fortino, P. Trunfio, *Internet of Things Based on Smart Objects: Technology, Middleware and Applications* (Springer International Publishing, 2014). https://doi.org/10.1007/978-3-319-00491-4
2. F. Bonomi, R. Milito, J. Zhu, S. Addepalli, Fog computing and its role in the Internet of Things, in *Proceedings of the First Workshop on Mobile Cloud Computing (MCC)* (2012), pp. 13–16
3. A. Guerrieri, V. Loscri, A. Rovella, G. Fortino, *Management of Cyber Physical Objects in the Future Internet of Things* (Springer International Publishing, 2016). https://doi.org/10.1007/978-3-319-26869-9
4. A. Giordano, G. Spezzano, A. Vinci, Rainbow: an intelligent platform for large-scale networked cyber-physical systems, in *UBICITEC* (2014), pp. 70–85
5. B. Ahlgren, C. Dannewitz, C. Imbrenda, D. Kutscher, B. Ohlman, A survey of information-centric networking, IEEE Commun. Mag. **50**(7) (2012)
6. M. Amadeo, C. Campolo, J. Quevedo, D. Corujo, A. Molinaro, A. Iera, R.L. Aguiar, A.V. Vasilakos, Information-centric networking for the Internet of Things: challenges and opportunities. IEEE Netw. **30**(2), 92–100 (2016)
7. M. Amadeo, O. Briante, C. Campolo, A. Molinaro, G. Ruggeri, Information centric networking for M2M communications: design and deployment. Comput. Commun. **89**, 105–116 (2016)
8. W. Shang, Y. Yu, R. Droms, L. Zhang, Challenges in IoT networking via tcp/ip architecture, tech. rep., NDN Project, Technical Report NDN-0038 (2016)
9. W. Shang, A. Bannis, T. Liang, Z. Wang, Y. Yu, A. Afanasyev, J. Thompson, J. Burke, B. Zhang, L. Zhang, Named data networking of things, in *2016 IEEE First International Conference on Internet-of-Things Design and Implementation (IoTDI)* (2016), pp. 117–128
10. W. Shang, Q. Ding, A. Marianantoni, J. Burke, L. Zhang, Securing building management systems using named data networking. IEEE Netw. **28**(3), 50–56 (2014)
11. O. Hahm, E. Baccelli, T.C. Schmidt, M. Wählisch, C. Adjih, L. Massoulié, Low-power Internet of Things with NDN and cooperative caching, in *Proceedings of 4th ACM Conference on Information-Centric Networking (ICN)* (2017)
12. J. Burke, P. Gasti, N. Nathan, G. Tsudik, Securing instrumented environments over content-centric networking: the case of lighting control and NDN, in *2013 IEEE Conference on Computer Communications Workshops (INFOCOM WKSHPS)* (IEEE, 2013), pp. 394–398
13. J. Burke, P. Gasti, N. Nathan, G. Tsudik, Secure sensing over named data networking, in *2014 IEEE 13th International Symposium on Network Computing and Applications (NCA)* (IEEE, 2014), pp. 175–180
14. W. Shang, Y. Yu, T. Liang, B. Zhang, L. Zhang, NDN-ACE: access control for constrained environments over named data networking, tech. rep., NDN Project. Technical Report NDN-0036, Revision 1 (2015)

15. M. Amadeo, C. Campolo, A. Iera, A. Molinaro, Information Centric Networking in IoT scenarios: the case of a smart home, in *2015 IEEE International Conference on Communications (ICC)* (IEEE, 2015), pp. 648–653
16. M. Amadeo, C. Campolo, A. Molinaro, Multi-source data retrieval in IoT via named data networking, in *Proceedings of the 1st International Conference on Information-Centric Networking* (ACM, 2014), pp. 67–76
17. M.A.M. Hail, M. Amadeo, A. Molinaro, S. Fischer, On the performance of caching and forwarding in information-centric networking for the IoT, in *International Conference on Wired/Wireless Internet Communication* (Springer, 2015), pp. 313–326
18. M.A. Hail, M. Amadeo, A. Molinaro, S. Fischer, Caching in named data networking for the wireless Internet of Things, in *2015 International Conference on Recent Advances in Internet of Things (RIoT)* (IEEE, 2015), pp. 1–6
19. E. Baccelli, C. Mehlis, O. Hahm, T.C. Schmidt, M. Wählisch, Information centric networking in the IoT: experiments with NDN in the wild, in *Proceedings of the 1st International Conference on Information-Centric Networking* (2014), pp. 77–86
20. G. Acs, M. Conti, P. Gasti, C. Ghali, G. Tsudik, Cache privacy in named-data networking, in *2013 IEEE 33rd International Conference on Distributed Computing Systems (ICDCS)* (2013), pp. 41–51
21. A. Carzaniga, M. Papalini, A.L. Wolf, Content-based publish/subscribe networking and information-centric networking, in *Proceedings of the ACM SIGCOMM Workshop on Information-Centric Networking* (ACM, 2011), pp. 56–61
22. A.Z. Alkar, U. Buhur, An Internet based wireless home automation system for multifunctional devices. IEEE Trans. Consum. Electron. **51**, 1169–1174 (2005)
23. J. Serra, D. Pubill, A. Antonopoulos, C. Verikoukis, Smart HVAC control in IoT: energy consumption minimization with user comfort constraints. Sci. World J. (2014)
24. P. Rashidi, D.J. Cook, Com: a method for mining and monitoring human activity patterns in home-based health monitoring systems. ACM Trans. Intell. Syst. Technol. **4**, 64:1–64:20 (2013)
25. N. Pavón-Pulido, J.A. López-Riquelme, J. Ferruz-Melero, M.A. Vega-Rodríguez, A.J. Barrios-León, A service robot for monitoring elderly people in the context of ambient assisted living. J. Ambient Intell. Smart Environ. **6**(6), 595–621 (2014)
26. F. Cicirelli, G. Fortino, A. Giordano, A. Guerrieri, G. Spezzano, A. Vinci, On the design of smart homes: a framework for activity recognition in home environment. J. Med. Syst. **40**, 200 (2016)
27. P. Richter, M. Toledano-Ayala, G.M. Soto-Zarazúa, E.A. Rivas-Araiza, A survey of hybridisation methods of GNSS and wireless LAN based positioning system. J. Ambient Intell. Smart Environ. **6**, 723–738 (2014)
28. L. Sang-hyun, J.-G. Lee, M. Kyung-il, Smart home security system using multiple ANFIS. Int. J. Smart Home **7**, 121–132 (2013)
29. D. Miorandi, S. Sicari, F. De Pellegrini, I. Chlamtac, Internet of Things. Ad Hoc Netw. **10**, 1497–1516 (2012)
30. G. Fortino, A. Guerrieri, W. Russo, Agent-oriented smart objects development, in *2012 IEEE 16th International Conference on Computer Supported Cooperative Work in Design (CSCWD)* (May, 2012), pp. 907–912
31. I. Bierhoff, A. van Berlo, J. Abascal, B. Allen, A. Civit, K. Fellbaum, E. Kemppainen, N. Bitterman, D. Freitas, K. Kristiansson, *Smart Home Environment* (COST, Brussels, 2007)
32. A. Savidis, C. Stephanidis, Distributed interface bits: dynamic dialogue composition from ambient computing resources. Pers. Ubiquitous Comput. **9**, 142–168 (2005)
33. J.E. Bardram, R.E. Kjær, M. Pedersen, Context-aware user authentication—supporting proximity-based login in pervasive computing, in *UbiComp 2003: Ubiquitous Computing* ed. by A. Dey, A. Schmidt, J. McCarthy, (eds.), vol. 2864 of Lecture Notes in Computer Science (Springer Berlin Heidelberg, 2003), pp. 107–123
34. M. Román, C. Hess, R. Cerqueira, A. Ranganathan, R.H. Campbell, K. Nahrstedt, A middleware infrastructure for active spaces. IEEE Pervasive Comput. **1**, 74–83 (2002)

35. R. Cerqueira, C. Cassino, R. Ierusalimschy, Dynamic component gluing across different componentware systems, in *Proceedings of the International Symposium on Distributed Objects and Applications, DOA '99* (Washington, DC, USA) (IEEE Computer Society, 1999), pp. 362–371
36. O. Evangelatos, K. Samarasinghe, J. Rolim, Syndesi: a framework for creating personalized smart environments using wireless sensor networks, in *Proceedings of the 2013 IEEE International Conference on Distributed Computing in Sensor Systems, DCOSS '13* (Washington, DC, USA) (IEEE Computer Society, 2013), pp. 325–330
37. C. Mastroianni, E. Cesario, A. Giordano, Efficient and scalable execution of smart city parallel applications, in *Concurrency and Computation: Practice and Experience*, Aug. 2017. Early view, http://dx.doi.org/10.1002/cpe.4258
38. A. Zanella, N. Bui, A. Castellani, L. Vangelista, M. Zorzi, Internet of things for smart cities. IEEE Internet of Things J. **1**, 22–32 (2014)
39. P. Garcia Lopez, A. Montresor, D. Epema, A. Datta, T. Higashino, A. Iamnitchi, M. Barcellos, P. Felber, E. Riviere, Edge-centric computing: vision and challenges. SIGCOMM Comput. Commun. Rev. **45**, 37–42 (2015)
40. F. Bonomi, R. Milito, J. Zhu, S. Addepalli, Fog computing and its role in the internet of things, in *Proceedings of the First Edition of the MCC Workshop on Mobile Cloud Computing*, MCC '12, (New York, NY, USA) (ACM, 2012), pp. 13–16
41. M. Yannuzzi, F. van Lingen, A. Jain, O.L. Parellada, M.M. Flores, D. Carrera, J.L. Perez, D. Montero, P. Chacin, A. Corsaro, A. Olive, A new era for cities with fog computing. IEEE Internet Comput. **21**, 54–67 (2017)
42. W. Shi, J. Cao, Q. Zhang, Y. Li, L. Xu, Edge computing: vision and challenges. IEEE Internet of Things J. **3**(5), 637–646 (2016)
43. M. Chiang, S. Ha, I. Chih-Lin, F. Risso, T. Zhang, Clarifying fog computing and networking: 10 questions and answers. IEEE Commun. Mag. **55**(4), 18–20 (2017)
44. M. Sifalakis, B. Kohler, C. Scherb, C. Tschudin, An information centric network for computing the distribution of computations, in *Proceedings of the 1st ACM International Conference on Information-centric Networking* (2014), pp. 137–146
45. Y. Ye, Y. Qiao, B. Lee, N. Murray, Piot: programmable IoT using information centric networking, in *2016 IEEE/IFIP Network Operations and Management Symposium (NOMS)* (2016), pp. 825–829
46. Q. Wang, B. Lee, N. Murray, Y. Qiao, Cs-man: computation service management for IoT in-network processing, in *IEEE Signals and Systems Conference (ISSC)* (2016), pp. 1–6
47. M. Król, I. Psaras, NFaaS: named function as a service, in *Proceedings of the 4th ACM Conference on Information-Centric Networking* (2017), pp. 134–144
48. C.-A. Sarros, A. Lertsinsrubtavee, C. Molina-Jimenez, K. Prasopoulos, S. Diamantopoulos, D. Vardalis, A. Sathiaseelan, Icn-based edge service deployment in challenged networks, in *Proceedings of the 4th ACM Conference on Information-Centric Networking* (ACM, 2017), pp. 210–211
49. S.S. Adhatarao, M. Arumaithurai, X. Fu, FOGG: a fog computing based gateway to integrate sensor networks to internet, in *IEEE Teletraffic Congress (ITC 29)*, vol. 2 (2017), pp. 42–47
50. M. Amadeo, A. Molinaro, S.Y. Paratore, A. Altomare, A. Giordano, C. Mastroianni, A cloud of things framework for smart home services based on information centric networking, in *2017 IEEE 14th International Conference on Networking, Sensing and Control (ICNSC)* (2017), pp. 245–250
51. F. Cicirelli, A. Guerrieri, G. Spezzano, A. Vinci, An edge-based platform for dynamic smart city applications. Future Gener. Comput. Syst. **76**, 106–118 (2017)
52. F. Cicirelli, A. Guerrieri, G. Spezzano, A. Vinci, O. Briante, A. Iera, G. Ruggeri, Edge computing and social internet of things for large-scale smart environments development. IEEE Internet of Things J. (2017)

Optimal Placement of Security Resources for the Internet of Things

Antonino Rullo, Edoardo Serra, Elisa Bertino and Jorge Lobo

Abstract In many Internet of Thing application domains security is a critical requirement, because malicious parties can undermine the effectiveness of IoT-based systems by compromising single components and/or communication channels. Thus, a security infrastructure is needed to ensure the proper functioning of such systems even under attack. However, it is also critical that security be at a reasonable resource and/or energy cost. This chapter deals with the problem of efficiently and effectively securing IoT networks by carefully allocating security resources in the network area. The problem is modeled according to game theory, and provide a Pareto-optimal solution, in which the cost of the security infrastructure and the probability of a successful attack are minimized. As in the context of smart urban ecosystems both static and mobile smart city applications can take place, two different formalizations are provided for the two scenarios. For static networks, the optimization problem is modeled as a mixed integer linear program, whereas for mobile scenarios, computational intelligent techniques are adopted for providing a good approximation of the optimal solution.

A. Rullo (✉)
DIMES Department, Universita della Calabria, 87036 Rende, Italy
e-mail: n.rullo@dimes.unical.it

E. Serra
Department of Computer Science, Boise State University, Boise, ID 83725, USA
e-mail: edoardoserra@boisestate.edu

E. Bertino
Lawson Computer Science Department, Purdue University, West Lafayette, IN 47907, USA
e-mail: bertino@purdue.edu

J. Lobo
ICREA and Department of Information and Communication Technologies, Universitat Pompeu Fabra, 08018 Barcelona, Spain
e-mail: jorge.lobo@upf.edu

F. Cicirelli et al. (eds.), *The Internet of Things for Smart Urban Ecosystems*, Internet of Things, https://doi.org/10.1007/978-3-319-96550-5_5

1 Introduction

The term *Internet of Things* (IoT) refers to the interconnection of small devices able to interact with each other and cooperate in order to accomplish common tasks. Many kinds of systems and technologies fall under the definition of IoT: identification and tracking technologies, wired and wireless sensor and actuator networks, enhanced communication protocols and distributed intelligence for smart objects, home automation, smart city applications [3]. It is critical that an effective security infrastructure be adopted, in order to ensure the integrity and reliability of exchanged data even under attack. Besides effectiveness, another important requirement for such an infrastructure is to carefully address security resource efficiency, as an infrastructure with very high demands in terms of additional resources for security would not be viable. Its cost would in fact outweigh the low cost of IoT technologies. Despite IoT networks can differ from each other in several aspects such as topology, size, location, etc., they also have some common characteristics that can be leveraged in the design process of a security infrastructure. They are prone to physical attacks and eavesdropping, since they are often unattended and typically communicate via wireless channels. They make use of (battery powered) devices with low computational power. They share known software and hardware vulnerabilities, and as a consequence, also many kinds of security systems, such as intrusion detection and prevention systems. Based on such common features, it is possible to design a general method that works for different IoT scenarios for designing an effective and efficient ad-hoc security resource allocation plan. By security resources we mean a combination of security tools like intrusion detection systems (IDS), intrusion prevention systems (IPS), special hardware, passive monitors of wireless traffic, etc., and by plan, the number and locations of such resources. Both efficiency and effectiveness depend on the choice of the security tools and how they are allocated in the system of interest. Among all possible security resource allocation plans, some are more efficient, as they require a lower amount of energy consumption and/or entail a cheaper cost in terms of additional equipment, while other plans are more effective, because harder to be evaded. As smart urban ecosystems can be implemented with either static (e.g. a smart city lightening system) or mobile (e.g. a vehicular network) IoT network devices, we make a distinction between static and mobile networks, for which the problem of computing the best allocation plan is addressed with different approaches. For static networks, the optimization problem is modeled as a mixed integer linear program, and an optimal solution is provided. For mobile networks instead, as device mobility confers different shapes to the monitored system, a security resource allocation plan must be designed in such a way to ensure a certain security level in the most of cases. To this end, computational intelligent techniques are adopted, that provide a good approximation of the optimal solution.

The rest of this chapter is organized as follows: in Sects. 2 and 3 a solution to the problem of computing the best resource allocation plan for static and mobile IoT networks, respectively, is provided; in Sect. 4 related works are discussed; finally, Sect. 5 concludes the chapter. Sections 2 and 3 are based on the content of [36, 37], respectively, where more detailed experimental results can be found. In this chapter we do not report any additional result with respect to [36, 37].

2 Static Networks

We start from the assumption that, given a set of available security tools, we can build a security resource as a combination of one or more tools, and we can estimate its installation cost and its average energy consumption at operating speed. We can thus estimate the efficiency of an allocation plan, based on the total energy consumption of the security infrastructure and the costs of its components. To measure the effectiveness of an allocation plan we use two metrics, namely *risk* and *criticality*. The former is defined as the maximum number of network nodes that are no longer protected when an attacker succeeds in taking down at least one security resource. The latter is a measure of how critical certain nodes are for the correct operation of the network.

Given a set of security resources $\mathscr{R}$ and a set of attacks to be faced $\mathscr{A}$, our method chooses the subset of $\mathscr{R}$ that best addresses the attacks in $\mathscr{A}$, and the set of locations where to place them, that ensure a certain security level at the minimum cost and energy consumption. We model the interaction between attacker and defender as a Stackelberg leadership model in which the leader (the defender) moves before the follower (the attacker) [42].

We compute a Pareto-optimal defender strategy (resource allocation plan) in two steps. First, among all possible allocation plans, we compute the subset of the Pareto-optimal plans, as the solutions of a three-objective optimization problem, which minimize *(i)* the total energy consumption, *(ii)* the installation cost, and *(iii)* the maximum criticality. In the second step, we select the plan, among those identified in the previous step, which minimizes the risk—as the solution of a single-objective optimization problem. The resulting defender strategy is efficiency-optimal because it entails the lowest energy consumption and the cheapest installation cost, while its effectiveness lies in the fact that the number of security resources an attacker needs to take down is maximized—and consequently, the probability of a successful attack is reduced.

2.1 Basic Concepts and Notation

Here we introduce the basic concepts and notations that help us to formalize the IoT environment and the security requisites for the defender.

Definition 1 (*security resource*) A *security resource* is a tuple $sr = \langle c, e, t, loc \rangle$, where:

- c is the cost;
- e is the energy consumption;
- t is the resource's type;
- loc is the location.[1]

The value of the attribute t (type) depends on the set of embedded security tools, so that security resources with different functions will be of different type.

Definition 2 (*network*) A *network* is an undirected graph $N = \langle V, E \rangle$, where:

- $V = \{(e, loc, cr)$: e is the node energy consumption, loc is the node location, and cr is the criticality value$\}$ is the set of network nodes;
- $E = \{\{v_1, v_2\} : v_1$ is in the communication range of v_2, and vice versa, and $v_1, v_2 \in V\}$ is the set of links.

We define a network as an undirected graph because we consider the *physical*-based topology, i.e. the one inferred by the communication range of network nodes, different from the *routing*-based topology, which is always a sub-graph of the physical one, whatever routing protocol is adopted, and whatever is the direction of the links. Thus, also devices directly connected to the Internet are considered part of a network if they fall in the communication range of other devices.

We use the notation $v.x$ and $sr.y$ to denote attributes x and y, of node v and and resource sr, respectively. Given a network N and a set of security resources $\mathscr{R}$, we define:

- $AP \subseteq \mathscr{R}$ as a resource allocation plan for N;
- $\mathscr{AP}$ as the set of all possible resource allocation plans for N;
- the *criticality* of a network node $n \in N$ as the measure of its relevance for the correct function of the network;
- function *crit*: $\mathscr{R} \times \mathscr{AP} \times N \rightarrow \mathbb{N}$ as the criticality associated with a security resource sr; given an allocation plan $AP \in \mathscr{AP}$ and a network N, $crit(sr, AP, N)$ returns the maximum criticality value over the set of nodes of N that are not any longer protected if sr stops working and no other security resource $sr' \in AP$ covers them;
- function *risk*: $\mathscr{R} \times \mathscr{AP} \times N \rightarrow \mathbb{N}$ as the risk associated with a resource sr; given an allocation plan $AP \in \mathscr{AP}$ and a network N, $risk(sr, AP, N)$ returns the number of nodes of N that are not any longer protected if sr stops working and no other security resource $sr' \in AP$ covers them.
- the function *Neighbors*: $V \rightarrow 2^V$, such that $Neighbors(v) = \{v' : \{v, v'\} \in E\}$;

---

[1]The attribute loc in Definition 1 helps us to simplify the formalization of the linear programs we show hereafter. The basic idea is the following: given that in the network area there are many locations where a security resource can be placed, for each resource sr^* we assume to have $sr_1, \ldots sr_n$ resources, one for each location where sr^* can be located.

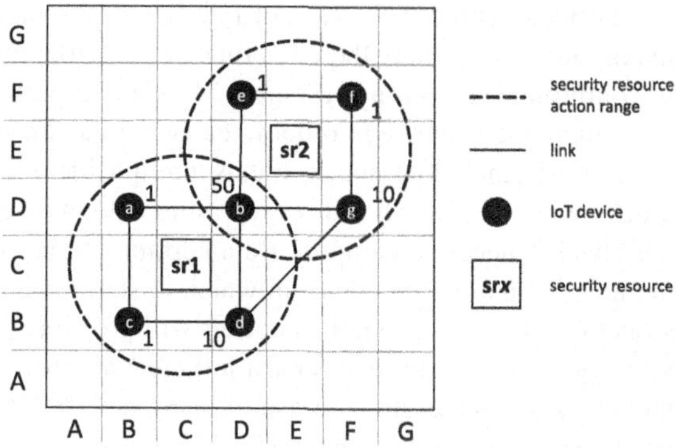

Fig. 1 An example network: the main square represents the network area; locations of the network area are represented by internals dotted squares

- the function $Edges$: $V \rightarrow 2^E$, such that $Edges(v) = \{\{v, v'\} : v' \in Neighbors(v)\}$;
- $\mathscr{L} = L_V \cup L_A$ as the set of locations, where L_V is the set of locations taken by each $v \in V$, and L_A is the set of all other available locations in the network area;
- the function $Res : 2^{\mathscr{L}} \rightarrow 2^{\mathscr{R}}$, such that, given a set of locations X, $Res(X) = \{sr : sr.loc \in X\}$, i.e., the resources that can be placed on locations in X;
- the *domain* of a security resource sr as the set of locations that are in the action range of sr, i.e., the portion of network area covered by sr;
- the function Dom: $2^{\mathscr{R}} \rightarrow 2^{\mathscr{L}}$, such that given a set of resources X, $Dom(X)$ returns the set of locations that belong to the domain of all $sr \in X$;
- $T = \{sr.t : sr \in \mathscr{R}\}$ as the set of security resources types.

Example 1 Figure 1 shows a network with 7 nodes protected by two security resources sr_1 and sr_2. The number associated to each node denotes the node's criticality value (here the criticality is proportional to the number of links that involve a node).

If one of the two security resources stops working, the set of nodes left without protection is $S_1 = \{a, c, d\}$ for sr_1, and $S_2 = \{e, f, g\}$ for sr_2. The node with maximum criticality in S_1 is node d (criticality=10), while in S_2 is node g (criticality=10). The criticality and risk values of the two security resources, thus, are: $crit(sr_1, AP, N) = crit(sr_2, AP, N) = 10$, and $risk(sr_1, AP, N) = risk(sr_2, AP, N) = |S_1| = |S_2| = 3$, with $AP = \{sr_1, sr_2\}$. Furthermore, we have that:

- $Neighbors(a) = \{c, b\}$, $Neighbors(e) = \{f, b\}, \ldots$;
- $Edges(a)$=$\{\{a, c\}, \{a, b\}\}$, $Edges(e)$=$\{\{e, b\}, \{e, f\}\}, \ldots$;
- $\mathscr{L} = \{\langle x_1, x_2\rangle : x_1, x_2 \in \{A, B, C, D, E, F, G\}\}$;
- $L_V = \{\langle B, B\rangle, \langle B, D\rangle, \langle D, B\rangle, \langle D, D\rangle, \langle D, F\rangle, \langle F, F\rangle \langle F, D\rangle\}$;
- $L_A = \mathscr{L} \backslash L_V$;
- $Res(\langle C, C\rangle) = \{sr_1\}$, $Res(\langle E, E\rangle) = \{sr_2\}$, $Res(\langle C, C\rangle, \langle E, E\rangle) = \{sr_1, sr_2\}$;
- $Dom(sr_1) = \{\langle x_1, x_2\rangle : x_1, x_2 \in \{B, C, D\}\}$, $Dom(sr_2) = \{\langle x_1, x_2\rangle : x_1, x_2 \in \{D, E, F\}\}$.

The idea behind the notion of risk lies in the fact that the minimization of the maximum risk restricts the operating range of the attacker, in case (s)he manages to compromise a security resource. Thus, (s)he probably will need to compromise more than one resource in order to carry out the attack. For instance, during a *sybil attack* [39] a malicious node presents multiple identities to other nodes. The attack is much more effective when those identities belong to real nodes of the network. Thus, the attacker needs to compromise a certain number of nodes, from which to steal the identity, before the one from which to start the attack. However, if the defender strategy is well designed, the attacker will probably have to damage more than one security resource in order to steal a sufficient number of identities, therefore delaying the attack and increasing the risk of being detected. Other types of attacks, instead, need just to compromise one node. This is the case of the *black hole* attack [39], in which the compromised node drops all the incoming packets. A black hole is much more effective when the attacker chooses to compromise a critical node, such as one with a high incoming traffic rate. In this case, a defender strategy must provide stronger coverage for the most critical nodes, i.e., those that are more relevant for the correct function of the network. An attacker will thus be forced to compromise more than one resource in order to attack the most critical nodes.

2.2 Definition of Secure Network

A formal definition of *secure network* that matches the techniques adopted by the security systems is crucial, since it gives the guidelines for the formalization of the linear constraints we will use in the Pareto analysis. We can classify security systems into two main categories: *detection* and *prevention* systems. Those categories correspond to two different security policies: *(i) node/link monitoring*, and *(ii) node/link hardening*. As an example of link monitoring we mention the IoT IDS by Raza et al. [34], which checks the link quality for the detection of network layer and routing attacks, and the Liteworp system [21] that uses guard nodes as a countermeasure for wormhole attacks in WSNs. We would like also to mention Dataguard [44] as an example of node monitoring based on a code attestation technique to check the presence of malicious code in the memory of network nodes. Titan [5] is an example of a security architecture for hardening tiny devices with a hardware-assisted dynamic root of trust. The encryption key management scheme by Eschenauer et al. [13], to harden links among nodes neighbors, is an example of link hardening. According to those two security policies, we derive the definition of secure network.

Definition 3 (*secure link*) A link *{v,v'}* is *secure* if at least one of the following conditions holds:

(σ) both nodes *v* and *v'* are in the communication range of the same watchdog;
(γ) both nodes *v* and *v'* can establish a secure communication channel.

Definition 4 (*secure node*) A node v is *secure* if at least one of the following conditions holds:

(α) v is tamper resistant;
(β) every link that involves v is secure.

In condition (α) the term *tamper resistant* means that the node is equipped with a security tool that makes it inaccessible to attackers wishing to compromise it. In other words, Definition 4 states that the correct functioning of a node can be guaranteed by avoiding malicious code injection, or by monitoring/strengthening the links connecting to the other nodes. In Definition 3, a *watchdog* [26] is a security resource that monitors network nodes behavior. Past approaches to intrusion detection use watchdogs for overhearing in/out-coming traffic of neighbors nodes, performing code attestation, checking the signal strength, etc. Conditions (α) and (γ) capture the policy adopted by prevention systems (node/link hardening), while conditions (β) and (σ) the policy adopted by detection systems (node/link monitoring).

We can now define a network $N = \langle V, E\rangle$ as secure if each node $v \in V$ is secure, according to Definition 4.

2.3 Players' Strategy

In this section, we define the concept of "strategy" for the defender and the attacker. Their interaction is modeled as a Stackelberg game [42]. In such a game the defender plays the leader and makes the first move by installing a security infrastructure AP. The attacker plays the follower by trying to compromise one or more security resources $sr \in AP$, so that the attack can be carried out on the nodes that are no longer protected by the damaged security resources.

2.3.1 Defender's Strategy

The defender strategy consists of a security resource allocation plan. Given that each security resource entails an installation cost and some energy consumption, the best resource allocation plan for the defender is the one that provides a reasonable balance between efficiency, in terms of energy consumption and cost, and effectiveness, in terms of maximum risk and maximum criticality. The best plan is computed in two steps. In the first step we perform a Pareto analysis which solves the optimization problem defined by the following equation:

$$\min_{AP \in \mathcal{AP}} \{ec(AP), tc(AP), \max_{sr \in AP} crit(sr, AP, N)\} \quad (1)$$

where $ec(AP)$ and $tc(AP)$ denote the total energy consumption and the total cost of the allocation plan AP, respectively, and $\max_{sr \in AP} crit(sr, AP, N)$ is the maximum criticality value over the set of nodes of N that are no longer protected when one resource in AP stops working. The Pareto analysis consists in computing a set of Pareto points $p = (ec, tc, cr)$, that we refer to as Pareto curve. Each point corresponds to a set of allocation plans, i.e., all the plans (strategies) that have an energy consumption, a total cost, and a maximum criticality equal to the values of ec, tc and cr, respectively. Then, we choose the point p^+ whose values best fit our efficiency and criticality requirements. In the second step we compute the best allocation plan (best defender strategy), by solving the optimization problem defined by the following equation:

$$\min_{AP \in AP^+} \{\max_{sr \in AP} risk(sr, AP, N)\} \tag{2}$$

where $AP^+ \subseteq \mathcal{AP}$ is the set of allocation plans that entail an energy consumption, a cost, and a maximum criticality as the values of the Pareto point p^+ chosen in the previous step; and $\max_{sr \in AP} risk(sr, AP, N)$ is the maximum number of nodes of N that remain unprotected when a security resource in AP stops working.

2.3.2 Attacker's Strategy

We assume that an attacker needs to compromise at least one security resource in order to carry out an attack. In fact, whatever is the attack, an attacker always needs to first open a breach in the security, before attacking the system. We address the worst case attacker, that is the one who knows the defenders strategy AP, and plays her/his best strategy, i.e., chooses to compromise the security resource that maximizes the risk (i.e., leaves unprotected the maximum number of nodes), the criticality (i.e., leaves unprotected the most critical nodes), or a combination of both. An attacker strategy is represented by a security resource $sr \in AP$. The best attacker strategy is defined as follows:

$$sr^* = \max_{sr \in AP} \alpha \cdot risk(sr, AP, N) + \beta \cdot crit(sr, AP, N) \tag{3}$$

where $\alpha + \beta = 1$.

2.4 Overview of the Pareto Analysis

Pareto analysis [27] is a classic optimization method used in situations in which there are multiple competing objectives that must somehow be satisfied simultaneously. The basic idea behind Pareto optimization of three competing objective functions ϕ_1, ϕ_2 and ϕ_3 subject to a set $\mathcal{C}$ of constraints is as follows. Suppose that $\sigma = (x_1, x_2, x_3)$ and $\sigma' = (x'_1, x'_2, x'_3)$ are two different solutions (resource allocation plans) and suppose that ϕ_1, ϕ_2 and ϕ_3 are minimization problems. We say that σ *dominates* σ', denoted $\sigma \rhd \sigma'$, iff:

$$(x_1 \leq x_1' \wedge x_2 \leq x_2' \wedge x_3 \leq x_3') \wedge (x_1 < x_1' \vee x_2 < x_2' \vee x_3 < x_3')$$

A solution σ is said to be *Pareto optimal* w.r.t. a set of minimization problems Φ, and constraints $\mathscr{C}$ if and only if there is no solution $\sigma' \neq \sigma$ such that $\sigma' \rhd \sigma$. We use Pareto analysis to solve the optimization problem defined by Eq. 1 in order to find a compromise strategy. The main point of this method is the computation of the Pareto curve (see [27]).

2.5 First Step: Pareto Analysis for the Defender

To perform our Pareto analysis, let

$$\mathscr{P} = \{(ec(AP), tc(AP), \max_{sr \in AP} crit(sr, AP, N)) | AP \in \mathscr{AP}\}$$

be the set of all possible values for our three objectives.

Definition 5 (*Pareto curve*) The *Pareto curve PC* for the three-objective optimization problem defined by Eq. 1 is the set $\{(a, b, c) \mid (a, b, c) \in \mathscr{P}$ and $\nexists (a', b', c')$ such that $(a', b', c') \rhd (a, b, c)\}$.

Algorithm 1

procedure COMPUTEPC(minCR,maxCR)
 for $cr = minCR - 1, \ldots, maxCR$ **do**
  $tc \leftarrow getTC(\infty, cr)$;
  while $(tc \neq null)$ **do**
   $ec \leftarrow getEC(tc, cr)$;
   $X \leftarrow X \cup (ec, tc, cr)$;
   $tc \leftarrow getTC(ec, cr)$;
  end while
 end for
 return Pareto-optimal points in X;
end procedure

where:

$$getEC(tc, cr) = \min_{\substack{AP \in \mathscr{AP} \\ tc(AP) \leq tc \\ \max_{sr \in AP} crit(sr, AP, N) \leq cr}} ec(AP)$$

$$getTC(ec, cr) = \min_{\substack{AP \in \mathscr{AP} \\ ec(AP) < ec \\ \max_{sr \in AP} crit(sr, AP, N) \leq cr}} tc(AP)$$

Algorithm 1 computes the Pareto curve of the three-objective optimization problem defined by Eq. 1. More specifically, by fixing each possible value of criticality cr (Line 2), the algorithm computes the set X of points $p = (ec, tc, cr)$ (Lines 3–8), where ec and tc are the total energy consumption and the total cost, respectively. The first step is to compute the minimum total cost without considering the energy consumption (Line 3). This optimization is performed by the function *getTC*, and the value is assigned to the variable tc. Then, given tc and cr, function *getEC* computes the minimum total energy consumption ec (Line 5), and the point (ec, tc, cr) is added to X (Line 6). Now the algorithm computes a different value of tc (Line 7). This value is obtained by the function *getTC*, which minimizes the total cost by imposing that the total energy consumption is strictly less than the one previously computed. If such value exists, the algorithm continues to generate more points by iterating in the while loop, otherwise it assigns $tc = null$ and exits the loop. Finally, in Line 10 the algorithm returns all Pareto (non-dominated) points of X. Notice that in Line 2, we start to enumerate the values of cr from $minCR - 1$, so that, as no node exists with $cr < minCR$, the plan being calculated with $cr = minCR - 1$ provides $risk = 0$, i.e., each network node is covered by at least two security resources. For allocation plans with $cr = 0$ it is not necessary to perform the second optimization step, since we already know that the risk will be also equal to 0. An example of Pareto curve can be found in Fig. 4. It can be noticed that no point (ec, c, cr) of the curve is dominated, according to the definition given in Sect. 2.4.

2.6 *Linear Constraints for the Pareto Analysis*

We formalize the set of basic constraints as follows.

2.6.1 Variables

- $x_1, \ldots, x_{|\mathscr{R}|}$; if resource sr_i belongs to AP, then $x_i = 1$, otherwise $x_i = 0$;
- $z_{11}, \ldots, z_{\lceil|\mathscr{L}||\mathscr{R}|\rceil}$; if location $loc_i \in Dom(sr_j)$ and $sr_j \in AP$, then $z_{ij} = 1$, otherwise $z_{ij} = 0$;
- $q_1, \ldots, q_{|V|}$; if $loc_i \in L_V$ belongs to the domain of no more then one security resource, then $q_i = 1$, otherwise $q_i = 0$;
- $l_1, \ldots, l_{|E|}$; if link e_i is secure, then $l_i = 1$, otherwise $l_i = 0$.

All variables are binary.

$$\sum_{sr_i \in Res(loc_j)} x_i \leq 1, \quad \forall loc_j \in \mathscr{L} \tag{1.4}$$

$$z_{ij} = x_j, \quad \forall sr_j \in \mathscr{R}, \forall loc_i \in Dom(sr_j) \tag{1.5}$$

$$\sum_{sr_j : loc_i \in Dom(sr_j)} z_{ij} \leq q_i + M \cdot (1 - q_i), \quad \forall loc_i \in L_V \tag{1.6}$$

$$\sum_{sr_j : loc_i \in Dom(sr_j)} z_{ij} \geq 2 - q_i, \quad \forall loc_i \in L_V \tag{1.7}$$

$$\sum_{sr_j \in Res(v_i.loc): sr_j.t \in T_\alpha} x_j + c_i \geq 1, \quad \forall v_i \in V \tag{1.8}$$

$$\sum_{e_k \in Edges(v_i)} l_k < c_i + |Neighbors(v_i)|, \quad \forall v_i \in V \tag{1.9}$$

$$\sum_{e_k \in Edges(v_i)} l_k \geq c_i \cdot |Neighbors(v_i)|, \quad \forall v_i \in V \tag{1.10}$$

$$l_k \geq a_k, \quad l_k \geq b_k, \quad l_k \leq a_k + b_k, \quad \forall e_k \in E \tag{1.11}$$

$$\sum_{sr_j : \{v.loc, v'.loc\} \in Dom(sr_j) \wedge sr_j.t \in T_\sigma} d_{kj} \geq a_k, \quad \forall \{v, v'\}_k \in E \tag{1.12}$$

$$a_k \geq d_{kj}, \quad \forall \{v, v'\}_k \in E, \forall sr_j : \{v.loc, v'.loc\} \in Dom(sr_j) \wedge sr_j.t \in T_\sigma \tag{1.13}$$

$$\sum_{loc_i \in \{v.loc, v'.loc\}} z_{ij} \geq 2d_{kj}, \quad \forall \{v, v'\}_k \in E, \forall sr_j : \{v.loc, v'.loc\} \in Dom(sr_j) \wedge sr_j.t \in T_\sigma \tag{1.14}$$

$$\sum_{loc_i \in \{v.loc, v'.loc\}} z_{ij} < d_{kj} + 2, \quad \forall \{v, v'\}_k \in E, \forall sr_j : \{v.loc, v'.loc\} \in Dom(sr_j) \wedge sr_j.t \in T_\sigma \tag{1.15}$$

$$\sum_{loc_i \in \{v.loc, v'.loc\}} z_{ij} \geq 2b_k, \quad \forall \{v, v'\}_k \in E, \forall sr_j : \{v.loc, v'.loc\} \in Dom(sr_j) \wedge sr_j.t \in T_\gamma \tag{1.16}$$

$$\sum_{loc_i \in \{v.loc, v'.loc\}} z_{ij} < b_k + 2, \quad \forall \{v, v'\}_k \in E, \forall sr_j : \{v.loc, v'.loc\} \in Dom(sr_j) \wedge sr_j.t \in T_\gamma \tag{1.17}$$

$$z_{ij} + q_i \leq 1, \quad \forall loc_i \in L_V, \forall sr_j \in \mathscr{R} : sr_j.t \in T_\alpha \wedge sr_j.loc = loc_i \tag{1.18}$$

$$\sum_{sr_j : loc_i \in Dom(sr_j)} z_{ij} \geq 1, \quad \forall loc_i \in \mathscr{L} \tag{1.19}$$

Fig. 2 Basic constraints

2.6.2 Basic Constraints

The set of basic constraints is shown in Fig. 2. C.1.4 says that no more than one security resource can be placed on the same location.[2] C.1.5 says that variable z_{ij} is assigned value 1 if $sr_j \in AP$ and $loc_i \in Dom(sr_j)$. C.1.6 and C.1.7 capture the fact that, if a location $loc_i \in L_V$ is guarded by no more than one security resource, then $q_i = 1$, otherwise $q_i = 0$. Here, M is a constant big enough to satisfy the constraints for all values of q_i and the *lhs* of the equation (any integer greater than $\mathscr{R}$). Variables q_i are used to compute the risk and the criticality associated with each resource $sr \in AP$. C.1.8 captures the definition of secure node, according to which a node is secure if conditions (α) or (β) of Definition 4 hold, i.e., expressions $\sum_{sr_j \in Res(v_i.loc): sr_j \in T_1} x_j$ or c_i, respectively, equal 1. The value of variable c_i is computed by constraints 1.9 and 1.10, and it has value equal to 1 if all links belonging to $Edges(v_i)$ are secure, otherwise it is equal to 0. The definition of secure link is captured by C.1.11, according to which a link is secure ($l_k = 1$) if conditions (σ) or (γ) of Definition 3 hold, i.e., variables a_k or b_k equal 1. The values of these variables are computed by C.1.12–1.15 and C.1.16–1.17, respectively.

C.1.18 says that for tamper resistant nodes the variable q_i is forced to 0, i.e., it does not affect the risk and criticality values of the allocation plan is being computed. Finally, C.1.19 says that each location has to belong to the domain of at least one

[2]C.1.4 does not imply any loss of generality since a security resource can embed more than one security tool.

security resource in AP. This last constraint is fundamental because it addresses the dynamic characteristic of IoT networks. In an IoT environment, as new devices can enter and leave the network area, it is important to monitor all locations where any new device can move over time. When this happens, we are sure to have at least one security resource monitoring its location. We emphasize that the concept of dynamic topology we refer to is different from the concept of mobility. In our network model it is expected that the nodes are static, and that the topology changes due to the (dis)appearance of nodes. In C.1.8 and C.1.14–1.18 the symbols T_α, T_σ and T_γ are subset of T, the set of security resource types. In particular, T_α is the set of types of security resources that make a node tamper resistant, T_σ is the set of types of security resources that behave as watchdogs, and T_γ is the set of types of security resources that make a node able to establish a security communication channel with its neighbors. In Sect. 2.9 we implement the proposed method for two real-case IoT scenarios where we show how to compute such sets.

2.6.3 Linear Programs

The basic constraints capture the dependencies between the variables in the integer linear programs shown in Fig. 3.

The objective function of the formulation in Fig. 3(top) says that we want to minimize the total energy consumption of the security infrastructure. Constraint 1.20 says that the total cost must be lower or equal to tc, and Constraint 1.21 says that the criticality value of the nodes protected by no more than one security resource must not exceed cr. The objective function of the linear program in Fig. 3(center) minimizes the total cost of the set of security resources. Constraint 1.22 imposes that the total energy consumption be strictly less than ec, while Constraint 1.23 is equal to Constraint 1.21.

2.7 *Second Step: Best Defender Strategy*

Once we have computed the Pareto curve PC, we can choose the point $p^+ = (ec^+, tc^+, cr^+) \in PC$ closest to our requirements of energy consumption, cost, and criticality. p^+ identifies the set AP^+ of different allocation plans that have an energy consumption equal to ec^+, a cost equal to tc^+, and a criticality value equal to cr^+. Among all plans in AP^+ we choose the one which minimizes the risk, i.e., the maximum number of unprotected nodes an attacker can exploit after compromising one security resource. To do that, we solve the optimization problem defined by Eq. 2, and formalized as shown in Fig. 3(bottom). The intuition behind constraint 1.24 is that instead of bounding the risk associated with each security resource $sr \in AP$ on the right-hand side of this constraint, we set it to an unknown value h and then require the objective function to minimize h. Here, M is a constant big enough to ensure the satisfaction of the equation for all values of q_i and x_j (any integer greater than $|V|$).

$$getEC(tc, cr) = \textbf{minimize} \sum_{sr_i \in \mathscr{R}} sr_i.e \cdot x_i$$

subject to: basic constraints, and

$$\sum_{sr_i \in \mathscr{R}} sr_i.c \cdot x_i \leq tc \tag{1.20}$$

$$v_i.cr \cdot q_i \leq cr \qquad \forall v_i \in V \tag{1.21}$$

$$getTC(ec, cr) = \textbf{minimize} \sum_{sr_i \in \mathscr{R}} sr_i.c \cdot x_i$$

subject to: basic constraints, and

$$\sum_{sr_i \in \mathscr{R}} sr_i.e \cdot x_i < ec \tag{1.22}$$

$$v_i.cr \cdot q_i \leq cr \qquad \forall v_i \in V \tag{1.23}$$

$$getR(ec, tc, cr) = \textbf{minimize}\ h$$

subject to: basic constraints, and

$$\Big(\sum_{loc_i \in \{Dom(sr_j) \cap L_V\}} q_i\Big) - (1 - x_j) \cdot M \leq h \qquad \forall sr_j \in \mathscr{R} \tag{1.24}$$

$$\sum_{sr_i \in \mathscr{R}} sr_i.e \cdot x_i \leq ec \tag{1.25}$$

$$\sum_{sr_i \in \mathscr{R}} sr_i.c \cdot x_i \leq tc \tag{1.26}$$

$$v_i.cr \cdot q_i \leq cr \qquad \forall v_i \in V \tag{1.27}$$

$$h \geq 0 \tag{1.28}$$

Fig. 3 Linear programs for computing the best defender strategy

Constraints 1.25, 1.26 and 1.27 impose that the allocation plan must be computed so to have an energy consumption, a cost and a criticality value not greater than *ec*, *tc* and *cr*, respectively.

2.8 *Discussion*

Our method computes the best defender strategy based on four objectives, that are energy consumption, cost, and criticality, mutually optimized in a first phase (see Sect. 2.5), and risk, minimized in a second phase (see Sect. 2.7). We chose to split the optimization process in those two phases for three main reasons. First, a Pareto

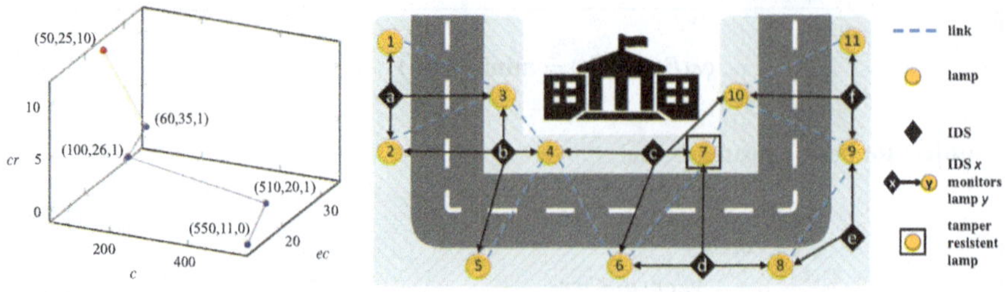

Fig. 4 Pareto curve (left) and the security reosurce allocation plans of Example 3. The Pareto points are expressend in the form (c, ec, cr)

analysis with more than three competing goals may require much more time, or alternatively, a much higher computing power in order to be performed. Second, a Pareto analysis involving less than three competing goals provides a much smaller set of solutions, thus limiting the administrator's decisional power. The reason is that the definition of dominated point becomes less restrictive when the number of dimensions decreases (see Sect. 2.4). The choice to optimize criticality in the first phase in place of risk is mainly a matter of convenience, since the range of criticality values is always much smaller than the range of risk values ($risk \in [0, |V|]$), and thus it is faster to enumerate (see Line 3 of Algorithm 1). By the way, the order in which the objectives are optimized is primarily a choice of the security officer, who might decide to employ the risk in the first phase in place of criticality because more important for the security of the network.

2.9 Examples of a Real Case Scenario

In this section we show how the proposed approach works for a real case IoT scenario, i.e., a smart street lighting system exposed to DoS attacks.

A DoS attack for such IoT scenario can be performed by either jamming the radio signal devices (lamps), such that they are not any longer able to communicate each other, or tampering with a device in a strategic position, in order to isolate part of the network. Suppose a system administrator willing to design the security plan before installing the lamps on the streets. (S)he can adopt two type of security resources, namely an IDS, for monitoring the links between lamps, and a tamper resistant lamp (TRL), which avoid a lamp to be compromised by an attacker. One IDS costs 10 monetary units (MU), consumes 5 energy units per hour (EUH), and can be deployed in the area delimited by the sidewalks. One TRL costs 50 MU, consumes 1 EUH, and has to be put in place of a normal lamp. The administrator has a security budget of 110 MU. Furthermore, (s)he wants to ensure a higher security level to the lamps close to the city council (lamps 3, 4, 7 and 10 of Fig. 4 (right)), thus (s)he assigns a criticality value $cr = 10$ to such lamps, and $cr = 1$ to the others.

The Pareto curve computed by Algorithm 1 is shown in Fig. 4(left). Two Pareto points outweigh the security budget (i.e., $(510, 20, 1)$ and $(550, 11, 0)$), thus they are not taken into account by the administrator. The other three points represent allocation plans capable of satisfying both the security and budget constraints. At this juncture, the administrator chooses the Pareto point on which perform the second step of our optimization method. For sake of completeness, we show in Fig. 4(right) the allocation plans obtained from the three Pareto points. The execution of the linear program *getR* on the point $(50, 25, 10)$ returns the plan consisting of IDSs *a, b, c, e, f*. With such plan, the nodes which would be left without security coverage in case at least one IDS stops working are two (lamps 6 and 7), thus risk = 2. For instance, if IDS *c* stops working (because compromised by the attacker, or for other reasons), lamps 6 and 7 would not be covered by any other IDS. Furthermore, lamps 6 and 7 have a criticality value of 1 and 10 respectively, thus the criticality associated to the plan is $max(1, 10) = 10$. $getR(60, 35, 1)$ returns the plan consisting of IDSs *a, b, c, d, e, f*, that has criticality = 1, because the lamps monitored by just one IDS have criticality = 1 (lamps 1, 5 and 11), and risk = 1, since no more than one lamp would remain unmonitored in case one IDS stops working. Finally, $getR(100, 26, 1)$ returns the plan consisting of IDSs *a, b, c, e, f*, and a TRL in place of lamp 7. Both criticality and risk equal 1 because the devices monitored by just one security resource are lamps 1, 5, 6, 8 and 11 (all with $cr = 1$), and no more than one lamp would remain unmonitored in case one IDS stops working.

3 Mobile Networks

The mobility of IoT devices makes the process of security provisioning more complicated than that described in the previous sections, where both the network topology and the number of network devices are assumed to be static. Mobile IoT networks, instead, continuously change topology because of mobility and/or (dis)appearance of devices. In such IoT environments, the geographical distribution and the number of connected devices are not stationary, and vary from spot to spot, according to the different activities taking place in the different areas. As an example of mobile IoT environment we can think of a university campus where thousands of people with several mobile connected devices continuously change their position, or also, of a vehicular network (VANET) that, although more homogeneous in the type of devices, assumes different shapes and different sizes according to the hour and day of the week. A security infrastructure must be able to address such a dynamic nature of IoT networks. To this end, a security resource allocation plan must take into account the numerous shapes the device mobility confers to the system of interest, and ensure a certain security level in the majority of cases.

We formalize a model for such scenarios, and provide a heuristic for computing allocation plans that minimize the risk of having IoT devices not monitored by any resource. We employ the key-concept of *shortfall* [4] as a risk measure. Shortfall is mostly used in economics to model the risk associated with an investment by combining in a single (risk) value the return of the investment in the worst scenarios together with the expected return. In fact, an investment might provide a return much more scarce than the expected one due to the significant changes that can affect the market. We adapt the concept of shortfall to model the risk associated with a security solution for which its effectiveness depends on how well the solution is able to address the continuous topology changes that affect the system of interest.

3.1 Defender Side

The goal of the defender is to passively monitor all wireless communications the devices moving in the area of interest do. To this end, (s)he needs to choose a set of security resources to monitor traffic and presence of devices, and decide where to deploy them in order to minimize the number of non-monitored devices. The geographical distribution and the number of connected devices in the area of interest are key parameters for the task of computing the best placement of security resources. Such distributions can be learned during a training period long enough to be a representative sample of the monitored eco-system. Following the two examples of IoT environments described above, in a campus area, most students follow patterns that repeat every day, and can be learned in a few months of an academic year: during the morning and the afternoon they are located at the classrooms; at lunch/dinner time they move to the dining areas; during the night they stay in the dorms or other in-campus accommodations. Such patterns can be learned by analyzing the network traces produced by the interaction between users' devices, or between access points and the devices. In a VANET, where cars move according to the hour and day of the week, such patterns can be learned from data collected by road side units (RSUs). Commuting patterns are also predictable by stochastic processes that capture local mobility decisions. Such processes help analytically derive commuting and mobility fluxes that require as input only information on the population distribution. The resulting model predicts mobility patterns in good agreement with mobility and transport patterns observed in a wide range of phenomena, from long-term migration patterns to communication volume between different regions [41].

In the next sections we show that for a security manager, the problem of computing the best allocation plan of security monitors for a changing IoT environment can be interpreted as the problem an investor faces when choosing an investment that maximizes the return, considering the continuous changes that affect the market.

3.2 Shortfall

In this section we briefly introduce the concept of shortfall, clarify the motivations that have lead us to adopt it in our formalization, and show how we manipulate it in order to obtain a more appropriate solution to our situation.

3.2.1 Definition

Shortfall is a risk measure used in economics which has conceptual, computational and practical advantages over other commonly used risk measures [4]. The shortfall, or more precisely, the *shortfall at level* α, measures how large losses, below the expected return, can be expected if the return of the investment drops below its α-quantile. Given all the possible market trends, we can compute the set of all possible returns of our investment. The shortfall at level α is the difference between the expected return and the average of the returns of the worst $\alpha\%$ of cases. Given an investment x and a value $\alpha \in (0, 1)$, the shortfall $s_\alpha(x)$ is defined as follows:

$$s_\alpha(x) = E[R] - E[R|R \leq q_\alpha(R)] \tag{4}$$

where $E[R]$ is the expected return of the investment, and $q_\alpha(X)$ is the α-quantile of a random variable X:

$$q_\alpha(X) = inf\{x|P(X \leq x) \geq \alpha\} \tag{5}$$

According to Levy and Kroll [24], for an investment x chosen to minimize $s_\alpha(x)$ for a fixed α and a given target mean μ_t, there is no other investment with the same mean which would be preferred to x, because less profitable in the worst $\alpha\%$ of cases. Thus, one is naturally led to minimize the quantity $s_\alpha(x)$ for some $\alpha \in (0, 1)$ as follows:

$$\begin{aligned} &\min && s_\alpha(x) \\ &\text{subject to } && E[R] = \mu_t \end{aligned} \tag{6}$$

3.2.2 From the Minimum Shortfall to the Best Choice

The solution we obtain by solving Problem 6 depends on the target mean μ_t we choose. In fact, given a target mean, we obtain the best investment (i.e., the one with minimum shortfall) among those with the same expected value, but we do not know whether there exists a better investment for different values of μ_t. In other words, there might exist a *dominating* investment, i.e., of a higher mean and a smaller shortfall, or alternatively, a *non-dominated* investment, i.e., of a higher (resp. lower) mean and a greater (resp. smaller) shortfall, which may be preferred by the investor. A possible solution is that of solving Problem 6 with different μ_t, and finally select the investment that best meets our needs. However, this solution is not efficient because it might compute some dominated investment, i.e., which would not be preferred to any other because of a lower expected value and a greater shortfall than those of the other ones. Consequently the time needed to compute such dominated solutions would be wasted. As alternative, we propose to adapt Problem 6 in such a way that it computes only non dominated solutions. This is possible by turning Problem 6 into a bi-objective optimization problem. First of all, we start considering the following equivalent problem:

$$\begin{aligned} \max \quad & E_\alpha[R] \\ \text{subject to } & E[R] = \mu_t \end{aligned} \tag{7}$$

where $E_\alpha[R] = E[R|R \leq q_\alpha(R)]$ is the mean of the worst $\alpha\%$ of cases, namely, the subtrahend in Eq. 4. Problem 7 is equivalent to Problem 6 because for a target mean μ_t, minimizing $E[R] - E_\alpha[R]$ is equivalent to maximizing $E_\alpha[R]$. Since we want to compute a set of non dominated investments based on their $E[R]$ and $E_\alpha[R]$ values, we place the expected value as an objective of the optimization problem, and we impose two inequality constraints on the values of $E[R]$ and $E_\alpha[R]$, in order to restrict the computation only to the cases we are interested in. The resulting problem is as follows:

$$\begin{aligned} \max \quad & E[R], E_\alpha[R] \\ \text{subject to } & E[R] \geq \mu_t \\ & E_\alpha[R] \geq \mu_\alpha \end{aligned} \tag{8}$$

Problem 8 is a less restricted version of Problem 6, where we split the two terms of the shortfall, $E[R]$ and $E_\alpha[R]$, and optimize them separately. The output is a Pareto frontier PF [27] whose points are non dominated investments x with values of $E[R]$ and $E_\alpha[R]$ below the desired thresholds. Formally, PF is defined as follows:

$$PF = \{(E[R], E_\alpha[R]) | \nexists (E'[R], E'_\alpha[R]) \in PF \text{ such that}$$
$$(E[R] \geq E'[R] \wedge E_\alpha[R] \geq E'_\alpha[R]) \wedge (E[R] > E'[R] \vee E_\alpha[R] > E'_\alpha[R])\}$$

An investor can thus conduct a cost-benefit analysis on the solutions of the Pareto frontier, and finally choose the one that best fits his/her requirements. In the next sections we show that the shortfall is well suited for describing the risk associated with an allocation plan, such that a certain level of security is ensured for every variation in the topology of the monitored environment.

3.3 Problem Definition

We define an IoT environment as a set of n geographic areas R_k, $k = 1, \ldots, n$, that we call *regions*. We divide each region in *locations*, i.e., space units where IoT devices or security resources could reside.

Definition 6 (*region*) A *region* R is a tuple $\langle L_R, P_R \rangle$, where:

- L_R is the set of locations of R;
- P_R is the probability distribution over a discrete random variable X that takes integer values in the interval $[0, \infty)$. $P_R(X = x)$ is the probability that there are x devices in R.

Definition 7 (*security resource*) A *security resource* sr is a tuple $\langle c, r, P_{sr} \rangle$, where:

- c is the cost;
- r (from radius) is the maximum action range;
- P_{sr} is the probability distribution over a discrete random variable Y that takes values in the interval $[0, r]$. $P_{sr}(Y = y)$ is the probability that sr is able to monitor a device located at a distance y.

Furthermore, we define $\mathscr{R} = \bigcup_{k=1}^{n} R_k$ as the set of all regions, and $\mathscr{L} = \{loc : loc \in \mathscr{R}\}$, as the set of all locations in $\mathscr{R}$. In the rest of the chapter we will use the notation $sr.c$ to denote the cost c of a security resource sr.

Security resources have an associated probability distribution, that describes how likely a device at a certain distance is seen by a resource. This choice is driven by the fact that the typical assumptions about all radios having circular range and perfect coverage in that range is far from real [22]. More realistic models take into account antenna height and orientation, terrain and obstacles, surface reflection and absorption, and so forth. It is often difficult in reality to estimate whether or not one has a functioning radio link between nodes, because signals fluctuate greatly due to mobility and fading as well as interference. Several signal attenuation models have thus been proposed [16, 32], and there are a few that can be used in concrete implementations of our model. For example, the Okumura model, the Hata model for urban areas, the Hata model for open areas, are models for outdoor attenuation; instead, the ITU model and the Log-distance path loss model are models for indoor attenuation. One of these models, or a combination of them, can be adopted for modeling the attenuation of a device according to its radio characteristics.

3.4 *From Economics to the IoT Domain*

The problem we face is that of computing the optimal security resource allocation plan for IoT scenarios characterized by a high degree of mobility. This problem can be reduced to Problem 8 if we consider an allocation plan along with its cost as an investment, and all the possible configurations a set of IoT devices may assume in the area of interest as all the possible market trends. A security manager (the investor) knows that the number of devices moving within a region R_k follows a certain probability distribution P_{R_k}, and that the resource allocation plan to choose must be able to provide a good security level not only in the average cases of P_{R_k}, but also in the rare ones. In fact, rare device configurations may result in high losses if not addressed by an adequate security infrastructure. We use $E[R]$ and $E_\alpha[R]$ as evaluation metrics of an allocation plan for an IoT environment, where the multitude of device configurations is wide because mobility. The optimal plan is the one that minimizes the expected number of IoT devices not reached by any security resource. We call this quantity the *risk* associated with the allocation plan (AP), denoted by $risk_{AP}$. We adopt the risk as a negative return of an allocation plan, such that the smaller the risk is, the more effective the plan. Being the risk a negative return, we need to adapt the definition of α-quantile given in Eq. 5 as follows:

$$q_\alpha(risk) = inf\{x|P(risk \geq x) \geq \alpha\} \tag{9}$$

Given a set of regions $\mathscr{R}$ (i.e., an IoT environment), and a fixed security budget b, we define an allocation plan for $\mathscr{R}$ as the set

$$AP = \{(sr_i, loc_j)| \sum_i sr_i.c \leq b, loc_j \in \mathscr{L}\} \tag{10}$$

Given the set of all possible device configurations, we can compute the set of all possible returns of an AP in terms of risk, and thus compute $E[risk_{AP}]$ and $E_\alpha[risk_{AP}]$. The adaptation of Problem 8 to our application domain is as follows:

$$\begin{aligned} \max_{AP \in \mathscr{A}\mathscr{P}} \quad & -E[risk_{AP}], -E_\alpha[risk_{AP}] \\ \text{subject to} \quad & \\ & E[risk_{AP}] \leq \mu_t \\ & E_\alpha[risk_{AP}] \leq \mu_\alpha \\ & \sum_{sr \in AP} sr.c \leq b \end{aligned} \tag{11}$$

where $\mathscr{A}\mathscr{P}$ is the set of all possible allocation plans. The minus sign before the two objectives means that we are actually minimizing the two measures, being the risk a negative return. For the same reason, μ_t and μ_α are upper bounds for $E[risk_{AP}]$ and $E_\alpha[risk_{AP}]$, respectively, and not lower bounds as in Problem 8. Note that without the constraint on the maximum budget, Problem 11 becomes meaningless because it would compute a unique allocation plan possibly with an unreasonable number of resources, providing risk 0 for any device configuration, which is surely the most effective solution, but also unlikely the most efficient because it would be very expensive in terms of cost.

Once a security manager has obtained the set of solutions to Problem 11, (s)he can choose the allocation plan that best fits his/her security requirements. However, there may be cases for which there is no solution, i.e., solving Problem 11 outputs an empty set. This can happen for two reason:

1. no allocation plan exists able to satisfy the constraints, because the security budget is too small to provide a value of $E[risk_{AP}]$ and $E_\alpha[risk_{AP}]$ below the desired thresholds;
2. the α-quantile is too small to admit $E_\alpha[risk_{AP}] \leq \mu_\alpha$.

In case (1), the security manager can try to increase the security budget until at least one solution is returned, or alternatively can turn Problem 11 in a three-objective optimization problem by placing the cost as a further objective as follows:

$$
\begin{aligned}
&\max_{AP \in \mathscr{A}\mathscr{P}} \; -E[risk_{AP}], -E_\alpha[risk_{AP}], -cost_{AP} \\
&\text{subject to} \\
&\qquad E[risk_{AP}] \leq \mu_t \\
&\qquad E_\alpha[risk_{AP}] \leq \mu_\alpha \\
&\qquad minCost \leq cost_{AP} \leq maxCost
\end{aligned}
\tag{12}
$$

where $cost_{AP} = \sum_{sr \in AP} sr.c$. This way, the security manager can have a comprehensive view of cost-benefit, by determining the minimum cost required to ensure a certain security level. However, this problem takes more time to solve than Problem 11, proportionally to the quantity $maxCost - minCost$.

In case (2), the security manager may decide to increase μ_α, or to tolerate a higher risk by having a larger quantile. In fact, α can be intended as a "measure of the risk tolerance": by increasing α, the value of $q_\alpha(risk)$ decreases, and Problem 11 computes $E_\alpha[risk]$ over a larger set of cases, which means that cases with high risk are more tolerated. On the contrary, when α decreases, $q_\alpha(risk)$ tends to the maximum value of risk, thus restricting the cases over which $E_\alpha[risk]$ is computed, meaning that we tolerate fewer cases with high risk.

3.5 Evaluation Algorithm

To solve Problem 11 an algorithm should enumerate all possible allocation plans $\mathscr{A}\mathscr{P}$, and evaluate each $AP \in \mathscr{A}\mathscr{P}$ over all possible configurations the IoT devices can take within the geographic area of interest $\mathscr{R}$. It easy to see that such an algorithm would take an unreasonable amount of time to solve the problem. In fact, for each region $R_k \in \mathscr{R}$, the number of possible configurations is the cardinality of the power set of L_{R_k}, and the size of $\mathscr{A}\mathscr{P}$ is upper-bounded by the number of all subsets of $\mathscr{L}$ of cardinality at most s, where s is the maximum number of security resources allowed by the security budget. This lead us to look for approximations in place of an exact solution. A standard tool for approximation is the use of genetic algorithms. There exist several genetic algorithms for computing a Pareto frontier, one of the most commonly used is NSGA-II [9], an evolutionary genetic algorithm able to find an approximation of the Pareto frontier for multi-objective optimization problems. The main difference between NSGA-II and other evolutionary genetic algorithms is the selection phase: NSGA-II has no unique fitness function but one for each objective. During the selection phase, the selected points are only the non dominated ones. When some constraints exist the selection phase will remove all points that do not satisfy the constraints.

In Problem 11 there are two objectives $-E[risk_{AP}]$ and $-E_\alpha[risk_{AP}]$. To compute their fitness at each iteration, the NSGA-II algorithm needs to evaluate the current AP over the set of all possible device configurations. An AP is implemented as an

Algorithm 2

```
procedure ComputeFitness(AP, n)
    →risk_AP = MonteCarlo(AP, n);
    q_α = α-quantile of →risk_AP;
    E[risk_AP] = 0;
    E_α[risk_AP] = 0;
    count = 0;
    for r ∈ →risk_AP do
        E[risk_AP]+ = r;
        if r ≥ q_α then
            E_α[risk_AP]+ = r;
            count + +;
        end if
    end for
    E[risk_AP] = E[risk_AP]/→risk_AP.length;
    E_α[risk_AP] = E_α[risk_AP]/count;
    return E[risk_AP], E_α[risk_AP];
end procedure
```

Algorithm 3

```
procedure MonteCarlo(AP, n)
    →risk_AP = [ ];
    while n > 0 do
        devicesLocations = ∅;
        for R_k ∈ ℛ do
            numberOfDevices = P_k.nextInt();
            while numberOfDevices > 0 do
                loc = a location ∈ R_k chosen randomly;
                devicesLocations.add(loc);
                numberOfDevices − −;
            end while
        end for
        risk = ComputeRisk(AP, devicesLocations);
        add risk to →risk_AP;
        n − −;
    end while
    return →risk_AP;
end procedure
```

individual of the population of the genetic algorithm, where each gene is the *id* of the location where a security resource has been placed. As stated before, for each region R_k the set of different configurations C_k is too big to be employed in the evaluation process without incurring in scalability problems. In its place we generate a subset of C_k with a Monte Carlo method [35], whose elements are generated according to the probability distribution P_{R_k} associated with the region R_k. More precisely, for a given region R_k, a device configuration is a set of locations chosen randomly, with cardinality equal to an integer randomly generated from P_{R_k}. Algorithms 2 and 3 illustrate the fitness function NSGA-II used for evaluating an allocation plan.

Algorithm 2 takes as input an allocation plan AP and an integer n. In Line 2, the MONTECARLO procedure returns a vector of length n with the risks associated with AP evaluated over n different device configurations. In Line 3, q_α is the value of the α-quantile of the vector $\overrightarrow{risk}_{AP}$. In Algorithm 3, the MONTECARLO procedure produces n different device configurations over which it evaluates AP. The number of devices for each region R_k is chosen randomly according to the probability distribution P_{R_k} (Line 6). In Line 13, the procedure COMPUTERISK computes the number of devices not monitored by any security resource $sr \in AP$ for each configuration, according to P_{sr} (the probability that sr is able to monitor a device located at a certain distance). The exact positions where to place the sensors are identified by each single gene of the individual corresponding to the chosen Pareto point.

3.6 Results Analysis

In this section we show the steps a security manager has to follow for computing the allocation plan that best fits his/her security requirements, according to his/her security budget. To this end, we report results related to the simulation of an area of 2000 m^2 and consisting of three regions divided in locations of 1 m^2. The details are shown in the following figure:

$Gamma(\alpha, \beta)$ is the Gamma probability distribution, adopted for simulating the devices distribution over each region.

Suppose the security manager has at his/her disposal a security budget of 30 monetary units, and only one type of security resource, based on the watchdog mechanism, with action range of 8 space units, and unitary cost. The security level (s)he wants to achieve is at most 30 unmonitored devices in no more than 10% of cases. This is formally translated as:

- $\forall$ security resource sr, $sr = \langle 1, 8, LDPL \rangle$;
- $\alpha = 0.1$;
- $b = 30$;
- $\mu_t = \infty$, $\mu_\alpha = 30$.

where $LDPL$ is the Log-distance path loss model [16], adopted as the signal attenuation model for the security resources. In order to compute the set of plans that satisfy his/her security constraints, the security manager solves Problem 11 with an NSGA-II algorithm, and set the Monte Carlo simulation with $n = 100K$ (see the MonteCarlo procedure shown in Algorithm 2) such that each plan is evaluated over a set of 100K different device configurations. Figure 5 shows the Pareto frontier (left), and the risk distribution of its plans (right). Looking at the Pareto frontier the security manager would choose the AP having the best combination of $E[risk]$ and $E_\alpha[risk]$. $AP\_30.1$ would be a better plan than $AP\_30.2$ and $AP\_30.3$ because $E[risk]$ is almost the same in the three cases and $E_\alpha[risk]$ is 25% lower. Looking at the risk distributions of Fig. 5 (right), (s)he can also base his/her decision on additional information and

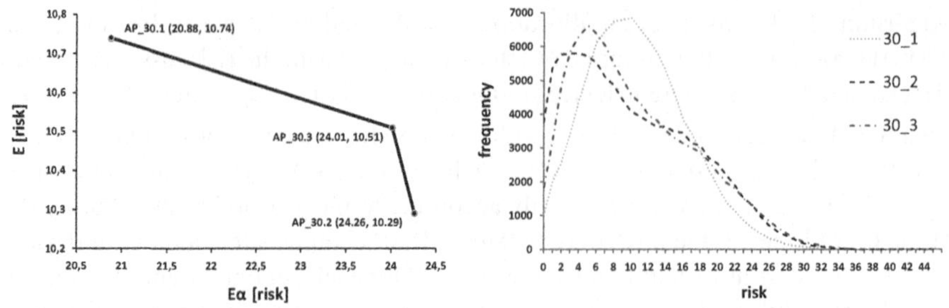

Fig. 5 Pareto frontier (left), and the risk distribution of its allocation plans (right)

conduct a more in-depth analysis. Over 100K different device configurations, each plan provides a minimum risk of 0 and a maximum risk of 45. *AP_30.2* is the plan that provides more cases with risk 0, thus a lower probability of an attack to happen in the network area. *AP_30.1*, although it provides the lowest number of cases with risk in the range bounded by the α-quantile, it also provides the lowest number of cases with risk 0, thus the highest probability to have an attack.[3] The trade-off between $E[risk]$, $E_\alpha[risk]$, and the probability of having an attack, is at the basis of the reasoning a security manager may want to follow to choose a plan. In this situation, a security manager interested in minimizing the cases with $risk > 0$ would choose *AP_30.2*, since *AP_30.3* provides similar values of $E[risk]$ and $E_\alpha[risk]$ Fig. 5 (left) but fewer cases with $risk = 0$ Fig. 5 (right).

Now suppose that the security manager wants to know how much to rise the budget to achieve higher security performances. In this case (s)he can solve Problem 12 and plot the values of $E[risk]$ and $E_\alpha[risk]$ along with the attack probability as shown in Fig. 6. It can be noticed that, although $E[risk]$ and $E_\alpha[risk]$ decrease of two orders of magnitude from $cost = 15$ to $cost = 40$, the probability of having an attack remains around 1, which means that there are very few cases (possibly none) with $risk = 0$. This way, the security manager would know (s)he needs a security budget greater than 40 for ensuring a higher level of security (Fig. 6).

4 Related Work

The problem of finding the optimal security resource allocation plan for IoT networks has been investigated in our recent papers [36, 37], where a solution is provided for static and mobile IoT networks scenarios, respectively. In these works, can be found further details on the type of security resources to be employed, on the heuristics for computing the node's criticality of the monitored network, more illustrative examples of real cases scenarios, a different formalization of the optimization problem for

[3]The probability to have an attack is computed as the ratio between the number of cases with $risk > 0$ and the total number of cases.

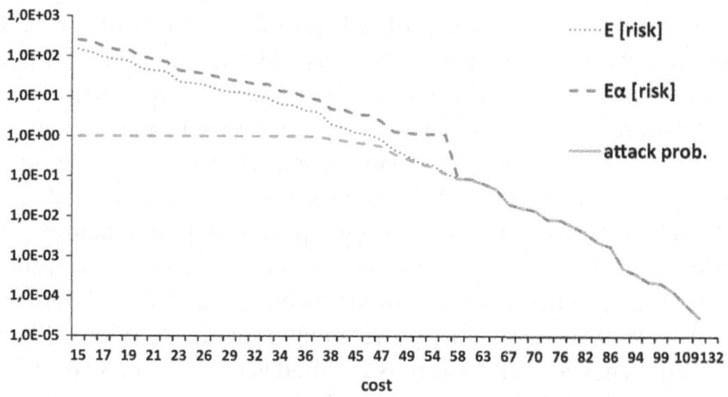

Fig. 6 The relation between cost, $E[risk]$, $E_\alpha[risk]$, and the probability to have an attack, of plans computed with $\alpha = 0.1$, $\mu_t = \infty$, $\mu_\alpha = 30$, $minCost = 15$ and $maxCost = 132$

large-scale networks, and a set of experiments that demonstrate the validity of the proposed approaches.

The problem of finding efficient security solutions with the help of game theory and Pareto analysis has been extensively considered for computer networks [1, 7, 10, 19, 31, 38, 46]. In Serra et al. [38] the interaction between attacker and defender is modeled as a Stackelberg game, and Pareto analysis is used to compute the best trade off between the cost of patching vulnerabilities and the cost due to the deactivation of vulnerable products within an enterprise system. In [31] the authors consider the problem of finding plans for patching vulnerabilities, that are tradeoffs between cost and risk, by using the Pareto analysis. In [10] the authors define a problem involving a game theory-based solution, but they do not mix it with the Pareto analysis and assume that the attacker does not know the strategy of the defender (we instead assume this is possible). The approach in [1] uses game-theory to study jamming attacks, while the approach in [19] uses Stackelberg games to model the interaction between defender and eavesdroppers. We like also to mention [7, 46] as typical examples of game theory and Pareto analysis applied to IoT scenarios. The approach in [46] investigates sensor networks in which an attacker can physically capture, replicate the nodes, and deploy sensors into a network, and then proceed to take over the network. A multi-player game is formalized in order to model the non-cooperative strategic behavior between the attackers and the network. The approach in [7] is based on a node clustering algorithm, with effective tax-based sub-carrier allocation tailored for wireless mesh networks with QoS support. Here, Pareto analysis is used for the optimal resource management. Other work focuses on security in the physical layer from eavesdropping and jamming attacks. In such previous approaches, players include attackers, non malicious users (that use the physical layer), and the layer itself (with its access control policy). These games are largely based on performance indexes of the physical layer, and the main goal is to optimize these performance indexes.

Solutions proposed in the context of IoT have focused on efficiency, due to the small "size" of network components, in terms of CPU, memory, and energy budget. Zhou and Chao [45] propose a media-aware security framework for facilitating various applications in IoT, and they provide a design rule and strategy to achieve a good trade-off between system's flexibility and efficiency. Raza et al. [33] propose an IPsec extension of 6LoWPAN, and show that IPsec is a feasible option for securing the IoT in terms of packet size, energy consumption, memory usage, and processing time. In [8] the authors propose a resource-aware self-adaptive network security provisioning scheme for the resource constraint Mobile Ad-hoc Networks (MANET), in order to avoid security provisioning Denial of Service (SPDoS) attack. In this work efficiency and effectiveness are modeled as two indexes, performance index (PI) and security index (SE), respectively. SI quantitatively reflects the security contribution of a secure protocol (set) to a MANET system: higher SI indicates a protocol (set) is more resistant to various attackers. PI quantitatively reflects network performance perspectives of a secure protocol (set): it is based on the QoS parameters (e.g., throughput, latency, jitter, power consumption, etc.); a higher PI indicates a secure protocol (set) of lower performance cost. In [28] Midi et al. propose Kalis, a knowledge-driven IDS for wide range of IoT systems. Kalis does not target individual protocols or applications, and adapts the detection strategy to the specific network features.

Many approaches have been proposed for computing the criticality of graph nodes. Recent approaches target the distributed evaluation and placement of the nodes most critical to network robustness, thus assessing node centrality in a distributed way [12]. Marsden [25] shows empirical evidence that localized centrality measures calculated for one-hop radius neighborhood are highly correlated to the global centrality measure. Kermarrec et al. [20] propose a new centrality measure, called second order centrality, defined in terms of the standard deviation of the time between visits of a perpetual random walk to each node.

Many algorithms are available for computing the Pareto frontier, one of the most used is NSGA-II [9]. NSGA-II is used by Dewri et al. [11] as optimization approach for minimizing the residual damage and the cost of security provisioning. They propose the use of workflow profiles to capture the contexts in which a communication channel is used in a pervasive environment. This is used to minimize the cost that the underlying business entity will have to incur in order to keep the workflow secure and running. Unlike us, they do not provide a formalization of the linear programs.

As alternative, simulation tools [15, 23, 43] can help understanding the relevance of a node for the correct function of the network. For instance, in a wireless sensor network scenario, where sources nodes collect data from the environment and send them to a sink through several edge nodes, a simulation can help understanding what are the most critical edge nodes, based on the amount of traffic involving them. Whereas the IoT system is already deployed, one can easily query network nodes about performances, in/out-coming traffic, etc.

We are not the first to use shortfall to characterize risk in a security context. Molloy et al. [30] adapt shortfall to take security decisions (such as access-control decisions or spam filtering decisions) under uncertainty when the benefit of doing so outweighs the need to absolutely guarantee that these decisions are correct. Molloy et al. have also put forward a more general vision on economic models for security [29]. We also would like to mention the risk-aware security solutions by Chen et Crampton incorporating the notion of risk into the RBAC model [6].

The problem of computing the best placement of network devices for mobile scenarios has been mostly addressed in the area of cellular networks. In such networks, the optimal placement of base stations (BSs) is crucial for the correct functioning of the communication system. The similarity with our scenarios is in the presence of mobile devices that have to be covered by the action range of static entities. However, the roaming problem, that has to be taken into account in the computation of the optimal BS placement, makes those solutions not suitable for the IoT scenario. We refer the reader to [18] for a more comprehensive discussion on this topic.

Signal attenuation models have also been widely investigated. Shen et al. propose an indoor wireless propagation model in WiFi radio-over-fiber network architecture for received signal strength (RSS) based localization in the IoT [40]. The proposed model adds attenuation terms of obstacles in each sub-space by dividing the room into several sub-spaces according to the obstacles' distribution. Alwajeeh et al. [2] propose an intelligent method to associate known models with spatial zones according to the electromagnetic interactions.

Finally, for a more comprehensive view of the management of smart physical objects and of IoT technologies in general, we refer the reader to [14, 17].

5 Conclusion

In this chapter, we have shown a game-theoretic model to answer the following question: *In an IoT scenario, given a set of security resources and a set of attacks to protect against, which resources should a security administrator choose, and how should (s)he allocate them in the network in order to ensure protection with the minimum cost and a certain degree of robustness against attacks?* To answer this question, we have provided a method for computing the best defender strategy, that gives one the possibility of choosing the resource allocation plan that best fits efficiency and effectiveness requirements. This leads to a problem involving Pareto optimization because efficiency and effectiveness are mutually competing goals. We show how to formulate the defender's problem as a linear optimization problem, and suggest a number of measures to formalize efficiency and effectiveness aspects of the defender's strategy.

We have made a distinction between static and mobile networks, for which the problem of computing the best allocation plan is addressed with different approaches. For static networks, the optimization problem is modeled as a mixed integer linear program, and an optimal solution is provided by using an approach with normal constraints. For mobile networks, instead, our proposed framework takes into consideration the possibility that the security resources are not able to monitor all the devices at all times, thus an evolutionary algorithm is adopted in order to provide a good approximation of the optimal solution.

References

1. E. Altman, K. Avrachenkov, A. Gamaev, Jamming in wireless networks: the case of several jammers, in *Proceedings of the First ICST International Conference on Game Theory for Networks* (2009)
2. T. Alwajeeh, P. Combeau, A. Bounceur, R. Vauzelle, Efficient method for associating radio propagation models with spatial partitioning for smart city applications, in *Proceedings of the International Conference on Internet of things and Cloud Computing* (ACM, 2016), p. 8
3. L. Atzori, A. Iera, G. Morabito, The internet of things: a survey. Comput. Netw. **54**(15), 2787–2805 (2010)
4. D. Bertsimas, G.J. Lauprete, A. Samarov, Shortfall as a risk measure: properties, optimization and applications. J. Econ. Dyn. Control **28**(7), 1353–1381 (2004)
5. F. Brasser, B. El Mahjoub, A.R. Sadeghi, C. Wachsmann, P. Koeberl, Tytan: Tiny trust anchor for tiny devices, in *DAC*, 2015 (IEEE, 2015), pp. 1–6
6. L. Chen, J. Crampton, Risk-aware role-based access control, in *International Workshop on Security and Trust Management* (Springer, 2011), pp. 140–156
7. H.T. Cheng, W. Zhuang, Pareto optimal resource management for wireless mesh networks with qos assurance: joint node clustering and subcarrier allocation. IEEE Trans. Wirel. Commun. **8**(3), 1573–1583 (2009)
8. C. Chigan, L. Li, Y. Ye, Resource-aware self-adaptive security provisioning in mobile ad hoc networks, in *IEEE Wireless Communications and Networking Conference* (2005)
9. K. Deb, A. Pratap, S. Agarwal, T. Meyarivan, A fast elitist multi-objective genetic algorithm: Nsga-ii. IEEE Trans. Evol. Comput. **6**, 182–197 (2000)
10. R. Dewri, I. Ray, N. Poolsappasit, D. Whitley, Optimal security hardening on attack tree models of networks: a cost-benefit analysis. Int. J. Inf. Secur. **11**(3), 167–188 (2012)
11. R. Dewri, I. Ray, I. Ray, D. Whitley, Security provisioning in pervasive environments using multi-objective optimization, in *ESORICS* (2008)
12. T.N. Dinh, Y. Xuan, M.T. Thai, E. Park, T. Znati, On approximation of new optimization methods for assessing network vulnerability, in *INFOCOM, Proceedings IEEE* (2010)
13. L. Eschenauer, V.D. Gligor, A key-management scheme for distributed sensor networks, in *Proceedings of the 9th ACM Conference on Computer and Communications Security* (ACM, 2002), pp. 41–47
14. G. Fortino, P. Trunfio, *Internet of Things Based on Smart Objects: Technology, Middleware and Applications* (Springer, 2014)
15. L. Girod, J. Elson, A. Cerpa, T. Stathopoulos, N. Ramanathan, D. Estrin, Emstar: a software environment for developing and deploying wireless sensor networks, in *USENIX* (2004)
16. J. Goldhirsh, W.J. Vogel, *Handbook of propagation effects for vehicular and personal mobile satellite systems*, vol. 1274 (NASA Reference Publication, 1998), pp. 40–67
17. A. Guerrieri, L. Valeria, R. Anna, F. Giancarlo, Management of Cyber Physical Objects in the Future Internet of *things* (Springer, 2016)
18. A. Guo, M. Haenggi, Spatial stochastic models and metrics for the structure of base stations in cellular networks. IEEE Trans. Wirel. Commun. **12**(11), 5800–5812 (2013)
19. Z. Han, N. Marina, M. Debbah, A. Hjørungnes, Physical layer security game: How to date a girl with her boyfriend on the same table, in *Proceedings of the First ICST International Conference on Game Theory for Networks* (2009)

20. A.M. Kermarrec, E. Le Merrer, B. Sericola, G. Trédan, Second order centrality: distributed assessment of nodes criticity in complex networks. Comput. Commun. **34**(5), 619–628 (2011)
21. I. Khalil, S. Bagchi, N.B. Shroff, Liteworp: a lightweight countermeasure for the wormhole attack in multihop wireless networks, in *International Conference on Dependable Systems and Networks, 2005*. DSN 2005. Proceedings (IEEE, 2005), pp. 612–621
22. D. Kotz, C. Newport, R.S. Gray, J. Liu, Y. Yuan, C. Elliott, Experimental evaluation of wireless simulation assumptions, in *Proceedings of the 7th ACM International Symposium on Modeling, Analysis and Simulation of Wireless and Mobile Systems* (ACM, 2004), pp. 78–82
23. P. Levis, N. Lee, M. Welsh, D. Culler, Tossim: accurate and scalable simulation of entire tinyos applications, in *Proceedings of 1st International Conference on Embedded Networked Sensor Systems* (ACM, 2003)
24. H. Levy, Y. Kroll, Ordering uncertain options with borrowing and lending. J. Finan. **33**(2), 553–574 (1978)
25. P.V. Marsden, Egocentric and sociocentric measures of network centrality. Soc. Netw. **24**(4), 407–422 (2002)
26. S. Marti, T.J. Giuli, K. Lai, M. Baker, Mitigating routing misbehavior in mobile ad hoc networks, in *Proceedings of the 6th Annual International Conference on Mobile Computing and Networking* (ACM, 2000), pp. 255–265
27. A. Messac, A. Ismail-Yahaya, C.A. Mattson, The normalized normal constraint method for generating the Pareto frontier. Struct. Multidiscip. Optim. **25**(2), 86–98 (2003)
28. D. Midi, A. Rullo, A. Mudgerikar, E. Bertino, Kalis: a system for knowledge-driven adaptable intrusion detection for the internet of things, in *IEEE 37th International Conference on Distributed Computing Systems (ICDCS)* (2017)
29. I. Molloy, P.C. Cheng, P. Rohatgi, Trading in risk: using markets to improve access control, in *Proceedings of the 2008 Workshop on New Security Paradigms* (ACM, 2009), pp. 107–125
30. I. Molloy, L. Dickens, C. Morisset, P.C. Cheng, J. Lobo, A. Russo, Risk-based security decisions under uncertainty, in *Proceedings of the Second ACM Conference on Data and Application Security and Privacy* (ACM, 2012), pp. 157–168
31. N. Poolsappasit, R. Dewri, I. Ray, Dynamic security risk management using bayesian attack graphs. IEEE Trans. Dependable Secur. Comput. **9**(1), 61–74 (2012)
32. T.S. Rappaport, et al., *Wireless Communications: Principles and Practice*, vol. 2 (Prentice Hall PTR, New Jersey, 1996)
33. S. Raza, S. Duquennoy, J. Höglund, U. Roedig, T. Voigt, Secure communication for the internet of things a comparison of link-layer security and IPsec for 6LoWPAN. Secur. Commun. Netw. **7**(12), 2654–2668 (2014)
34. S. Raza, L. Wallgren, T. Voigt, Svelte: real-time intrusion detection in the internet of things, in *Ad Hoc Networks* (2013)
35. C. Robert, G. Casella, *Monte Carlo Statistical Methods* (Springer Science & Business Media, 2013)
36. A. Rullo, D. Midi, E. Serra, E. Bertino, Pareto optimal security resource allocation for Internet of Things. ACM Trans. Priv. Secur. (TOPS) **20**(4), 15 (2017)
37. A. Rullo, E. Serra, E. Bertino, J. Lobo, Shortfall-based optimal placement of security resources for mobile IoT scenarios, in *European Symposium on Research in Computer Security* (Springer, 2017), pp. 419–436
38. E. Serra, S. Jajodia, A. Pugliese, A. Rullo, V. Subrahmanian, Pareto-optimal adversarial defense of enterprise systems. ACM Trans. Inf. Syst. Secur. (TISSEC) **17**(3), 11 (2015)
39. K. Sharma, M. Ghose, Wireless sensor networks: an overview on its security threats. IJCA (Special Issue on Mobile Ad-hoc Networks MANETs), 42–45 (2010)
40. X. Shen, K. Xu, X. Sun, J. Wu, J. Lin, Optimized indoor wireless propagation model in wifi-rof network architecture for rss-based localization in the Internet of Things, in *Microwave Photonics, 2011 International Topical Meeting on & Microwave Photonics Conference, 2011 Asia-Pacific, MWP/APMP* (IEEE, 2011), pp. 274–277
41. F. Simini, M.C. González, A. Maritan, A.L. Barabási, A universal model for mobility and migration patterns. Nature **484**(7392), 96–100 (2012)

42. H. von Stackelberg, D. Bazin, R. Hill, L. Urch, Market Structure and Equilibrium (Springer, 2010)
43. A. Varga et al., The omnet++ discrete event simulation system, in *Proceedings of the European Simulation Multiconference (ESM2001)* (2001)
44. D. Zhang, D. Liu, Dataguard: dynamic data attestation in wireless sensor networks, in *DSN, 2010* (IEEE, 2010)
45. L. Zhou, H.C. Chao, Multimedia traffic security architecture for the Internet of Things. IEEE Netw. **25**(3), 35–40 (2011)
46. Q. Zhu, L. Bushnell, T. Basar, Game-theoretic analysis of node capture and cloning attack with multiple attackers in wireless sensor networks, in *CDC* (IEEE, 2012), pp. 3404–3411

Embedding Internet-of-Things in Large-Scale Socio-technical Systems: A Community-Oriented Design in Future Smart Grids

Yilin Huang, Giacomo Poderi, Sanja Šćepanović, Hanna Hasselqvist, Martijn Warnier and Frances Brazier

Abstract In traditional engineering, technologies are viewed as the core of the engineering design, in a physical world with a large number of diverse technological artefacts. The real world, however, also includes a huge number of social components—people, communities, institutions, regulations and everything that exists in the human mind—that have shaped and been shaped by the technological components. Smart urban ecosystems are examples of large-scale Socio-Technical Systems (STS) that rely on technologies, in particular on the Internet-of-Things (IoT), within a complex social context where the technologies are embedded. Designing applications that embed both social complexity and IoT in large-scale STS requires a Socio-Technical (ST) approach, which has not yet entered the mainstream of design practice. This chapter reviews the literature and presents our experience of adopting an ST approach to the design of a community-oriented smart grid application. It discusses the challenges, process and outcomes of this apporach, and provides a set of lessons learned derived from this experience that are also deemed relevant to the design of other smart urban ecosystems.

Y. Huang (✉) · M. Warnier · F. Brazier
Section Systems Engineering and Simulation, Faculty of Technology, Policy and Management, Delft University of Technology, Delft, The Netherlands
e-mail: y.huang@tudelft.nl

M. Warnier
e-mail: m.e.warnier@tudelft.nl

F. Brazier
e-mail: f.m.brazier@tudelft.nl

G. Poderi
Department of Computer Science, IT University of Copenhagen, Copenhagen, Denmark
e-mail: gipo@itu.dk

S. Šćepanović
Department of Computer Science, Aalto University, Helsinki, Finland
e-mail: sanja.scepanovic@aalto.fi

H. Hasselqvist
Department of Media Technology and Interaction Design, KTH Royal Institute of Technology, Stockholm, Sweden
e-mail: hannaha@kth.se

F. Cicirelli et al. (eds.), *The Internet of Things for Smart Urban Ecosystems*, Internet of Things, https://doi.org/10.1007/978-3-319-96550-5_6

1 Introduction

The traditional science and engineering philosophy is dominated by technological determinism, the idea that technology determines societal development [35, 45, 52]. Within this reductionist view, technologies are core to the engineering design, where the physical world consists of a large number of diverse technological artefacts. The plausibility of this view is challenged by the Socio-Technical Systems (STS) view [56] that argues that technological and social development form a "seamless web" where there is no room for technological determinism or autonomy of technological systems [14]. The latter view is premised on the interdependent and deeply linked relationships among the features of technological artefacts or systems and social systems (i.e. the mutual constitution) [45], since the man-made world also comprises a huge number of social components—people, communities, institutions, regulations, policies and everything that exists in the human mind—that have shaped and been shaped by technological components [22, 56]. In this view, engineering design is identified as a process through which technologies materialize into products, a process that substantively shapes and reshapes our lives and societies and vice versa [33]. This focus on Socio-Technical (ST) interconnectedness becomes even more visible in new emerging technologies [33].

Smart cities, for example, use technologies such as Internet-of-Things (IoT) within a large complex social context in which they are embedded to facilitate coordination of fragmented urban sub-systems and to improve urban life experience [17]. The rise of IoT has important ST implications for people, organizations and society. Although connecting devices is technically possible, little is known about the implications [50]. An ST perspective can be insightful when looking at dynamic technological development and when considering sustainable development [50]. Although STS have been studied for decades, ST approaches are relatively new to the design and systems engineering communities [2, 38, 45]. Such approaches are not widely practised despite growing interests [2].

This chapter reviews the literature and presents our experience of adopting an ST approach in designing a community-oriented smart grid application called *YouPower*. It discusses the challenges, process and outcomes of this design experience, and provides a set of lessons learned that are also deemed relevant to the design of other smart urban ecosystems.

2 Designing Large-Scale Socio-technical Systems

STS are systems arising through encompassing people communicating with people whose interactions are mediated (at least partially) by technology rather than only in the natural world [59]. The term "socio-technical" embodies both a research perspective and a subject matter [34]. Facing a complex system, researchers from different disciplines often examine the system from their own perspectives. Engineers, for example, see hardware systems, computer scientists see information systems, psychologists see cognitive systems, sociologist see social systems—all these views are valid [61]. Figure 1 uses the notion of system levels to illustrate this difference of perspectives in STS [59, 61]. Notably, the levels in Fig. 1 are not different systems nor partitions of systems, but overlapping views of the same system corresponding to the engineering, computing, psychological and sociological perspectives [59]. The top and bottom of the levels are open-ended, as social groups can coalesce into larger entities such as organizations, cities, nations and beyond [62], while physics and hardware can be studied in micro, nano and smaller scales. The system boundary and the boundaries of those views are not necessarily clear-cut (hence drawn as dashed lines). An STS view is one that incorporates and meaningfully interconnects all levels of considerations: the upper two levels (Group/Community and Personal/Individual) together being social and the lower two (Information/Software and Physical/Hardware) technical. Each upper level can be seen as "arising" or "emerging" from the lower levels. For example, personal cognitions "emerge" from information exchanges supported by software, which "arises" from hardware [59]. The higher a level of view, the higher its degree of abstraction, and the less deterministic and predictive it becomes. With the levels of these different perspectives in mind, the STS view can be articulated as the recognition of three fundamental properties as follows [45].

First, the mutual constitution of people and technologies. This mutual constitution (by the social and the technological) generates complex and dynamic interactions among technological capacities, social norms, histories, situated context, human choices, actions and so on. In STS, social interactions are enabled or supported by technological means. The two adapt to one another, which is referred to as mutual adaptations.

Second, the contextual embeddedness of the mutuality. The context of a sociotechnical system is not taken as static or delineable. There are dynamic situational and temporal conditions that influence the mutual adaptations throughout the course of design, development, deployment, uses and even retirement phases of systems of interest.

Third, the importance of collective action. Collective action refers to the joint pursuit of one or more shared (potentially conflicting) goals by two or more interested parties such as problem owners, shareholders, users and communities affected

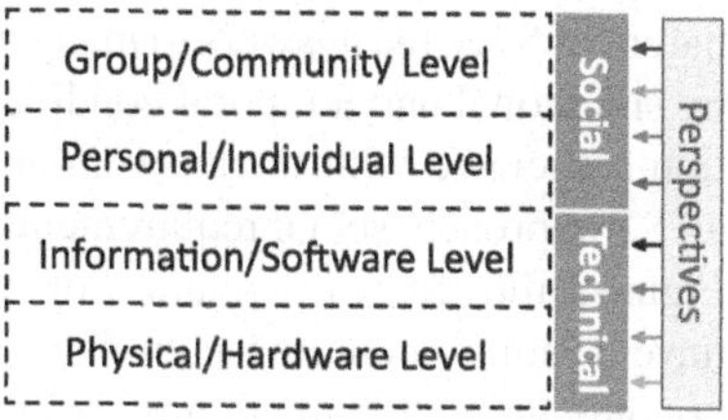

Fig. 1 Levels of STS viewing from different perspectives: the levels are not different systems but overlapping views of the same system [59, 61]

(without implying positive or negative outcomes). It shapes and is shaped by both the context and the technological components.

Researchers who hold an STS view investigate more than just the technological (sub-)system or just the social (sub-)system or even the two side by side, but also the phenomena that emerge when the two interact [34]. An ST approach tries to abstain from oversimplifications that seek a single or dominant cause of change, but studies the complexity, dynamic and uncertainty in the networks of institution, people and technological artefacts in the process of technologically involved change [45]. The levels of perspectives and the three fundamental properties of STS aforementioned help researchers to organize, categorize and allocate their inquiries and knowledge.

What does an STS view mean to design in particular? The rest of this section discusses the impact of an STS view on (I) the understanding of design problems, and (II) the design process and design artefacts.

Understanding the Design Problems or Situation Designing STS is becoming increasingly challenging partly due to the increasing systems complexity and scale. Large-scale STS often are not designed as a whole by one team in one project, but are incrementally "piece by piece" transformed and evolved from many generations of "legacy" systems. Designers and engineers are therefore faced with ill-structured or wicked problems that do not allow to straightforward determine what systems boundaries to choose, what issues to address and what aspects to consider regarding the design [4].

An STS view by definition advocates a systemic approach towards understanding including but not limited to information acquisition, diagnosis and analysis. Developing an understanding of the design problems or situation entails firstly looking into the roles, responsibilities, powers, interests and requirements of the stakeholders involved [11]. As will be discussed later in this section, iterations in a design process deepens this understanding. Pragmatically, a designer can start with upper level (more abstract) views and dive into the lower level (less abstract) ones. At each level, a designer investigates questions such as what are the corresponding goals to achieve (or problems to tackle) [11, 58] and associated requirements to fulfil [62], which social/technical elements (or components) are important to each level of views, how do the elements operate/behave individually, how do they interact within and across the levels, and what are the possible outcomes of the interactions and in what context [2]. Table 1 provides a set of such questions categorized by the three STS properties and associated to the levels of focus. The questions are by no means exhaustive but serve as examples to orient ways of thinking during design. Given the nature of STS, the answers to many of such questions are context specific, influenced by situational and temporal conditions [2, 38]. This means the contextual information associated with the answers also need to be well studied and documented. In an ST approach, social requirements must become part of the technical design [60]. Figure 2 illustrates the relation of requirements at different levels [62]. Each level unveils requirements that cumulate level by level. The requirements at one level affect not only that level but all those below it [62]. For example, a communal requirement may add new requirements at the personal level which in turn affects software and

Table 1 Examples of questions to investigate categorized by STS properties and associated to levels of focus

Properties	Levels of focus	Examples of questions to investigate
Mutual constitution	All	Which elements (or components) are important at each level?[a]
		How do the elements behave and interact?
		What are the possible outcomes of the interactions?
		What are the goals, constrains and requirements, if any, of the elements?
Contextual embeddedness	Group/community	What are the situational and temporal conditions where the behaviours and interactions take place?
	Personal/individual	What are the influences of the situational and temporal conditions on the outcomes of the behaviours and interactions?
		How those situational conditions may change over time?
Collective action	Group/community	What are the community (or institutional) goals, constrains and requirements?
	Personal/individual	How are the community (or institutional) goals, constrains and requirements aligned with the individual goals, constrains and requirements?
		What is the group and individual attitude towards the community (or institutional) goals or collective action?
General[b]	All	What is the level of resolution to use when describing and analysing the system?
		What is the set of values that underpin the design thinking about the system?
		What are the criteria and metric of evaluating whether and to what extent the desired goals are achieved and maintained?

[a]Elements can also be categorized by weighted scale, e.g. from *important* (must be included in the study), to *can be relevant* (can be included in the study), to *not relevant* (can be excluded from the study)
[b]It concerns all three properties above

hardware requirements. When a technical design fails to fulfil requirements derived from the personal or social level, there is a deficit between what society needs and what technology does—this is when a "ST gap" emerges [60].

As mentioned earlier, large-scale STS are often "systems of evolution" rather than "systems of revolution" [2, 38]. Significant changes in a system should be accompanied by a well designed and managed change process where feedback is returned for analysis and adaptation [2]. For this, a good understanding of the existing system and work/operation processes is necessary to design and plan the change

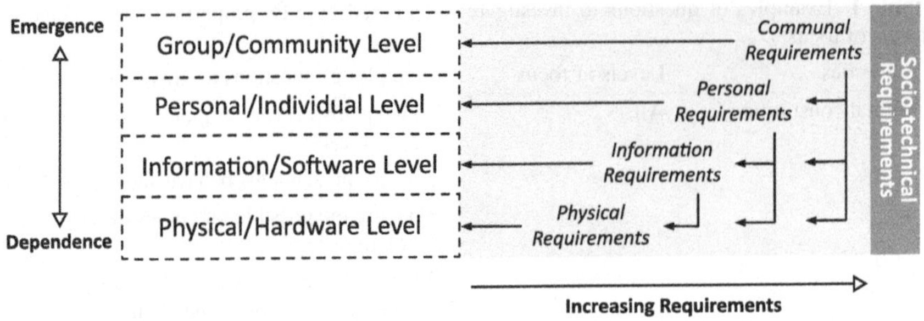

Fig. 2 Levels of socio-technical requirements [62]

process. Many core difficulties in complex projects stem from implementation of the design in the real world [38]. Designers therefore need to address the possible impediments to implementation (and change process) already from the beginning, and they must play an active role in implementation, and develop solutions through small incremental steps [38].

Design Process and Design Artefacts The design process of STS is often conceived and implemented as a participatory decision-making process where problem owners, shareholders, users, developers and other stakeholders are actively involved to represent their interests and negotiate agreements. Designers should be working *in* the context of an STS as an insider, not outside of the system as a bystander, with the intention of changing or improving some part of that system [4].

The evolutionary nature of STS means that what matters more in the design is the design process itself rather than the "final status" of the system [50]. When an STS keeps evolving and exhibits emergent behaviour [36], any designed "final status" soon becomes a transitional state. An important goal of a design process is to make the design relevant to the evolving context where the technology is utilized [50]. This is not a pure technological inquiry but an ST one that demands human-centred design, progressing by iteration and "muddling through" [38].

The interdisciplinary nature of STS calls for interdisciplinary teams. Although this need has been widely accepted, working in an interdisciplinary team remains a persisting challenge. It is group work of the most challenging sort, especially when those involved are in fields far apart intellectually as well as physically [5]. Despite efforts at creating teams across disciplines in the design process, interdisciplinary integration is often poor and disciplinary borders have been largely maintained [2]. Some common issues include [2, 5, 38]: (1) difficulties concerning the logistics of group interactions at management level; (2) failures in understanding and communication due to methodological, disciplinary, language, cultural and value differences; (3) personal challenges related to gaining trust and respect of others working in different disciplines, and (4) institutional impediments related to incentives and priorities given to disciplinary versus interdisciplinary work. One discipline has to understand (at least at an conceptual level) and appreciate what the other disciplines can do in order to ask them to deliver something that assists the analysis and design during the development process [2].

The design artefacts can be aligned to achieve specific goals or effects across all four levels of views (shown in Fig. 1) through which designers wish to intervene in STS. They can be, for example, hardware, software artefacts, a new idea of human-computer interaction design, rules for behaviour, policies, social programs, and any combination of them. Good solutions are often balanced "satisficing" solutions between different requirements that will be acceptable to and used by end users as well as delivering the expected benefits to stakeholders [2, 38]. As mentioned earlier, designers should not stop at the design stage but play an active role in implementation, developing "evolving" contextual solutions through iteration [38]. Acontextual and detemporalized general solutions are actually self-limiting [45]. In addition, the solutions should be accompanied by a thoughtful change process that is concerned with, among others, sensitising stakeholders for awareness and constructive engagement taking into account social and organizational issues [2].

3 CIVIS: A Community-Oriented Design in Future Smart Grids

This section presents the EU CIVIS project[1] as an illustration of designing STS. The project took place under EU's interest to address the societal challenges of energy efficiency. The vision of smart grids and the use of IoT and ICT are the main drivers for the project's ambition to reconfigure the relationships among traditional and emerging actors—producers, distributors, retailers, prosumers and cooperatives—in the energy value chain. In the following, we review the literature about IoT with regard to smart grids and STS, provide an overview of the design situation, and then discuss the collaborative design process and the main outcomes.

3.1 Internet-of-Things and Smart Grids as Socio-technical Systems

The International Telecommunication Union defines IoT as the worldwide network of interconnected objects uniquely addressable based on standard communication protocols—a definition focusing on the technological aspect of IoT. Since IoT is expected to have a massive impact on society and wider cultural milieu, its ultimate status should accordingly be a human-centred STS although how the IoT landscape will look like in the future is yet uncertain [1, 20, 37, 50, 54].

A key application domain of IoT is envisioned for smart grids [50]. IoT technologies can collect energy and environmental data, and form high-speed real-time

[1] http://cordis.europa.eu/project/rcn/110429_en.html.

bidirectional connections among consumers, utilities and the electrical grid [63]. Improved data collection and communications can support decision making and in turn improve the overall efficiency of the grid. IoT is also an integral technology in future smart homes, smart buildings and smart cities [15, 46, 64, 65] where IoT devices are expected to cooperate, actively share energy, and participate in energy management [31, 39]. In addition to object-object interaction, the IoT design must also consider human-object, human-environment and human-human interactions [20, 21]. As an ST ensemble, IoT and smart grids should be embedded into society to build new communities of empowered users with an emphasis on contextual design, so that the technologies will be adapted to different psychological, social, legal, policy factors considering actual adoption possibilities (in contrast to designing intrusive technology) [37, 50].

For more than two decades, energy transition has shifted the energy domain towards decentralization and distributed renewable sources [43, 53]. This transition can be attributed to several intertwined facts: (1) the increasing awareness of the inherent complexity among energy systems, societies and the environment [7, 55], (2) the widespread diffusion of new enhanced technologies, such as IoT, and their hybridization with modern ICT [42, 47], (3) the pursuit of national and supranational energy policies promoting low carbon emission, energy efficiency and sustainability [12], and (4) the emergence of new actors such as energy cooperatives and energy communities in the energy value chain [57], and the transformation of traditional actors such as housing associations and amateur energy managers [23]. Under these conditions, many new challenges and possibilities emerge, particularity from an ST perspective [50].

3.2 An Overview of the CIVIS Project

For the CIVIS project, an ST approach was in prospect by design from onset in the project goal and team composition. The goal in large was to provide ICT support for social participation in smart grids to manage communities and support energy services in the domestic sector. The project team had the ambition to increase citizens' energy awareness, promote environmental values, improve citizens' know-how about sustainable consumption, and to facilitate citizens to improve energy consumption behaviours in their everyday life together with local communities [26–28]. The research attention was oriented towards the potentials and challenges of citizens' collective actions, pro-social values and sense of community. The composition of the project consortium included a diversity of disciplinary profiles—electrical engineers, computer scientists, HCI designers and sociologists—that was necessary for tackling ST challenges in the project from multiple perspectives.

Another overarching goal of the CIVIS project was to integrate the core features of CIVIS design and its underlying infrastructure into rather different contexts, to meet diverse needs and expectations as well as to serve various types of users. This is why the pilot sites of CIVIS—two sites hosted in Italy and two in Sweden—were also deemed as sources of collaborative design and development rather than merely passive recipients of technologies to be tested.

In the two Italian pilot sites,[2] the focus (at the community level) was cooperative-owned electricity provision to local houses. Two electricity cooperatives, that produce and sell 100% renewable energy to their associate members, together with two samples of recruited associate member households were the main stakeholders. The regional distribution system operator (DSO), the institutional representatives of the two municipalities, and two local cultural associations participated as stakeholders in different phases of the project, by providing knowledge and support for technical aspects related to energy and households engagement. The CIVIS design in Italy needed to support energy communities in demand-side management.[3]

In the two Swedish pilot sites,[4] the focus (at the community level) was housing cooperative's energy management in apartment buildings and town-houses. One site included apartment buildings owned by housing cooperatives.[5] Recruited households from the cooperatives, and the cooperatives' board members were key stakeholders. The other site was a townhouse area where the local residents' association and some of its member households participated to CIVIS. The design in Sweden needed to support knowledge sharing about energy management practices at building and household levels.

The project was structured around three main areas of interest—energy, ICT, and social innovation—and was organized in three broad phases that roughly overlapped with the three project years. Each phase ensured a close interaction with the local realities and context of the pilot sites: (I) an exploratory phase, aligned and refined CIVIS' objectives with the local context, (II) a real-world prototyping phase, concerned with the design and development of the platform (from data monitoring devices to the front-end applications), and (III) a piloting phase, for the full scale deployment of the platform in the pilot sites and assessment.

3.3 Collaborative Design Process

The CIVIS design process was theory-driven, human-centred, collaborative and iterative. A literature review was carried out early in the project and later updated regarding energy intervention strategies and social smart grid applications for the promotion of environmental behaviour change. This provided a broad set of initial design ideas which had been iteratively assessed, expanded, refined and improved throughout the

---

[2]Two municipalities of Storo and San Lorenzo in Trento, Northwest Italy.

[3]For example, moving peaks of electricity demand towards peaks of local energy production or, in other words, improving the self-consumption capabilities of the electric cooperatives and their associate members.

[4]The neighbourhoods of Hammarby Sjöstad and Fårdala in the Stockholm area.

[5]In Sweden, those who buy an apartment must join a corresponding *housing cooperative* that owns and maintains the estates. The members of a cooperative annually elect a board that makes energy related decisions on behalf of the members.

design process with the collaboration and participation of stakeholders affected. The rationale behind this approach rested on the conviction that applying a human-centred and collaborative design process to the development of large STS has positive theoretical, practical and ethical implications [3, 18] by, for instance, increasing users engagement, usability and integration into existing local conditions [6, 13, 40]. During the three project years, the process unfolded as a complex and articulated network of meetings and artefacts which strived to align the interests of different stakeholders involved, from project partners to local stakeholders and end-users. The project team organized brainstorming sessions and design workshops, and run exploratory and evaluation focus groups with end-users in the pilot sites. Due to limited space, the main aspects of the process are summarized as follows. Interested readers can refer to [41] for more detail on how the process shaped the main outcomes of CIVIS.

User Stories User stories [30] were used and adapted it to the ST context of the project acrossed CIVIS both horizontally (to the scope of the work packages) and vertically (to the needs of the two countries). Each user story identified a realistic scenario, a main scope of energy intervention, supporting ICT tools, and central social dynamics. During the 3 years, user stories were drafted, refined, merged, abandoned and finalized as part of our constant work of alignment and negotiation. They were discussed in internal workshops, round-tables with stakeholders, and focus-groups with participant end-users; circulated to software engineers and platform designers; publicly presented for feedback and used as frames for collaborative workshops. They represented evolving artefacts that were consolidated in formal versions at the end of each year during the project.

Stakeholder Meetings Stakeholder meetings were held primarily at the level of pilot sites involving CIVIS key technical stakeholders and key local energy stakeholders. Meetings were held quarterly, although at the project's onset and during the most intense design phase, they occurred more frequently. These meetings proved helpful for agreeing on the project overarching objectives at the local levels, but also for understanding the feasibility and rationality of the choices for the social and technical aspects of the platform. For instance, identification and selection of the energy monitoring devices (to be installed in participants' households to enable the collection of energy data) required long discussions and negotiation. The suitability of these devices could not be assessed at a technical level only (regarding cost/efficiency, type of data, reliability and protocols). The typology of end-users and housing conditions[6] also played an important role.

Focus Groups Focus groups involved potential and actual participating household members, recruited for the project, and they were run as collective discussions. Usually they lasted around 2 hours and included between six to eight discussants. In case of the exploratory meetings, the scope of the discussion was intentionally broad

[6]In Italy, participants were older and less tech-savy, living in independent, large houses; while in Sweden participants were relatively young and more tech-savy, but living in smaller apartments in residential buildings.

(a) (b)

Fig. 3 **a** First project plenary meeting where local stakeholders took part; **b** Stakeholder meeting among technical project partners and local stakeholders to discuss demand-side management

Fig. 4 An initial moment of an exploratory focus group in Italy

and was aimed at revealing possible latent needs or expectations, as well as discussing explicit ones. More importantly these were used to get first-hand knowledge about the social and cultural environment for which the platform was to be deployed. In contrast, the evaluation discussions had more specific focus and involved concrete artefacts (e.g. an interface mock-up or app prototype) as a basis. For instance, exploratory meetings helped some of the features initially thought to be welcomed by end-users, such as "sharing" of energy performances or measurements typical of social network platforms, into due perspective. In our context, it was both difficult to grasp the meaning of such a feature, but it also raised concerns with respect to privacy. At the same time, intermediate evaluation activities allowed us to spot limitations of our data visualization (e.g. oversimplifications of energy data through a certain type of charts), and of the engagement and participatory process itself[7] (e.g. expectation of more frequent interactions with the project) (Figs. 3, 4, 5).

Design workshops These workshops involved concrete hands-on activities done primarily with participant household members. Occasionally a few workshops took place among project partners or had a broader target. Different workshop method-

[7]A study of the end-users appreciation of the engagement and participatory process in the Italian pilot sites is published in [10].

(a) (b)

Fig. 5 **a** Beginning of group activities in one of the first workshops held in Italy and focusing on user requirements; **b** One of the group outcomes for mapping energy consumption habits at home

ologies (e.g. brainstorming, future scenarios, collages, usage simulation) were used to suit diverse needs in the different phases of CIVIS. End-user requirements[8] were identified for the platform front-end as well as for the interface layout. For instance, for the module of *Action suggestions*, the workshops were relevant for adjusting the various tips for energy conservation to the local contexts of use. These were in fact quite different between the two countries, and certain tips had no meaning when delivered to one or another country or they needed a different rationale for their presentation.

In general, continual alignment took place at a high level of abstraction mainly due to the use of user stories as key boundary object among stakeholders, expertise and local contexts. At a more concrete level, a set of platform features were prototyped in simple mock-ups and also used as a basis for discussion. These underwent iterative rapid prototyping which produced wireframes as better visual guides that could be more effectively communicated to end-users. Prior and after each iteration, exploratory activities on how to proceed and evaluation sessions for their outcomes took place in different venues and with different stakeholders. Table 2 provides a brief overview on the relationships among the various activities of the collaborative design process and their influence on CIVIS platform design viewed through the perspective of an ST approach.

3.4 Main Outcomes of the Design Process

The main outcomes of the CIVIS collaborative design process include (1) an open source social smart grid application called YouPower [25], and (2) community engagement approaches that were implemented during the change process of the project [10, 24], both contextualized to the local situations.

[8]A preliminary analysis of these emerging requirements in the Italian pilots is presented in [9].

Table 2 A simplified view of the relationships among type of activities the stakeholders involved in the collaborative design process and their influence on CIVIS platform design—viewed through the perspective of a socio-technical approach. Aspects that had a specific link with one of the two pilot countries (Italy ITA Sweden SWE) are reported in the table

Type of activity main stakeholders involved	Social levels	Technical levels
Stakeholder meetings Project partners, institutional local energy stakeholder	• Endorsement and preparation of recruitment strategy for participant households • Refinement and public endorsement of participatory energy budgeting (ITA)	• Definition of main energy targets: demand-side management (ITA), energy knowledge sharing (SWE) • Refinement of energy optimization models and feasibility of a time-of-use signal for demand-side management • Selection of optimal energy monitoring devices: CurrentCost (ITA), Smappee (SWE) • Understanding of DSO energy data structure and availability • Understanding of existing energy/ICT infrastructure • Availability to invest "energy bonus" for participatory energy budgeting (ITA)
Focus groups Recruited household members	• Understanding local context: strength of local community groups and associations reinforcing the idea to promote joint actions through the platform • Emerging concerns about privacy • Emerging concerns about anonymity related to energy data comparisons (ITA) • Exploration of "ICT literacy"	• Exploration and rejection of social networking features • Refinement of the understanding of ICT devices availability, type and use
Co-design workshops Recruited household members, institutional local energy stakeholder, project partners	• Requested interface features: real time and historical data for PV production (ITA) • Definition of PEB policy documents (ITA) • Collaborative content-generation sessions for Housing Cooperative module (SWE)	• Exploration and rejection of social networking features • Refinement of the understanding of ICT devices availability, type and use
	• Assessment and usability feedback on all module mock-ups, leading to improvement of interface designs	

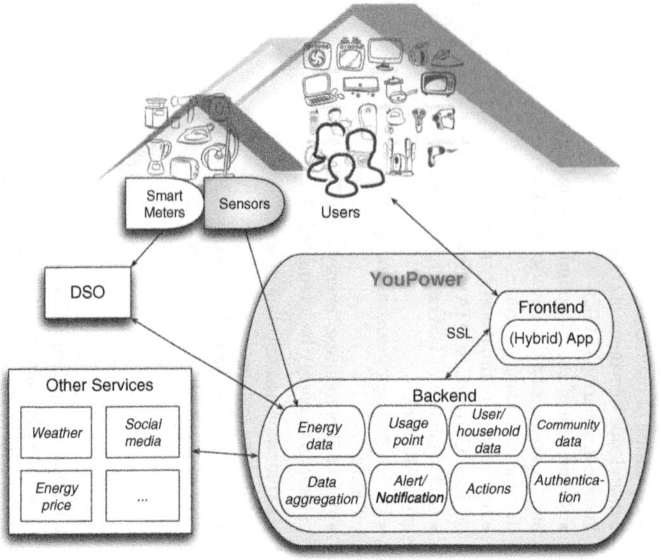

Fig. 6 The CIVIS project platform overview. DSO (Distribution System Operators); SSL (Secure Sockets Layer)

3.4.1 YouPower: An Open Source Social Smart Grid Application

Combining smart sensing and web technologies among others, YouPower is designed as a social smart grid application (developed by the CIVIS project as a hybrid mobile app) that can connect users to friends, families and local communities to learn and take energy actions that are relevant to them together [25, 29]. The app encourages an energy-friendly lifestyle and can be linked to users' energy consumption and production data for quasi real-time and historical prosumption information. The CIVIS platform as a whole (shown in Fig. 6) is mainly composed of (I) the *energy sensor level services* mainly dealing with energy data collection, and (II) the *energy data level and social level services* mainly dealing with energy data analytics as well as user, household and community management among others.

Energy Sensor Level Services The CIVIS project installed hardware (smart plugs and sensors) and software required for appliance-level energy data collection. The hardware/software choices differ in the two sites due to the local context. For example, *Smappee*[9] for 40 households in Stockholm, and *CurrentCost*[10] for 79 households in Trento. Trento also installed Amperometric clamps for PV production measures. Household-level energy data of the pilot sites in both countries is measured by smart meters and provided by local DSOs.

[9]http://www.smappee.com.

[10]http://currentcost.com.

Energy Data Level and Social Level Services These services are provided by the YouPower app and its back-end. The design consists of three self-contained composable modules: (1) *House Cooperatives* (contextualized and deployed to the Stockholm pilot sites); (2) *Demand-Side Management* (contextualized and deployed to the Trento pilot sites); and (3) *Action Suggestions* (contextualized and deployed to all pilot sites). They are discussed in the following paragraphs.

3.4.2 Housing Cooperatives

This module is designed for the community of housing cooperatives[11] in the Stockholm pilot sites [23]. Similar housing ownership and management models exist in a number of EU and non-EU countries, which allow potential wider application of the design. A housing cooperative annually elects a board which manages cooperative properties and decides on energy contracts, maintains energy systems, and proposes investments in energy efficient technologies. Since board members are volunteers who may have limited knowledge of energy or building management, this module aims to support board members in energy management, in particular energy reduction actions. Cooperative members can also use the app to follow energy decisions and works of the cooperative. Additionally, the app can be of interest by building management companies working with housing cooperatives. The information presented in the app is visible for these user groups and shared between housing cooperatives. This openness of energy data is key to facilitating users in sharing experiences relevant for taking energy reduction actions.

Linking Energy Data to Energy Reduction Actions The design links energy data with energy reduction actions taken (Fig. 7a) at cooperative levels, making the impact of energy actions visible to users. Energy use is divided into district heating and hot water, as well as facilities electricity in apartment buildings. Users can switch between the views per month or per year to show overall changes. Users with editing rights, typically board members, can add energy reduction actions that the cooperative has taken, e.g., improvement of ventilation, lighting or heating systems, and related cost. Trusted energy or building management companies can be given editing rights to add energy reduction actions taken on behalf of the cooperative. Actions taken are depicted per month and are listed below the graph. Clicking on an action provides more details. To make the impact of actions visible, users can compare the energy use of the viewed months to that of a previous year. This can be used e.g. by a cooperative to explore what energy reduction actions to take in the future by learning actions taken by other cooperatives and what the effects were in relation to costs.

Comparing Housing Cooperatives The cooperatives that are registered for the app are displayed in a map or list view (Fig. 7b). Their icons are colour coded (from red to green) based on each cooperative's energy performance, i.e. from high to low energy use per heated area, scaled according to the Swedish energy declaration for build-

[11] *Bostadsrättsförening* or *Brf* in Swedish.

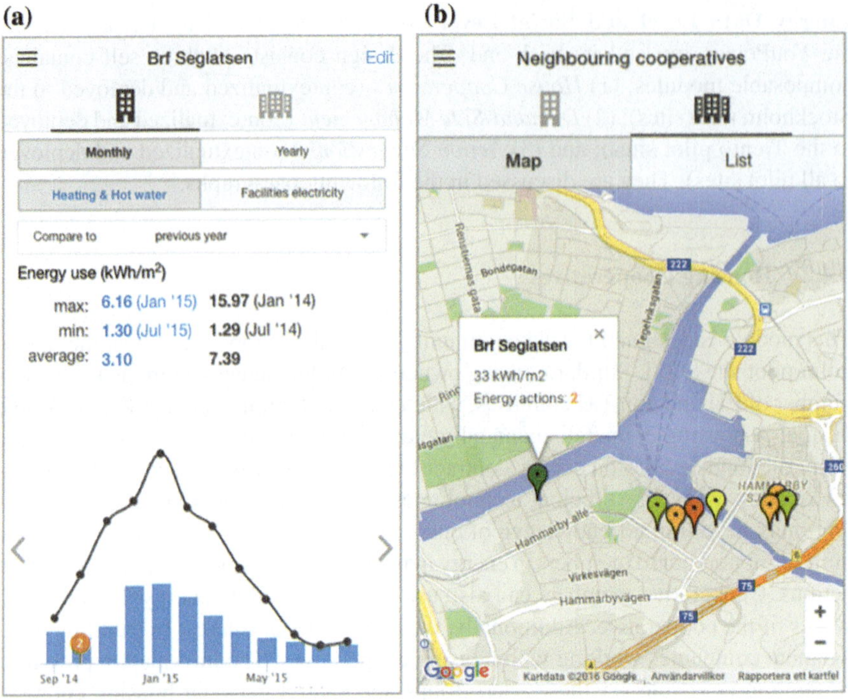

Fig. 7 **a** Heating and hot water use graph. Blue bars show the current year's use per month; the black line shows that of previous year. Energy reduction actions taken are mapped to the time of action and listed below (not shown); **b** Map view of participating housing cooperatives. The energy performance of cooperatives is indicated by colour and in numbers

ings.[12] Users can also see the energy performance as a number (in kWh/m^2), and the information about energy reduction actions of the cooperatives. During stakeholder studies, energy managers in cooperative boards stressed the importance of knowing the difference between cooperatives in order to understand the difference in their energy performance. Thus, the design also includes information about cooperatives (Fig. 8) such as the number of apartments and heated areas in a cooperative, a building's construction year, and types of ventilations (e.g. with or without heat recovery). Users can compare a cooperative's energy use per month or per year to another cooperative or to the neighborhood average. The electricity use is also displayed per area (kWh/m^2) to make it comparable.

Sharing Experiences A cooperative interested in taking an action may wish to know more, e.g. which contractor was chosen for an investment and why or how to get buy-in from cooperative members. The design provides commenting functions for each action added, where users can post questions and exchange experiences. The

[12]http://www.boverket.se/sv/byggande/energideklaration/energideklarationens-innehall-och-sammanfattning/sammanfattningen-med-energiklasser/energiklasser-fran-ag/.

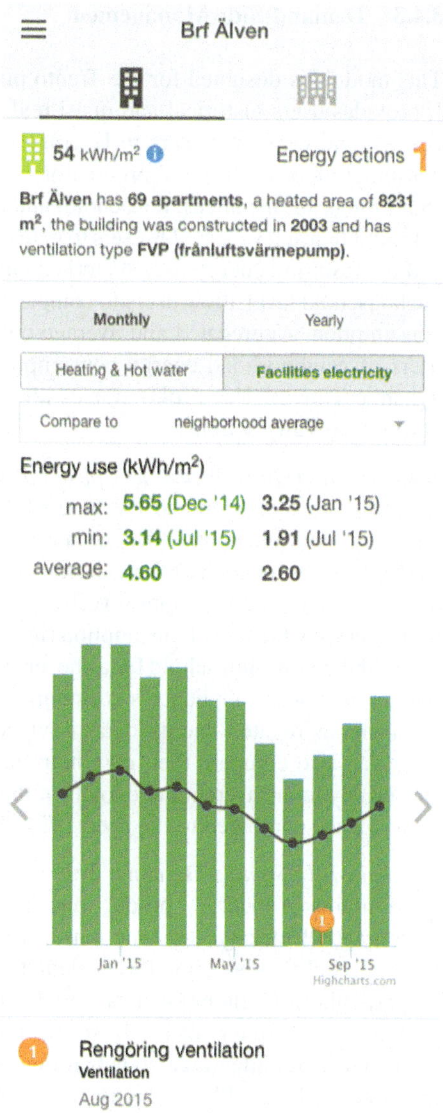

Fig. 8 Facilities electricity use graph. Information about housing cooperatives and actions is displayed at the top. Green bars show the housing cooperative's current year's use per month; the black line shows the average use of all housing cooperatives

cooperatives can also add email addresses of their contact persons, which are visible on each cooperative's app page. Sharing experiences certainly also happens outside of the digital world, e.g. during meetings of cooperative boards or with local energy networks. The app aims to support discussions and knowledge exchange also in such situations, where someone can easily demonstrate the impact of an energy investment with smart phones.

3.4.3 Demand-Side Management

This module is designed for the Trento pilot sites and can have wider application. It provides users historical and quasi real-time consumption and production information, and facilitates users to leverage load elasticity in order to maximize self-consumption of rooftop PV productions. Energy data is displayed at appliances (if smart plugs are installed), household, and electricity consortia levels. Consumption at the appliance level enables users to gain deeper understanding of their daily actions and the resulting energy use. Historical and current consumption and production at the household level allow users to compare those two and potentially maximize self-consumption. Aggregated and average consumption at the consortia level informs users of neighborhood energy consumption and allows comparisons. In addition, dynamic Time-of-Use (ToU) signals are displayed to assist users in load shifting during their daily actions.

Historical and Quasi Real-time Consumption and Production At the household level, electricity consumption and PV production levels (in W and Wh) are displayed in quasi real-time and updated for the latest 6 min[13] (Fig. 9a). This information can also be displayed as a bar chart for a chosen period (in the past) to provide an aggregated daily overview of consumption versus production. When smart plugs are installed, users can view the daily consumption (in Wh) of the corresponding connected appliances of their own household for a chosen period. This helps them gain better insights into the individual appliance's consumption level and its daily or seasonal patterns. With the aggregated energy data provided by the two local electricity consortia, users can also compare their own households' hourly consumption profiles over a chosen day to the averages and totals of the consortia to gain a sense of their relative performance compared to the peers (Fig. 9b).

Dynamic ToU Signals Dynamic ToU signals are provided to facilitate users' self-consumption of local PV productions. They give clear indications to encourage or discourage electricity consumption at a certain moment based on the forecasted local renewable production level calculated with open weather forecast information (in particular solar radiation data) and the local rooftop PV production capacity. The signals are at 3 h intervals for the forthcoming 30 h (Fig. 10a), and are updated every 24 h. A green smiley face signals a time slot suitable for self-consumption where the forecasted local PV production exceeds the current local consumption, while an orange frowny face signals otherwise. On a weekly basis, users receive a summary of the proportion of their own household consumption that took place under green or orange ToU signals to allow them to reflect on their levels of self-consumption (Fig. 10b). The same information is also provided at the consortia level to enable peer comparison.

[13]For technical reasons such as households' data transfer connections and processing time, there can be up to 2-min delay between the time of actual power measurement and the data displayed.

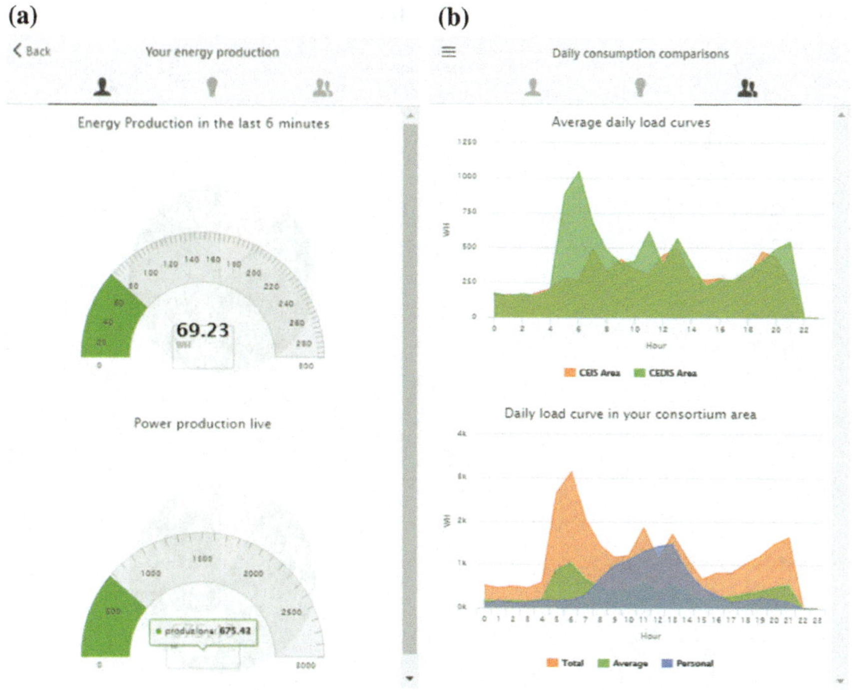

Fig. 9 **a** Quasi real-time meters for household PV production; **b** A household's hourly consumption profile over a chosen day compared to the averages and totals of the consortia

3.4.4 Action Suggestions

This module aims to facilitate all household members to take part in energy conservation in their busy daily life. About fifty action suggestions are composed to provide users practical and accurate information about energy conservation. They include one-time actions such as "Use energy efficient cooktops", routine actions such as "Line dry, air dry clothes whenever you can", as well as in-between actions (reminders) such as "Defrost your fridge regularly (in x days)". Some suggestions may seem obvious and trivial, but as indicated by literature, people often has an attitude-behaviour gap when it comes to environmental issues. The goal is to facilitate the behaviour change process to bridge the attitude-behaviour gap, making energy conservation new habits integrated in everyday household practices.

Free Choice and Self-monitoring of Energy Conservation Actions Actions are not meant to prescribe what users should do but to present different ideas of what they can do (and how) in household practices. Users can freely choose whether (and when) to take an action and possibly reschedule and repeat the action according to the needs and interests in their own context (Fig. 11). After all, users are experts of their own reality. They also have an overview of their current, pending, and completed actions.

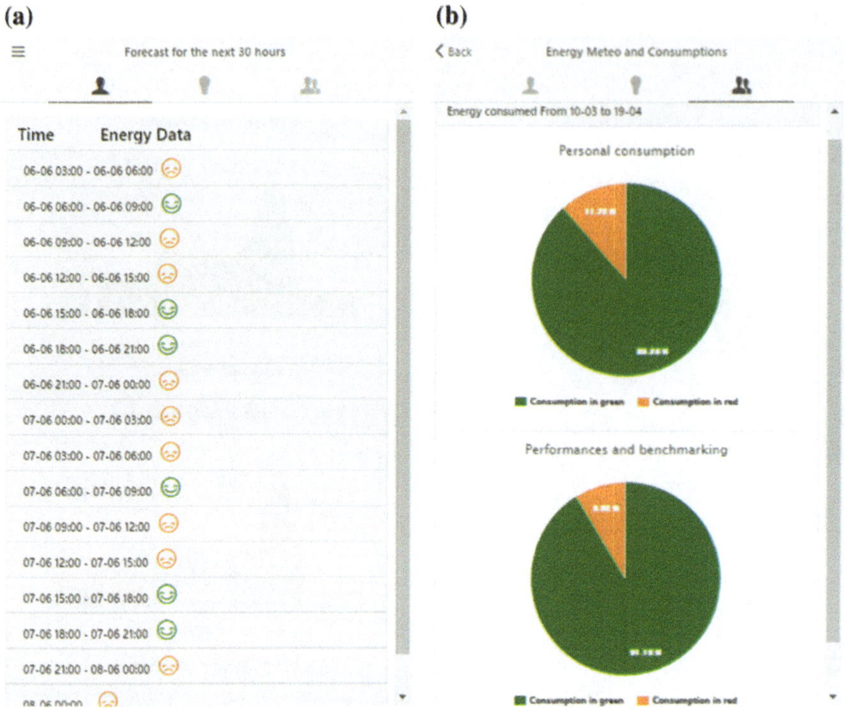

Fig. 10 **a** Dynamic ToU signals at 3 h intervals for the forthcoming 30 h; **b** A household's hourly consumption profile over a chosen day compared to the averages and totals of the consortia

A new action is suggested when one is completed. When an action is scheduled, its reminder is triggered by time. Users' own choices of actions and the action processes facilitate the sense of autonomy which enhances and maintains motivation [44].

Promoting Motivation and Engagement The design uses a number of elements to promote users' motivation and engagement. Suggestions are tailored to the local context by local partners and focus groups. Each action is accompanied by a short explanation, the entailed effort and impact (on a five-point scale) and the number of users taking this action. The design encourages users to take small steps (and not to have too many actions at a time) and gives positive performance feedback. In addition, users can invite household members, view and join the energy conservation actions of the whole household. Users can also login with Facebook, like, comment, share actions, give feedback and invite friends. Users are awarded with points (displayed as Green Leaves) once they complete an action, or provide feedback or comments.

3.4.5 Community Engagement Approaches

Another main outcome of the design process, which also reflects the potential richness of designing for large-scale STS, rests at the level of community engagement. The

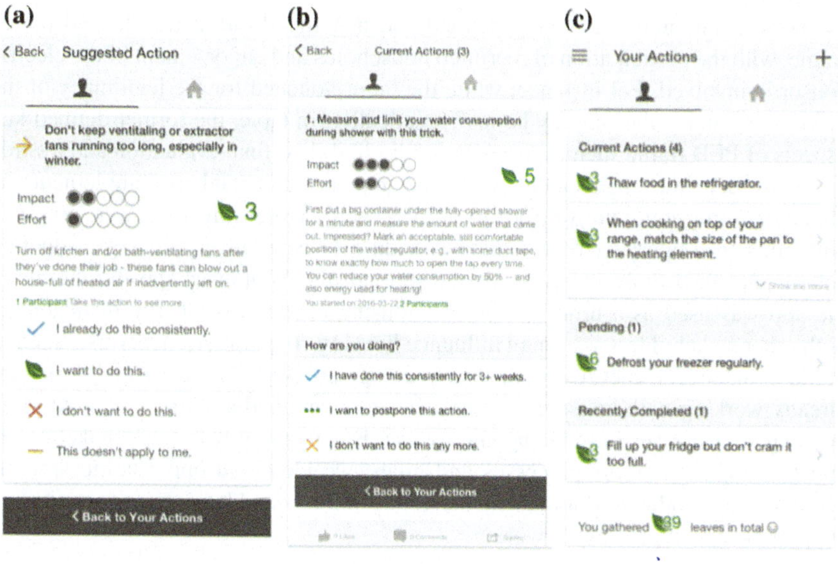

Fig. 11 **a** Action suggestion; **b** Action in progress; **c** User actions

ambition to foster energy behaviour change at the collective level of communities (or neighbourhoods), instead of simply aiming for technology adoption at individual level, made it clear the need to design for engagement. In the two national contexts, two different engagement processes accompanied the deployment and testing of the technology. They tried to stimulate the emergence of the social dynamics connected to the change of energy behaviour. (Note that the collaborative design process discussed in Sect. 3.3 also contributed to engagement.)

In Italy, a full fledged process named *Participatory Energy Budgeting* (PEB) [8, 10] was run with the twofold goal of subsidizing people's efforts in demand-side management and empowering them to handle their achievements in a collective and transparent way. PEB is a policy frame that relies on a call for tender that defines: the energy budget to be administered; the criteria and procedures to submit proposals for funds request; the procedures to evaluate, select and award the winning proposals; and a roadmap for the process development. Grounded on the community funds model of participatory budgeting [16, 51], PEB promoted engagement and allowed collective decision making around the management and allocation of "energy bonus", which could be collected through the collective effort of shifting electric energy demand towards local production peaks. The PEB and the demand-side management module of YouPower were thought and designed to act in synergy and "reinforce" each other. The more people consumed energy during peaks of local production—foretold and displayed with "green smileys" in YouPower—the more the energy bonuses grew. PEB can be considered a main outcome of the design process for the Italian sites, because the notion of collectively managing energy savings emerged during the first exploratory focus groups, and throughout the fist two project years. During the project, this notion has been refined and negotiated into a full-fledged policy frame, with the participation of recruited households and endorsement of the electric consortia involved. For instance, while the latter vouched for the legitimacy of the process and made the "energy bonus" practically available, the former defined key aspects of PEB frame such as the criteria for eligibility, final evaluation and award.

In Sweden, the engagement work and app design aimed to complement the already existing community efforts to address energy issues. Meetings were arranged with housing cooperative representatives to discuss experiences of energy reduction actions and how those could be shared through the Youpower app. Furthermore, the app was used as a probe to discuss housing cooperative energy management with other stakeholders who may influence housing cooperative energy use, such as building managers, energy providers, and energy advisers. These stakeholders were already working with housing cooperatives and many had ambitions of supporting housing cooperatives in reducing energy use. By engaging with these stakeholders and learning about their processes and goals, we identified opportunities for the Youpower app to be used jointly by these stakeholders and housing cooperatives to support energy improvement work.

4 Discussion and Conclusion

Collaboratively designed with the stakeholders from different pilot sites, the main outcomes of CIVIS addressed the goals and context of the project at different ST levels. They include the CIVIS platform that consisted of YouPower, an open source social smart grid application, and the corresponding hardware and software installation for energy data collection at participating households from the pilot sites. The deployment was accompanied by community engagement approaches to ensure that the stakeholders were well aware of the key issues and results the project was aiming at, and to develop positive attitude and encourage active participation.

At the Italian pilot sites, self-consumption of local renewable production was promoted at household and consortium levels, while at the Swedish pilot sites, knowledge sharing about housing cooperatives energy management practices was supported among cooperatives' board members and across different cooperatives. To bridge the attitude-behaviour gap of people's environmental values (and attitudes) and their actual behaviour in energy consumption [32, 48, 49], the platform also provided a set of features that could facilitate users' behaviour change process towards sustainable consumption that was implementable in their daily life along their existing practices. A number of lessons learned from the CIVIS design experience that could also be relevant to the design of other smart urban ecosystems beyond the particular case of CIVIS project are discussed below.

First, despite the many advantages already discussed previously, implementing a collaborative participatory design process is highly challenging in practice with an interdisciplinary team in an international setting. The design and development team, together with stakeholders involved, have various professional and cultural backgrounds, possibly speak diverse languages, hold disparate values and principles, work in different styles, not to mention the personal and organizational interests they may withhold. Misunderstandings on terminologies, methodologies and actions may go unnoticed and accumulate until it is very challenging or even critical to mediate the diverging opinions. The full awareness of such issues, frequent and efficient communications, positive and constructive attitude, plus open-mindedness are the keys to make the development process effective and enjoyable.

Second, the relevance, importance and challenge of setting up an engagement strategy or change process for the potential users of the new or modified system should not be underestimated. Engaging people in changing behaviour has much to do with understanding local contexts, people's heterogeneous attitudes, and local cultures. It also needs careful planning and execution. Develop a clear engagement strategy starting from the beginning of the project and let the professionals with the proper skills in this area to interact with the stakeholders.

Third, with respect to STS design and engagement strategies, users and other stakeholders should be provided with accurate and actionable information about how to achieve target behaviour. At the CIVIS pilot sites, for example, people expressed the desire and need to want to do more for a sustainable future. They liked the idea of receiving relevant and contextual suggestions and tips for action. Given the heterogeneity of potential stakeholder groups, understanding them and their interests and needs remains a crucial and challenging part of design that requires careful confrontation with stakeholders directly.

Fourth, consumers' intrinsic motivation for engagement needs to be fostered. Users need to be allowed to freely practice and adapt their course of action. This facilitates the sense of competence and autonomy that promotes and enhances motivation for behaviour change [44]. For example, people in the pilot sites are skeptical about how much monetary gains they can actually have by using less energy in households, but they are driven by intrinsic motives as well as altruistic and environmental values for energy saving. The social and community-oriented features as those designed in the CIVIS project articulate those values.

With the explosive growth of smart devices and smart everything, the coming wave of IoT and the hyperconnected world will soon bring the society into a smart environment where computing is pervasive [19, 50]. Will this smartness bring its inventors and the natural world into a sustainable future? This chapter advocates the potential fruitfulness of IoT and smart urban ecosystems that do not mainly rely on the technological side. Designers and engineers need to indispensably take a human-centred ST approach in developing a smart sustainable future.

References

1. L. Atzori, A. Iera, G. Morabito, From "smart objects" to "social objects": the next evolutionary step of the internet of things. IEEE Commun. Mag. **52**(1), 97–105 (2014)
2. G. Baxter, I. Sommerville, Socio-technical systems: from design methods to systems engineering. Interact. Comput. **23**(1), 4–17 (2011)

3. K. Bødker, F. Kensing, J. Simonsen, *Participatory IT Design: Designing for Business and Workplace Realities* (MIT Press, Cambridge, MA, 2004)
4. P.W. Bots, Design in socio-technical system development: three angles in a common framework. J. Design Res. **5**(3), 382–396 (2007)
5. G.D. Brewer, The challenges of interdisciplinarity. Policy Sci. **32**, 327–337 (1999)
6. H. Brynjarsdottir, M. Håkansson, J. Pierce, E. Baumer, C. DiSalvo, P. Sengers, Sustainably unpersuaded: how persuasion narrows our vision of sustainability, in *Proceedings of the SIGCHI Conference on Human Factors in Computing Systems*, CHI' 12 (ACM, New York, NY, USA, 2012), pp. 947–956
7. H. Bulkeley, V.C. Broto, G. Edwards, Bringing climate change to the city: towards low carbon urbanism? Local Environ. **17**(5), 545–551 (2012)
8. A. Capaccioli, G. Poderi, M. Bettega, V. D'Andrea, Exploring alternative participatory budgeting approaches as means for citizens engagement: the case of energy, in *2016 IEEE International Smart Cities Conference (ISC2)*, pp. 1–4, Sept. 2016
9. A. Capaccioli, G. Poderi, M. Bettega, V. D'Andrea, Participatory infrastructuring of community energy, in *Proceedings of the 14th Participatory Design Conference: Short Papers, Interactive Exhibitions, Workshops (PDC'16)*, vol. 2 (ACM, New York, NY, USA, 2016), pp. 9–12
10. A. Capaccioli, G. Poderi, M. Bettega, V. D'Andrea, Exploring participatory energy budgeting as a policy instrument to foster energy justice. Energy Policy **107**, 621–630 (2017)
11. P. Checkland, *Systems Thinking, Systems Practice* (Wiley, 1981)
12. M. da Graa Carvalho, EU energy and climate change strategy. Energy **40**(1), 19–22 (2012)
13. H. Dick, H. Eden, G. Fischer, J. Zietz, Empowering users to become designers: using meta-design environments to enable and motivate sustainable energy decisions, in *Proceedings of the 12th Participatory Design Conference: Exploratory Papers, Workshop Descriptions, Industry Cases* (PDC' 12), vol. 2 (ACM, New York, NY, USA, 2012), pp. 49–52
14. L. Fleischhacker, E. Agazzi, Chapter commentaries: the non-linearity of the development of technology and the techno-scientific system, in *Right, Wrong and Science The Ethical Dimensions of the Techno-Scientific Enterprise*. Monographs-in-Debate (Brill, 2004), pp. 301–310
15. G. Fortino, P. Trunfio (eds.), *Internet of Things Based on Smart Objects: Technology, Middleware and Applications* (Springer International Publishing, 2014)
16. E. Ganuza, G. Baiocchi, The power of ambiguity: how participatory budgeting travels the globe. J. Public Delib. **8**(2) (2012)
17. A. Glasmeier, S. Christopherson, Thinking about smart cities. Camb. J. Reg. Econ. Soc. **8**, 3–12 (2015)
18. J. Greenbaum, K. Halskov, PD a personal statement. Commun. ACM **36**(6), 47 (1993)
19. J. Gubbi, R. Buyya, S. Marusic, M. Palaniswami, Internet of Things (iot): a vision, architectural elements, and future directions. Future Gener. Comput. Syst. **29**(7), 1645–1660 (2013)
20. B. Guo, Z. Yu, X. Zhou, D. Zhang, Opportunistic iot: exploring the social side of the internet of things, in *2012 IEEE 16th International Conference on Computer Supported Cooperative Work in Design (CSCWD)*, pp. 925–929 (IEEE, 2012)
21. B. Guo, D. Zhang, Z. Wang, Z. Yu, X. Zhou, Opportunistic iot: exploring the harmonious interaction between human and the Internet of Things. J. Netw. Comput. Appl. **36**(6), 1531–1539 (2013)
22. Y.N. Harari, *Sapiens: A Brief History of Humankind* (Harvill Secker, 2014)
23. H. Hasselqvist, C. Bogdan, F. Kis, Linking data to action: designing for amateur energy management, in *Proceedings of the 2016 ACM Conference on Designing Interactive Systems*, pp. 473–483 (2016)
24. H. Hasselqvist, C. Bogdan, M. Romero, O. Shafqat, Supporting energy management as a cooperative amateur activity. CHI **2015**, 1483–1488 (2015)
25. Y. Huang, H. Hasselqvist, G. Poderi, S. Sćepanović, F. Kis, C. Bogdan, M. Warnier, F. Brazier, Youpower: an open source platform for community-oriented smart grid user engagement, in *2017 IEEE 14th International Conference on Networking, Sensing and Control (ICNSC)*, pp. 1–6, May 2017

26. Y. Huang, D. Miorandi, D3.1 simulation model of integrated energy system. Technical report, EU FP7 CIVIS Project, 2014. Deliverable 3.1
27. Y. Huang, D. Miorandi, H. Hasselqvist, M. Warnier, S. Scepanoic, R. Eskola, D3.2 intergrated energy system. Technical report, EU FP7 CIVIS Project, 2015. Deliverable 3.2
28. Y. Huang, G. Poderi, L. Yishagerew, H. Hasselqvist, A. Massaro, S. Scepanovic, H. Ensing, F. Cuscito, D3.3 final field tested integrated energy system. Technical report, EU FP7 CIVIS Project, 2016. Deliverable 3.3
29. Y. Huang, M. Warnier, F. Brazier, D. Miorandi, Social networking for smart grid users—a preliminary modeling and simulation study, in *Proceedings of 2015 IEEE 12th International Conference on Networking, Sensing and Control*, pp. 438–443 (2015)
30. A. Kankainen, K. Vaajakallio, V. Kantola, T. Mattelmki, Storytelling Groupa co-design method for service design. Behav. Inf. Technol. **31**(3), 221–230 (2012)
31. S. Karnouskos, The cooperative internet of things enabled smart grid, in *Proceedings of the 14th IEEE International Symposium on Consumer Electronics (ISCE2010), June*, pp. 07–10 (2010)
32. A. Kollmuss, J. Agyeman, Mind the gap: why do people act environmentally and what are the barriers to pro-environmental behavior? Environ. Educ. Res. **8**(3), 239–260 (2002)
33. P. Kroes, P.E. Vermaas, A. Light, S.A. Moore, Chapter design in engineering and architecture: towards an integrated philosophical understanding, in *Philosophy and Design: From Engineering to Architecture* (Springer, Dordrecht, 2008), pp. 1–17
34. A.S. Lee, Mis quarterlys editorial policies and practices. MIS Q. iii–vii (2001)
35. C.C. Mody, Chapter 5 Small, but determined: technological determinism in nanoscience, in *Nanotechnology Challenges: Implications for Philosophy, Ethics, and Society* (World Scientific, 2006), pp. 95–130
36. I. Nikolić, Co-evolutionary method for modelling large-scale socio-technical systems evolution. PhD thesis, Delft University of Technology, 2009
37. H. Ning, Z. Wang, Future internet of things architecture: like mankind neural system or social organization framework? IEEE Commun. Lett. **15**(4), 461–463 (2011)
38. D.A. Norman, P.J. Stappers, DesignX: complex sociotechnical systems. She Ji: J. Design Econ. Innov. **1**(2), 83–106 (2015)
39. J. Padget, H. Riat, M. Warnier, F. Brazier, S. Natarajan, An agent-based infrastructure for energy profile capture and management, in *1st International Workshop on Agent Technologies for Energy Systems, 9th International Conference on Autonomous Agents and Multiagent Systems*, Toronto, Canada, 2010
40. J. Pierce, E. Paulos, Beyond energy monitors: interaction, energy, and emerging energy systems, in *CHI'12* (ACM, 2012), pp. 665–674
41. G. Poderi, M. Bettega, A. Capaccioli, V. DAndrea, Disentangling participation through time and interaction spacesthe case of IT design for energy demand management. CoDesign, 0(0), 1–15 (2017)
42. G.A. Putrus, E. Bentley, R. Binns, T. Jiang, D. Johnston, Smart grids: energising the future. Int. J. Environ. Stud. **70**(5), 691–701 (2013)
43. J. Rifkin, *The Third Industrial Revolution: How Lateral Power is Transforming Energy, the Economy, and the World* (Palgrave Macmillan, New York, NY, USA, 2011)
44. R.M. Ryan, E.L. Deci, Intrinsic and extrinsic motivations: classic definitions and new directions. Contemp. Educ. Psychol. **25**(1), 54–67 (2000)
45. S. Sawyer, M.H. Jarrahi, Chapter 5 Sociotechnical approaches to the study of information systems, in *Computing Handbook: Information systems and information technology*, (Taylor & Francis, 3rd edn., 2014)
46. M. Schatten, Smart residential buildings as learning agent organizations in the internet of things. Bus. Syst. Res. J. **5**(1), 34–46 (2014)
47. L. Schick, B.R. Winthereik, Innovating relations—or why smart grid is not too complex for the public. Sci. Technol. Stud. **26**(3), 82–102 (2013)
48. P. Schultz, Chapter Knowledge, information, and household recycling: examining the knowledge-deficit model of behavior change, in *New Tools for Environmental Protection:*

Education, Information, and Voluntary Measures (National Academy Press, Washington DC, 2002), pp. 67–82
49. P.W. Schultz, Strategies for promoting proenvironmental behavior—lots of tools but few instructions. Eur. Psychol. **19**(2), 107–117 (2014)
50. D. Shin, A socio-technical framework for internet-of-things design: a human-centered design for the internet of things. Telemat. Inform. **31**(4), 519–531 (2014)
51. Y. Sintomer, C. Herzberg, A. RCke, Participatory budgeting in Europe: potentials and challenges: participatory budgeting in Europe. Int. J. Urban Reg. Res. **32**(1), 164–178 (2008)
52. M.R. Smith, L. Marx (eds.), *Does Technology Drive History? The Dilemma of Technological Determinism* (MIT Press, 1994)
53. B.K. Sovacool, How long will it take? Conceptualizing the temporal dynamics of energy transitions. Energy Res. Soc. Sci. **13**, 202–215 (2016)
54. M. Tomasini, B. Mahmood, F. Zambonelli, A. Brayner, R. Menezes, On the effect of human mobility to the design of metropolitan mobile opportunistic networks of sensors. Pervasive Mob. Comput. **38**, 215–232 (2017)
55. F. Umbach, Global energy security and the implications for the EU. Energy Policy **38**(3), 1229–1240 (2010)
56. K.H. van Dam, I. Nikolic, Z. Lukszo (eds.), *Agent-Based Modelling of Socio-technical Systems* (Springer Science & Business Media, 2012)
57. E. Viardot, T. Wierenga, B. Friedrich, The role of cooperatives in overcoming the barriers to adoption of renewable energy. Energy Policy **63**, 756–764 (2013)
58. P.E. Waterson, M.T.O. Gray, C.W. Clegg, A sociotechnical method for designing work systems. Hum. Factors **44**, 376–391 (2002)
59. B. Whitworth, Chapter 66 A brief introduction to sociotechnical systems, in *Encyclopedia of Information Science and Technology* (IGI Global, 2nd edn., 2009), pp. 394–400
60. B. Whitworth. *The Social Design of Technical Systems: Building Technologies for Communities* (The Interaction Design Foundation, 2014)
61. B. Whitworth, A. Ahmad, *The Encyclopedia of Human-Computer Interaction*, chapter 24. Socio-Technical System Design (The Interaction Design Foundation, 2nd edn., 2013)
62. B. Whitworth, A. De Moor (eds.), *Handbook of Research on Socio-technical Design and Social Networking Systems* (IGI, 2009)
63. M. Yun, B. Yuxin, Research on the architecture and key technology of internet of things (iot) applied on smart grid, in *2010 International Conference on Advances in Energy Engineering (ICAEE)*, pp. 69–72 (IEEE, 2010)
64. A. Zanella, N. Bui, A. Castellani, L. Vangelista, M. Zorzi, Internet of things for smart cities. IEEE Internet of Things J. **1**(1), 22–32 (2014)
65. S. Zygiaris, Smart city reference model: assisting planners to conceptualize the building of smart city innovation ecosystems. J. Knowl. Econ. **4**(2), 217–231 (2013)

Aggregation Techniques for the Internet of Things: An Overview

Barbara Guidi and Laura Ricci

Abstract Internet of Thing (IoT) can be generally defined as a network connecting millions of smart objects, most of them equipped with sensors. Since sensors are devices generating a huge amount of data, the transmission of raw data to the edge nodes and then to higher level cloud nodes may give rise to transmission delays and energy consumption. Furthermore, sensors are characterized by limited resources. For all these reasons, aggregation techniques are required to reduce the size of data to be transmitted and stored, while maintaining a reasonable level of approximation. In this paper, we propose an overview of a set of aggregation techniques which may be exploited in IoT. We present a set of techniques, ranging from Space Filling Curves, to Q-digest, Wavelets, Gossip aggregation, and Compressive Sensing. We also show how these techniques are exploited in IoT applications.

1 Introduction

The Internet of Things (IoT) is a new paradigm based on the vision of a worldwide network of intercommunicating devices. Despite many definitions of the Internet of Things exist, it can be described as a network of devices interacting with each other enabling data collection and exchange [16, 21]. The basic idea is the presence of a variety of objects ranging from wearable devices such as smart-watches to sensors and mobile phones which are able to interact with each other and cooperate with their neighbours to reach common goals. In the IoT, billions of objects can be found through various types of actuators and sensors, which are connected to the Internet via Wireless Sensor Network (WSN) [1].

Devices involved in IoT systems generate a huge amount of data, and despite the heterogeneity of the IoT devices, the data in IoT applications such as smart

B. Guidi (✉) · L. Ricci
Department of Computer Science, University of Pisa, Largo Bruno Pontecorvo, 3, Pisa, Italy
e-mail: guidi@di.unipi.it

L. Ricci
e-mail: laura.ricci@unipi.it

F. Cicirelli et al. (eds.), *The Internet of Things for Smart Urban Ecosystems*, Internet of Things, https://doi.org/10.1007/978-3-319-96550-5_7

home, smart city, and smart energy services can be easily combined, correlated, compared and merged to match the people's needs and requirements [35]. However, these small devices have limited resources, such as limited batteries and energy supplies [3]. Therefore, it is necessary to improve the lifetime of the devices by optimizing both the data collection and the analysis of the huge amount of data. For this reason, IoT systems require data aggregation techniques. Data aggregation appears as a key element in reducing the amount of traffic in WSNs and achieving energy conservation. The fundamental purpose of the a data aggregation strategy is to aggregate and collect data in an effective way in order to improve energy consumption and to avoid traffic bottleneck [41]. Current data aggregation techniques are in origin proposed for WSNs. During the last years, some surveys have been published [24, 36, 38], specifically focusing on aggregation techniques for WSNs.

In this paper we propose an overview of a set of complex data aggregation techniques applied to the IoT. This paper differs from other works because we describe the mathematical background of the proposed techniques and we propose an overview of how current systems exploit these techniques. The data aggregation techniques presented in this paper are: Space Filling Curves, the Quantile Digest, the Wavelet Transform, the Compressive Sensing, and Gossip Aggregation.

The paper is organized as follows. In Sect. 2 we propose an overview of the data aggregation process and its importance in IoT by introducing also two well known network structures: Structured Networks and Unstructured Networks. In Sect. 3 we introduce a set of data compression techniques, and in Sect. 4 we show how gossip algorithms can be exploited to aggregate data. Finally, in Sect. 5 we present the conclusions.

2 Data Aggregation

Data aggregation is an important phase in IoT because of the number of devices involved in the IoT system and the huge amount of data produced. Indeed, data aggregation is considered one of the fundamental processing procedures for saving the energy and to limit the amount of required resources. Sensors included in IoT networks have limited computational power, limited memory and battery power. Usually, there is the need to find a way to combine data and aggregate them. This would optimize streaming data flows and edge analytics resulting in faster time and greater overall value. In detail, data aggregation can be defined as the process of integrating and summarizing data. Data needs to be aggregated to be used and analyzed, and during the last years several approaches have been applied. The aim of data aggregation is multiple because it is useful to reduce the amount of space used to store data, to remove redundant transmissions of data, to decrease the traffic load, and to save energy of the nodes. Classic aggregation techniques applied in Wireless Sensor Networks (WSNs) are usually applied in the IoT context, and specific complex aggregation functions are used to aggregate data.

In this paper, we propose an overview of the main data aggregation techniques in IoT. In particular, we focus our attention on a set of complex aggregation functions that produce a kind of digest, which summarizes the data distribution. Digests essentially consists on the compression of all inputs into a fixed size data structure, using probabilistic methods, and losing some information. We present the Space Filling Curves, the Quantile Digest, the Wavelet Transform, the Compressive Sensing, and the Gossip Aggregation. Before introducing the complex data aggregation functions, we propose an overview of the main network structures generally used in IoT systems: Structured or Unstructured networks.

2.1 Structured Networks

In structured networks both a well defined network topology and a routing scheme are defined. This implies that data aggregation algorithms need to take into account the structure of the underlying topology.

A hierarchical communication structure is mostly used, in particular in Wireless Sensor Networks (WSNs). In a *Hierarchical-based scheme* the structured network is a tree where the intermediate nodes perform the data aggregation process [12]. Usually, a bottom-up approach is exploited, where the data aggregation process starts from leaf nodes till the root, and each node has a parent node which is used to move aggregate data to the upper levels. Figure 1 shows an example of a Hierarchical-based architecture implemented by a tree, presented in [36].

Another important scheme is the cluster-based one, where the network is separated into a set of clusters, and each cluster consists of many sensor nodes. Set of nodes with the same attributes or characteristics are put in the same group or cluster. One header node is selected for each cluster, which is also called as cluster-head, as show in Fig. 2, presented in [36]. The clustering structure results in reducing the bandwidth overhead as a total number of packets to be transmitted [36].

Cluster-based data aggregation algorithms are one of the most popular because they have the advantages of high flexibility and reliability.

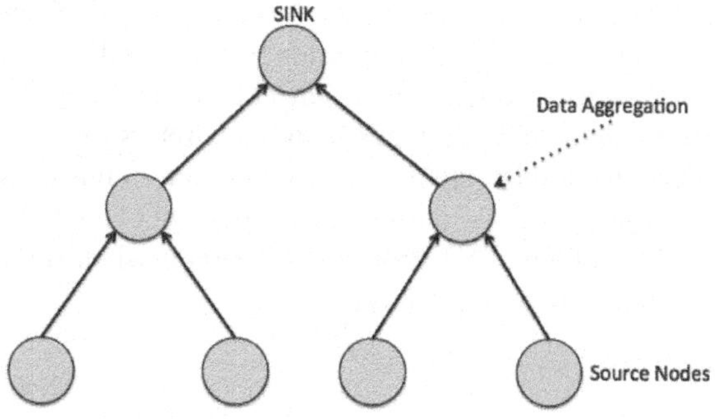

Fig. 1 An example of a tree-based architecture

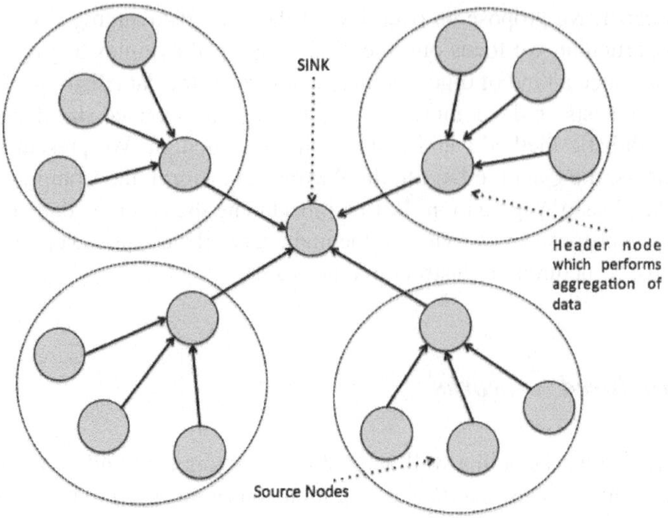

Fig. 2 An example of a cluster-based architecture

In Structured Networks, simple aggregation functions such as MAX, MIN, SUM, and COUNT can be computed easily and efficiently [29]. Indeed, nodes first auto-organize into a spanning tree and then starting from the leaves, the aggregation propagates upwards to the root. When a node receives the aggregates from its children, it computes the aggregation of these with its own data, returning the aggregation of all the data in the nodes subtree, and forwards it to its parent. Several complex data aggregation mechanisms have been proposed. In the following sections, we describe a set of data compression techniques used for data aggregation in IoT, and we describe how they are used.

2.2 *Unstructured Networks*

An unstructured sensor network is a collection of sensor nodes deployed in ad-hoc manner into a region. Connection between nodes are generally generated at random.

Network maintenance, which includes detecting failures and managing connectivity are quite difficult in unstructured wireless sensor networks. Also, due to random deployment there are uncovered areas left in unstructured WSN. Moreover, several problems may arise during the deployment of an unstructured sensor network due to the lack of a reliable collision detection mechanism, and the lack of the knowledge about the network topology.

3 Complex Data Aggregation Techniques in IoT

In this section we introduce an overview of a set of complex data aggregation approaches used in IoT. In detail, we present the Space-filling Curves (Hilbert Curve and Z-Curve), the Quantile Digest, the Compressive Sensing, the Wavelet Transform, and finally the Gossip aggregation approach.

3.1 Space-Filling Curves

The mathematical definition of a space filling curve [34] refers to a curve whose range contains the entire 2-dimensional unit square (or more generally an n-dimensional unit hypercube). Space filling curves were initially discovered by Giuseppe Peano, and for this reason they are known as Peano curves [34].

There are several space filling curves which differ according to the mapping function. In Fig. 3 we show three of them: the z-curve, the Hilbert-curve, and finally the gray code curve.

A significant amount of interest has been demonstrated for the Hilbert curve, proposed by Hilbert [19] as a common method to reduce multi-dimensional problem into a one-dimensional problem. A Hilbert curve (also known as a Hilbert space-filling curve) is a continuous fractal space-filling curve proposed as a variant of the Peano curves.

Given a domain I, I can be mapped continuously onto the unit-square Ω. If I is partitioned into four congruent subintervals, it should be possible to partition Ω into four congruent sub-squares, such that each subinterval will be mapped continuously onto one of the sub-squares. This process generates a partition of I into 2^{2n} congruent subintervals and of Ω into 2^{2n} congruent sub-squares for $n = 1, 2, 3, \ldots$. The sub-squares of each partition can be ordered in such a manner that consecutive subintervals correspond to consecutive sub-squares with a common edge, and if a square corresponds to an interval, then its sub-squares correspond to the subintervals of that interval. The Hilbert curve is a continuous and surjective mapping [39].

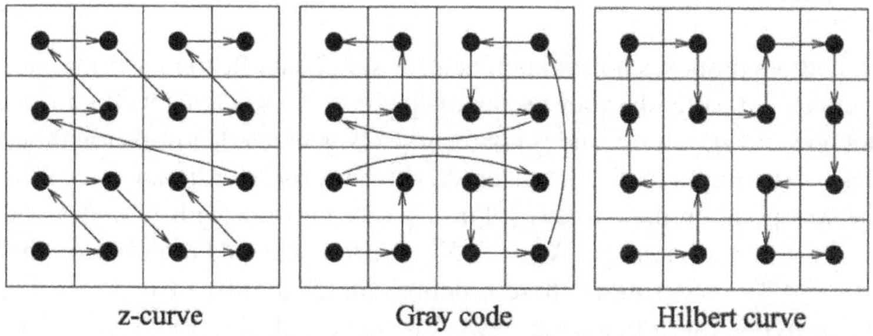

Fig. 3 Illustration of three different space-filling curves: z-curve, Gray code, and Hilbert curve

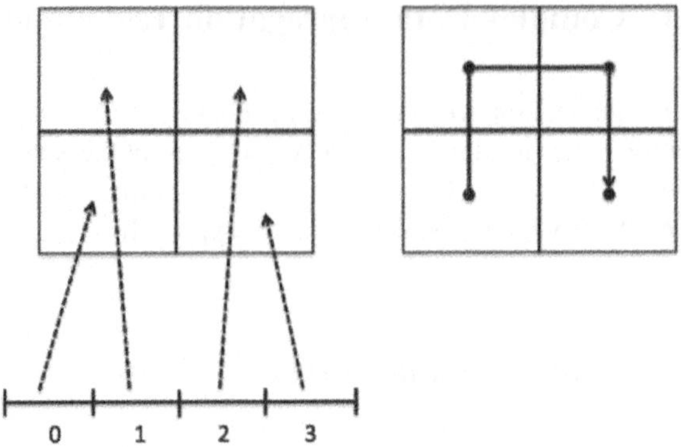

Fig. 4 First order Hilbert curve

The construction of a Hilbert curve needs to specify the order of curve which means the number of steps or iterations. The first step of the process concerns the one-dimensional interval [0, 1] and the square $[0, 1]^2$ which are divided into four congruent quarters. Each sub-interval is then mapped to a different sub-square in such a way that sub-squares mapped to from adjacent sub-intervals share a common edge. The sub-squares are thus ordered. The line passing through the center points of the sub-squares is referred to as a first order Hilbert curve (Fig. 4). The process of division of intervals and squares is repeated by taking into account the order of the curve. A curve of order k, where $k > 1$ is constructed by replacing each point on a curve of order $(k - 1)$ by a scaled-down curve of order 1 [27].

Another important curve is the Z-curves which maps multidimensional data to one dimension preserving locality of the data points. It was introduced in 1966 by Guy Macdonald Morton [31]. The z-value of a point in multi-dimensions is simply calculated by interleaving the binary representations of its coordinate values. Once the data are sorted into this ordering, any one-dimensional data structure can be used (binary search trees, B-trees, skip lists, or hash tables). The resulting ordering can equivalently be described as the order one would get from a depth-first traversal of a quadtree.

3.1.1 Space Filling Curves in IoT

In [33], authors proposes a new approach for serial data fusion by using the Space Filling curves, in particular they use a sweep curve. This is not a well known curve, does not preserve proximity and is not recursively generated, however authors adopt it because of its simplicity. Each sensor (X_i) in the two-dimensional space is mapped to its nearest point on the curve (t_i). For a sensor network with N nodes, the locations of sensor nodes, say, $X_1, X_2, \ldots, X_N$, are thus mapped to points on the curve $t_1, t_2, \ldots, t_N$. The ordering of these t_i defines an ideal order of nodes that are traversed. The hypothesis is that if the network is connected, sensor nodes are distributed somewhat uniformly in the two dimensional space, and the curve is drawn with an appropriate density for the network. Two consecutive nodes i_j and i_{j+1} in the traversal order are also in close proximity in the physical space. The traversal starts from a specific node by assuming that each node knows its neighbours, including their identifications and coordinates. This information can be collected statically via one time beaconing or hello messages, or periodically, in case of dynamic changes in the topology. When visiting a sensor node, say P, the agent orders all unvisited neighbours of P by their curve indices (t_i). The curve indices can be computed by knowing their locations. This process is repeated, and if the agent finds itself at a node P that does not have any unvisited neighbour, it backtracks to the node it came from, and similarly looks for unvisited neighbours, backtracking further if there is none. The traversal terminates when the appropriate termination condition is satisfied. Note that the traversal is similar to a depth-first traversal of an unknown graph. The only difference is that when visiting an unvisited neighbour, the neighbour to visit is not arbitrarily picked when there are more than one possibilities, but is picked based on its index on the curve. This imposes an a priori order for thc traversal that helps minimizing the number of transmissions necessary to complete the traversal.

In [2] a secure data aggregation scheme based on the space filling curves is proposed. The proposed method encrypts the sensor data by elliptic-curve based seed exchange algorithm and Hilbert-curve based data transformation. To process a users query, a parent node aggregates its data and all data received from its children nodes. Then, the parent node transforms the aggregated result into two-dimensional encrypted data by using the Hilbert curve. Authors assume that each sensor node transforms one-dimensional sensed value into two-dimensional data. The one-dimensional value is the aggregated value after applying the seed exchange algorithm for each node group. The two-dimensional data is a coordinate of the aggregated value along with the Hilbert curve in $2^n \times 2^n$ metrics. A key is set by considering both the level and the direction of Hilbert curve. Formally, aggregated data can be encrypted by using two dimensional data (x, y) into a tuple of $< key(d, l), x, y >$, where l is a level and d is a direction. During the data aggregation phase, each node sends the encrypted data to its parent node. Then, the parent node analyses the received encrypted data, and if the curve direction and level of its child node are different from its own ones, they should be changed into the curve direction and level of its parent ones. In this way, a sink node aggregates all of the encrypted data from the hierarchy of nodes. For example, in Fig. 5, a node 5 receives encrypted data $< key(B, 2), 1, 1 >$ and $< key(T, 2), 3, 2 >$ from its child nodes 8 and 9, respectively. The encrypted data from child nodes should be changed into $< key(R, 2), 2, 1 >$ and $< key(R, 2), 2, 0 >$ by following the curve direction and the level of the node 5. Then, node 5 aggregates their data and sends aggregated data $< key(R, 2), 3, 2 >$ to the parent node. The proposed data aggregation method outperforms the current methods in terms of privacy and energy preservation according to the performance analysis.

Space Filling Curves have been used in Dragon [6], a hierarchically structured system, which obtains a derived key from the multi-dimensional attributes of each node of the system. Each point of the curve corresponds to a derived key obtained by the values of the attributes in the multi-dimensional space. A node in the system

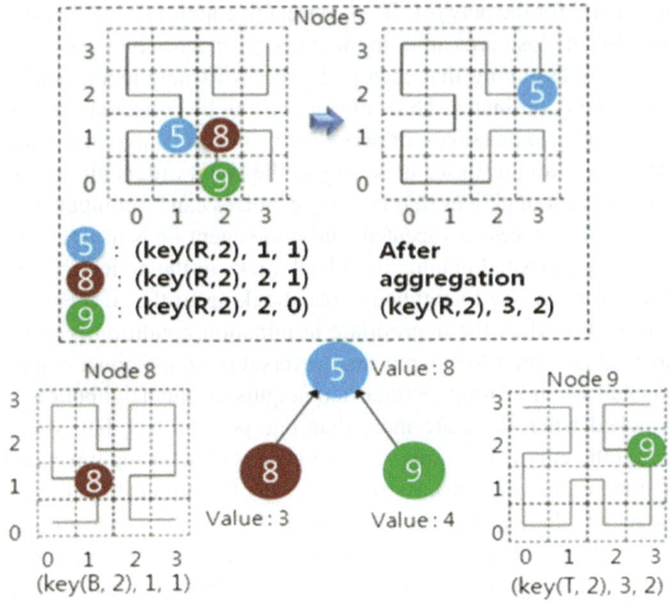

Fig. 5 Example of data aggregation

computes its derived key by using an algorithm which takes the binary representation of the values of each attribute and interleaves bits taken cyclically to construct the single derived key. The resulting value is transmitted to the internal nodes of the tree, which aggregate data. Among several Space Filling Curves, the Z-order space filling curve is used because of its good level of locality, [31] and in particular because the Z-curve visits the quadrants resulting from the recursive definition of the n-dimensional space, according to the order defined by their identifiers, at any recursion level.

3.2 *The Quantile Digest (Q_Digests)*

The q_digest is used to reduce the space of a set of values by maintaining a fixed error margin. The q_digest structure automatically groups values into variable sized buckets of almost equal weights and consists of a set of buckets of different sizes and their associated counts. Consider a network of n sensor devices, where all devices are sensing in a common modality and each sensor generates an integer value in the range $[1, \sigma]$, where σ is the maximum possible sensed value. Let define a binary tree T whose nodes correspond to sub-interval of the initial range and with height $log\sigma$. For instance, root is paired with the range $[1, \sigma]$, and its two children with ranges $[1, \sigma/2]$ and $[\sigma/2 + 1, \sigma]$. All leaf nodes have a range of width 1 (single values), instead the ranges paired with the internal nodes of the tree are the union of the ranges

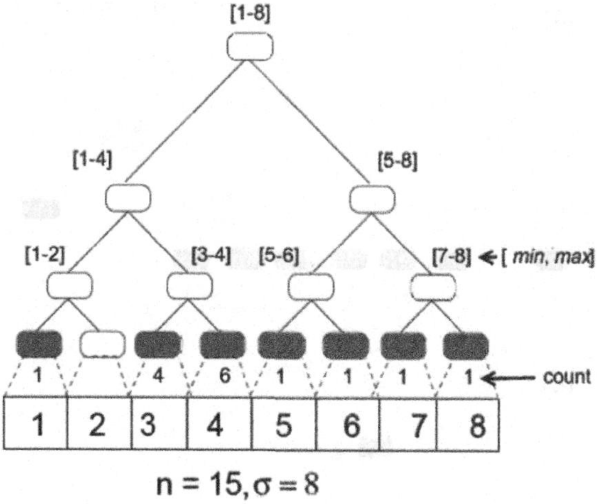

Fig. 6 Example of a Q_digest tree

of their children. Each node has a frequency counter *count*, which is the number of values inside the range. Figure 6 shows an example of q_digest in which the domain range has a dimension of 8 and the total number of values to reduce is equal to 15. The number of instances of each value is defined by $count(v)$. The value of n is defined as $n = \sum count(v), v \in LeafSet$. When the tree is defined, the q_digest function consists of the selection of a set of nodes (hence of intervals) in the tree, in order to describe the distribution of values minimizing the loss of information.

As we can easily image, the best scenario is to maintain all the single values stored the leaf nodes. However, IoT systems contain huge amounts of sensor nodes and to optimize the memory, a compression phase of the tree is required. For instance, each sensor may have a separate q_digest reflecting a summary of data available on that sensor. The size of the q_digest is determined by the compression parameter k. Given the compression parameter k, a node v is in q_digest if and only if it satisfies the following digest property:

$$count(v) \leq \lfloor n/k \rfloor \tag{1}$$

$$count(v) + count(v_p) + count(v_s) > \lfloor n/k \rfloor \tag{2}$$

where v_p is the parent and v_s is the sibling of v. The first constraint (1) asserts that unless it is a leaf node, no node should have a high count. The second constraint (2) says that we should not have a node and its children with low counts. The only exception to these properties are the root and leaf nodes. Indeed, if a leaf node frequency is larger than n/k, then it belongs to the q_digest. Also the root can violate these properties, in particular the property (2), because it doesn't have parents or siblings, and still belong to the q_digest. To construct the q_digest it is necessary to merge and to reduce the number of ranges in a bottom-up way. The compression algorithm verifies that the q_digest properties are respected. In the case at least one of them is violated, the compression phase resets the counters of the children nodes

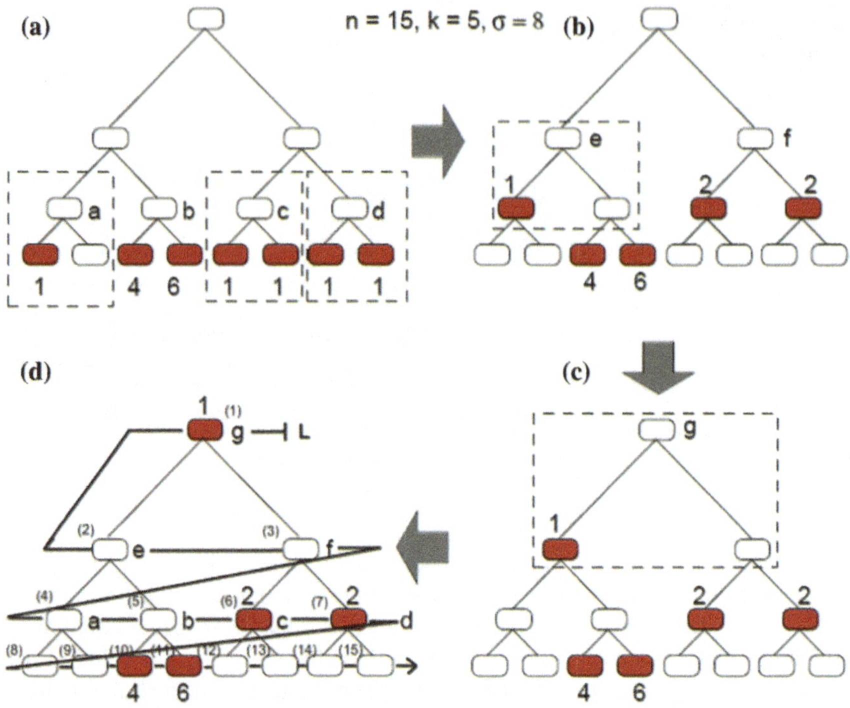

Fig. 7 Example of the compression of the Q_digest tree

and assigns to the parent node the value given by the sum of the previous counters values.

Consider the example shown in Fig. 7 where we consider a set of n = 15 values in the range [1, 8]. The leaf nodes represent the values 1, 2, ..., 8 and the numbers under the nodes is the frequency count. By considering a compression factor k = 5, $\lfloor n/k \rfloor = 3$, so we can note that a, c, and d violate the property (2). The algorithm compresses these nodes by combining their children (Fig. 7b). By analysing the tree, the node e violates the property (2), and the algorithm compresses the node e (Fig. 7c). The property is still violated by the node g and the compression is still applied to the tree by obtaining the final q_digest (Fig. 7d).

We can consider the nodes of the tree T as a list L (Fig. 7d), concatenated and enumerated from right to left, and from the top to the bottom. At the end, the q_digest is composed by a set of tuples (nodeid(v), count(v)), in which the frequency counter of each node is > 0 and the q_digest is represented as:

$$q\_digest(L) = (1, 1), (6, 2), (7, 2), (10, 4), (11, 6) \tag{3}$$

We can see how the compression value k influences the aggregation of nodes. In particular through the properties of the q_digest, we can maintain the adaptive compression. For sake of readiness, data values which occur frequently are preserved in the digest, while less frequently values are put together in larger ranges resulting in information loss.

3.2.1 Merging of Q_digests

The merging operation is used to join two different q_digest to obtain a single tree. The level i of the new tree contains all nodes at the same level i of the two original trees. The algorithm follows the rule:

- if a node v is present in both the original trees, at the same level i, the node v is added only one time to the new tree at the level i, and the counter is equal to the sum of the counters of v in the two original trees.
- if a node v is present only in one of the two original trees, at the level i, v is added to the new tree at the level i without changes.

When each node is added to the new tree, the compression algorithm is executed at each level of the new tree, from the leaf nodes to the root. The compression algorithm is necessary because, even if the nodes in each single original tree respect the properties, it is possible that the union of the two trees has changed the counters of some nodes and the properties could be not respected.

An example of the merging of two trees is proposed in Fig. 8. As we can see, the new tree has a number of resources equals to the sum of the resources contained into the two original trees, and the threshold $\lfloor n/k \rfloor$ is changed as a consequence.

3.2.2 Q_Digest in IoT

In [40] a data aggregation scheme based on q_digest is proposed. After computing the q_digest, each sensor transmits it to its parent. To represent a q_digest tree in a compact fashion, the nodes are numbered from 1 to $2\sigma - 1$ in a level by level order. Consider the definition of n and σ given in the previous section, each node $v \in T$ can transmit the q_digest as a set of tuples of the following form $< nodeid(v), count(v) >$, requiring a total of $(log(2\sigma) + logn)$ bits for each tuple. The proposed approach uses the q_digest for managing the quantile queries: given a fraction $q \in (0, 1)$, find the value whose rank in sorted sequence of the n values is qn. To find the qth quantile nodes are sorted in increasing max values and this returns a list L which represents the post-order traversal of list nodes. Once the q_digest is computed, it is used to provide approximate answers to a variety of queries:

- Inverse Quantile. Given a value x, determine its rank in the sorted sequence of the input values, traversing L from beginning to end.
- Range Query. Find the number of values in a given range [low, high] by performing two inverse quantile queries to find the ranks of low and high, and take their difference.

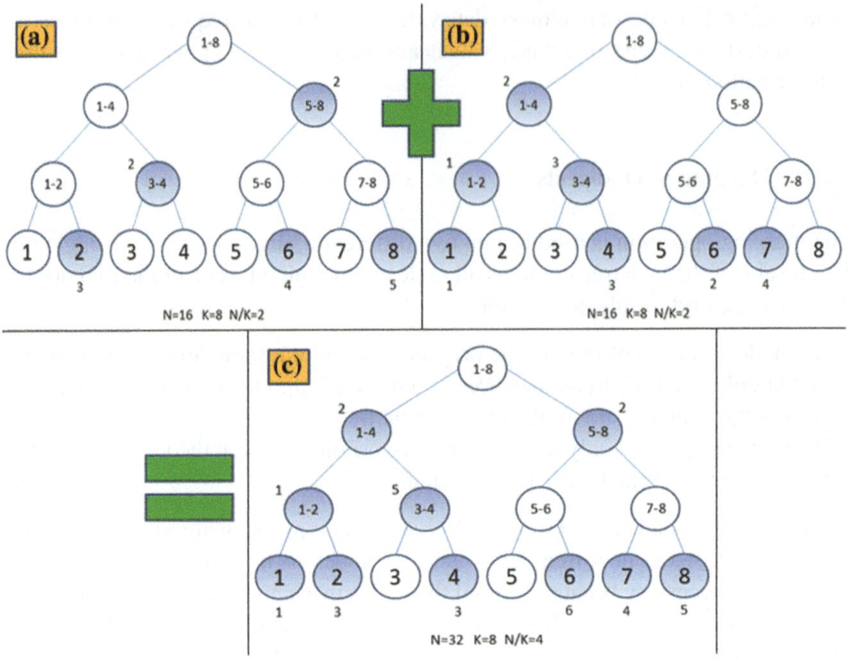

Fig. 8 Example of the merging of two Q_digest trees

- Consensus Query. Given a fraction $s \in (0, 1)$, find all the values which are reported by more than sn sensors. These values are called *Frequent items*.

Another approach which uses q_digest is DRAGON, proposed in [6], a tree-based overlay targeting for fog computing in IoT. In detail, it is a distributed search-able data structure organized as a binary aggregation tree. In DRAGON each node publishes an object characterized by a set of attributes. The attributes are linearized through a space-filling curve and the derived key is published at the leaves of the tree. Each internal node of the tree stores a digest which summarizes the data contained in the relative sub-tree. The digests are updated by a bottom-up visit of the tree, starting from the leaf nodes and modifying the digest on the path to the root. Q_digests in the internal nodes of the tree are used to support the resolution of range queries, by directing the search process toward the leaves which probably have values satisfying the query. Figure 9 proposes an overview of the DRAGON tree-based overlay with four nodes: P1 with identifier 100, P2 with identifier 001, P3 with identifier 111, and P4 with identifier 101. Grey areas highlight the mapping of tree nodes on real nodes. Moreover, dotted stroke nodes are not mapped. The table under each node represent the Routing Table.

In [20] another approach to compute q_digest in sensor networks is proposed. Authors propose a flat approach in which each node i independently samples each of its data values with some probability p. For each sampled value a, it computes

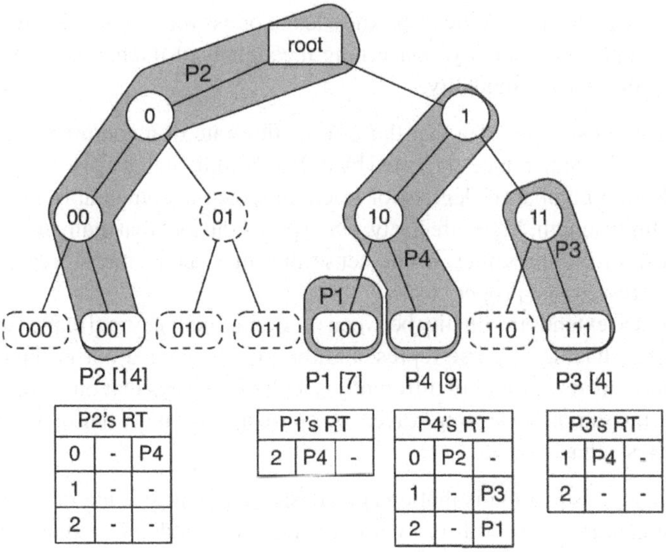

Fig. 9 DRAGON tree-based overlay

its local rank $r(a, i)$. Then, it sends all the sampled values and their local ranks to the base station. Moreover, authors propose a tree-based approach by considering the issue that if each node sends its message through its ancestors in the routing tree to the base station without any data reduction, an intermediate node might see too much traffic going through. The idea is to merge these messages to reduce their size. The unit of the merging operation is a sample s taken from a ground set $D(s)$. Let $n(s)$ denote the size of the ground set $D(s)$, and $n(s)$ is stored with s. The sample s is a *small sample* if $n(s) < n/\sqrt{k}$ and a large sample if $n(s)n/\sqrt{k}$. When a node has received a number of samples from its children, together with one of its own, it will first check if the ground sets of all the small samples have a combined size of at least $n/\sqrt{k}$. If so, it will merge all the small samples into a large sample. Next, it will repeatedly merge two large samples of the same class into one in the next class, until no two large samples are in the same class. As a result, there will be at most one large sample per class left.

3.3 *Compressive Sensing*

The Compressive Sensing (CS) method focuses on compressing data during the gathering procedure, so reducing the amount of data (or the number of packets) to achieve aggregation. CS theory asserts that it can be possible to recover certain signals and images from far fewer samples or measurements than those used by traditional methods. CS [14] is based on the sampling theory that leverages compressibility without relying on any specific prior knowledge or assumption on signals. CS relies on two principles [5]: sparsity, concerning the signals of interest, and incoherence, concerning the sensing modality.

- Sparsity expresses the idea that the information rate of a continuous time signal may be much smaller than suggested by its bandwidth, or that a discrete-time signal depends on a number of degrees of freedom which is comparably much smaller than its finite length. More precisely, CS exploits the fact that many natural signals are sparse or compressible, in the sense that they have concise representations when expressed in a proper basis.
- Incoherence extends the duality between time and frequency and expresses the idea that objects having a sparse representation must be spread out in the domain in which they are acquired. Put differently, incoherence says that unlike the signal of interest, the sampling/sensing waveforms have an extremely dense representation in the chosen basis.

The crucial observation is that one can design efficient sensing or sampling protocols that capture the useful information content embedded in a sparse signal and condense it into a small amount of data. By a sparse representation, we can say that for a signal of length N, we can represent it with $K \ll N$ non-zero coefficients. The signal is well-approximated by a signal with only K non-zero coefficients. To be more formal [15], let X be a signal of size n, $X = [x_1, \ldots, x_n]^T$, and X is K-sparse in the space ϕ, i.e. ϕX contains at most K significant coefficients. ϕ is of size $n \times n$ and degenerates to the identity matrix if X is sparse itself. Let Y be a measurement vector of size k, $Y = [y_1, \ldots, y_k]^T$, $k \ll n$, and Y is formed by the projection of the K-sparse coefficients ϕX on the random matrix A of size $k \times n$, i.e. $Y = A\phi X$. The entries of A are drawn from either Gaussian or Bernoulli distribution. The CS theory asserts that one can recover X from Y, if $k \gg O(K \log(n/K)$, by solving for the noiseless case

$$min \parallel \phi X \parallel_1 \quad s.t. Y = A\phi X \tag{4}$$

or for the noisy case

$$min \parallel \phi X \parallel_1 \quad s.t. \parallel Y = A\phi X \parallel_2^2 \ll \epsilon \tag{5}$$

For applications of IoT or WSNs, CS is promising as it can reduce the number of data transmission from n to k.

3.4 Compressive Sensing in IoT

A novel compressive sensing approach capable to exploit both spatial relevance and temporal smoothness of sensory data is proposed in [47]. Authors exploit the Treelet transform [28] to derive proper sparsification bases. This transform is data-driven and extends localized multi-scale analysis to unordered data. Indeed, Treelet transform is adopted as sparse transformation tool to derive transformations from training data. After, the main process of the CS-based scheme is that it needs training rounds to collect training set. The approach collects training sets in training rounds, and adopt the information discovered from training set for the next μ monitoring rounds before collecting the next training set.

In [15], authors propose a data aggregation method based on CS. The idea is to include the integration of CS with the routing protocol and the choice of the sparse space ϕ which is generally specific to the application. Authors assume that each sensor broadcasts messages which can only be received by sensors in a certain range and the radio transmission is under certain packet loss rates. The routing is basically performed by broadcasting and aggregation. In particular, the data collection procedure is divided into time slots of equal size within each of which sensor data are relayed to the sink by a multi-round routing algorithm. In each round, each sensor aggregates messages heard in the previous round using CS and broadcasts the aggregated data.

In [46] a data aggregation scheme based on compressed domain is presented and implemented to overcome the problem of excessively high amount of data transmission and excessive energy consumption in wireless sensor networks. The sensor network is configured to generate multiple layers of different types of cluster-based structure for the transition data collection. The leaf nodes only transmit raw data, other data collection clusters do compressed samples, and then, send the compressed measurements to up level. When the mother collection clusters received measured value they use CoSaMP algorithm [32] to recover the original data. In the initial stage, $N_1(1) - 1$ sensors in each region only send their initial data to the cluster head $c_1(1)$ to reduce the redundancy of data fusion. Cluster head $c_1^{(1)}$ compresses these data into $M_1 = K \log N_1^{(1)}$ random measurements. In level $i \geq 2$, the k-th cluster head $c_i^{(k)}$ receives $M_{i-1}^{(k)}$ random measurements from the cluster head $c_{i-1}^{(k)}$, and then runs the compressive sensing data recovery algorithm to reconstruct redundant data. After collecting all the data from children nodes, cluster head selects $M_i^{(k)}$ signal random measurements and sends them to the parent cluster head on level $i + 1$, until it arrives at the cluster head in the top level.

In [45] an energy-efficient data aggregation technique based on a multilevel hierarchical clustering architecture and hybrid compressive sensing (HDACS) is proposed. The idea of this approach is to configure sensor nodes in such a way of one sink node being targeted by all sensors, and several nodes, arranged in a way to yield a hierarchy of multilevel clusters, are designated for the intermediate data collection. The amount of data that needs to be transmitted by each sensor is determined by the local cluster size at different levels rather than the entire network. The multilevel hierarchy is more efficient than other configurations when applied to data aggregation. In this approach, the hierarchy is constructed based on the geographical location and area. At level one, all the clusters are defined by identical regions each with area s. With the increase in level i, where $i > 2$, cluster l is configured by merging the areas of n clusters at level $i - 1$. This construction process guarantees the same area for all the clusters at the same level.

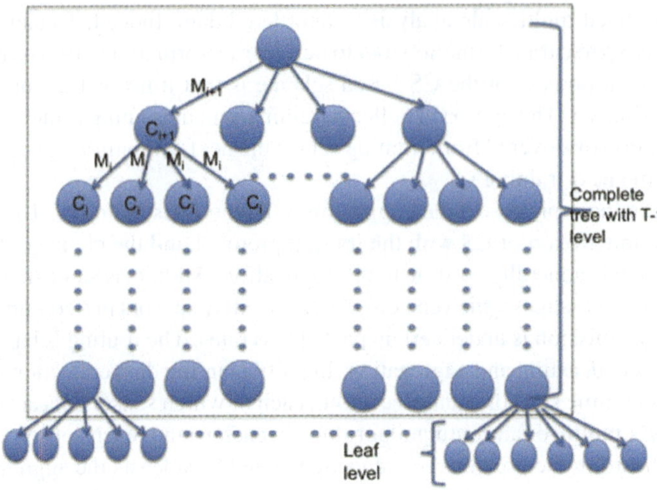

Fig. 10 HDACS: CS data aggregation architecture

Figure 10 shows the hierarchical tree where the logical relationship among clusters at multiple levels in the hierarchy is visible. The hierarchy consists of n nodes at level i for $i \geq 2$, and the degree of the hierarchical tree is n. When a cluster head receives CS random measurements from its children cluster, it executes the CS recovery algorithm to first estimate the transformed signal and then retrieve the original data through inverse transformation.

3.5 Wavelets

Wavelet transformation are used in the signal processing because it expresses a signal as a linear combination of shifted and scaled versions of a finite duration waveform. In the field of data aggregation, wavelets are considered a useful tool for range query processing and for the data representation in histograms. Histograms are common used in partitioning an attribute values domain into a set of buckets by considering the cumulative number of observations. Thanks to the wavelet decomposition, it is possible to obtain histograms based on the data distribution and the estimation of a range query is treated by using a subset of wavelet coefficients for approximating data distribution. The Haar Wavelet [43], into its discrete form, is related to a mathematical operation called Haar Transform, which can be considered a prototype for all the wavelets transforms.

A discrete signal is a temporal function whose values correspond to a set of discrete instants of time represented into the domain of the whole numbers. A discrete signal is represented as $f = (f_1, f_2, \ldots, f_N)$ where N is a positive integer number indicating the length of f. Values of f are the real numbers $f_1, f_2, \ldots, f_N$, which are computed by applying an analogical signal to the instant of time $t_1, t_2, \ldots, t_N$. For this reason, the values of f are:

$$f_1 = g(t_1),\ f_2 = g(t_2), \ldots, f_N = g(t_N). \tag{6}$$

As all the wavelet transforms, the Haar Wavelet decomposes a discrete signal into two sub-signals with a half size: trend and fluctuation. The sub-signal trend $a^1 = (a_1, a_2, \ldots, a_{N/2})$ is obtained by computing the average of the first couple of values in f: $(f_1 + f_2)/2$ and by multiplying the obtained value by $\sqrt{2}$. At the same way, the next value a^2 is computed by executing $(f_3 + f_4)/\sqrt{2}$. The formula to obtain a generic value a^m is:

$$a^m = \frac{f_{2m-1} + f_{2m}}{\sqrt{2}} \tag{7}$$

with $m = 1, 2, \ldots, N/2$.

The other sub-signal is called the *first fluctuation*. The first fluctuation of the signal f, which is denoted by $d^1 = (d_1, d_2, \ldots, d_{N/2})$ is computed by executing the difference between values. For sake of readiness, the first value d_1 is obtained from $(f_1 - f_2)/\sqrt{2}$. At the same way, all the other values are computed and the formula to compute a general value is:

$$d^m = \frac{f_{2m-1} - f_{2m}}{\sqrt{2}} \tag{8}$$

with $m = 1, 2, \ldots, N/2$.

The Haar Transform is obtained by executing a set of phases, or levels. The first level is the mapping H_1 defined as:

$$f \xrightarrow{H_1} (a^1|d^1)$$

from a discrete signal f to its first trend a^1 and first fluctuation d^1. The mapping H_1 has an inverse which rebuilds the discrete signal f starting from the trend and the fluctuation $(a^1 \mid d^1)$ through the following formula:

$$f = \left(\frac{a_1 + d_1}{\sqrt{2}}, \frac{a_1 - d_1}{\sqrt{2}}, \ldots, \frac{a_{N/2} + d_{N/2}}{\sqrt{2}}, \frac{a_{N/2} - d_{N/2}}{\sqrt{2}} \right) \tag{9}$$

For sake of readiness, in Fig. 11 we propose an example of the Haar wavelet of 1-level.

There are several advantages using the Haar Transform because the Haar Transform has the Small Fluctuations property, which says that the magnitudes of the values of the fluctuation sub-signal are often significantly smaller than the magnitudes of the values of the original signal. This property is important because through the Haar wavelet, data can be aggregate or transmitted with a low number of bits.

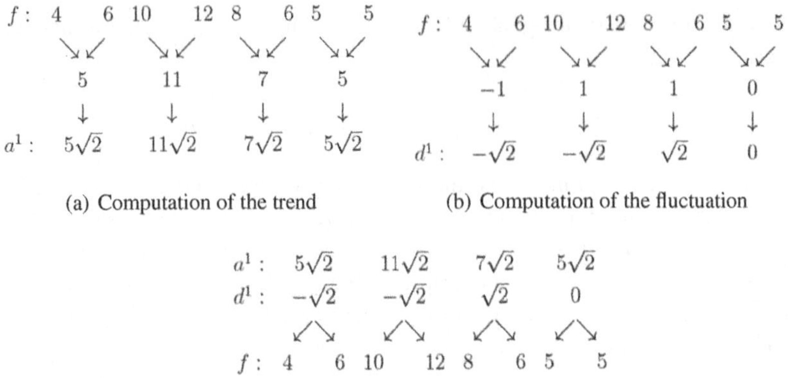

(a) Computation of the trend

(b) Computation of the fluctuation

(c) Computation of the inverse of H_1

Fig. 11 Haar Transform 1-level

Furthermore, the Haar Transform has the property of energy compression. Indeed, the energy of the trend sub-signal a^1 maintains a high percentage of energy of the original signal $(a^1 \mid d^1)$.

The wavelet decomposition is very computationally efficient, requiring only $O(N)$ to compute for a tuple of N values. Furthermore, there is no loss of information, i.e., the original histogram can be regenerated.

3.5.1 Wavelet in IoT

Wavelets have been used in database [30, 42] to represent the information through histograms. Histograms permit to synthetise the data distribution, and a range of a query can be easily estimated (the number of records which satisfy the query). For this reason, in the last years, this technique has been taken into account in both IoT and WSN, for data aggregation. In [17], wavelets are used for data aggregation in a hierarchical system. The construction of the histogram is composed by two distinct phases:

1. *Wavelet Decomposition*: the wavelet decomposition is applied obtaining a set N of wavelet coefficients. The aim of the wavelet decomposition phase is to obtain a histogram which helps to represent the data to different detail levels.
2. *Pruning*: in this phase only the most significant m wavelet coefficients are considered. The choice of the m selected coefficients depends on the pruning algorithm.

Suppose that an attribute can assume the following values:

$$[0, 0, 2, 3, 3, 4, 4, 5, 5, 6, 7, 7]$$

Table 1 Wavelet decomposition

Resolution	Average	Detail Coefficients
8	[2, 2, 3, 5, 7, 9, 10, 12]	
4	[2, 4, 8, 11]	[0, −1, −1, −1]
2	$[3, 9\frac{1}{2}]$	$[-1, -1\frac{1}{2}]$
1	$[6\frac{1}{4}]$	$[-3\frac{1}{4}]$

The approach proposed in [17] performs the decomposition on a data distribution by computing the average of the values, pairwise, and obtaining a new array of values. Some information is lost during this phase, and to recover the missed values, the Haar wavelets store the pairwise differences of the original values (divided by 2) as *detail coefficients*.

By repeating this process recursively on the averages, the full decomposition is obtained. An example of the approach is shown in Table 1. A critical step related to the usage of wavelet happens when two histograms need to be merged. The proposed approaches applies a merge algorithm.

Another approach based on wavelet is proposed in [4]. The paper proposes a novel data aggregation scheme based on wavelet-entropy (DAWE). The idea of DAWE is the data aggregation using wavelet-entropy discriminance in cluster members and cluster heads to reduce transmitting packets. The approach uses LEACH [18] as clustering algorithm. Every cluster member fuses its raw data with wavelet entropy first, and then transmits fused data to its cluster head. In detail, the decomposition of data is a 4-level decomposition. Then, the wavelet variance of real data is computed and normalized to obtain energy sequence and the wavelet entropy based-on energy distribution synchronously. The distribution of the energy sequence is used to replace the probability distribution of the data. Entropy is used on every level of every group of data to obtain aggregated data by weighted average method. To obtain the original frequency, the wavelet entropy on every level is computed, and then weight values of wavelet entropy are distributed to high frequency coefficient of wavelet transform according to the inverse wavelet transform formulation.

There are several works in the literature that propose a modified version of distributed wavelets transform using the lifting scheme [11]. The lifting scheme is an alternative method to compute bi-orthogonal wavelets. It allows a faster implementation of the wavelet transform along with a full in-place calculation of the coefficients.

In [9] a clustering model based on lifting wavelets is proposed. The approach divides all the sensor nodes into two categories: candidate members and normal nodes according to their residual energy levels. After the clustering process, a distributed fast lifting wavelet compression technique is used to aggregate data at each cluster head and recover data at the sink. The wavelet coefficients are compressed and encoded to reduce the amount of coefficients. The process of lifting wavelet transformation is divided into three steps:

- **Splitting process**: The received data at the cluster head is stored as a two-dimensional matrix which can be decomposed into four sub-bands by row- or column-lifting wavelet. The four sub-bands are: one low frequency sub-band and three high frequency sub-bands. For each row of data matrix, the data at a time slot, expressed as $x_m(t)$, can be divided into even sequence $xe_m(t)$ and odd sequence $xo_m(t)$ according to the sampling interval, that is,

$$xe_m(t) = x(2\tau) \quad and \quad xo_m(t) = x(2\tau + 1). \tag{10}$$

  where $1 \leq 2\tau t$.
- **Predicting process**: To maintain signals at even time slots, the prediction operator $P[]$ is performed on the signals at even time slots to predict the even data signal. The difference $d(t)$ is the details of data and reflects the high frequency part of original signal, defined as:

$$d(t) = xo_m(t) - P[xe_m(t)] \tag{11}$$

- **Updating process**: By using the difference $d(t)$, a new update operator $U[]$ is performed,

$$c(t) = xe_m(t) + U[d(t)] \tag{12}$$

  where $c(t)$ is an approximation signal, which denotes the low frequency part of original signal. After the row-lifting wavelet transformation, column-lifting wavelet transformation is executed in the similar way, as shown in Fig. 12.

Fig. 12 Transformation process of lift wavelets

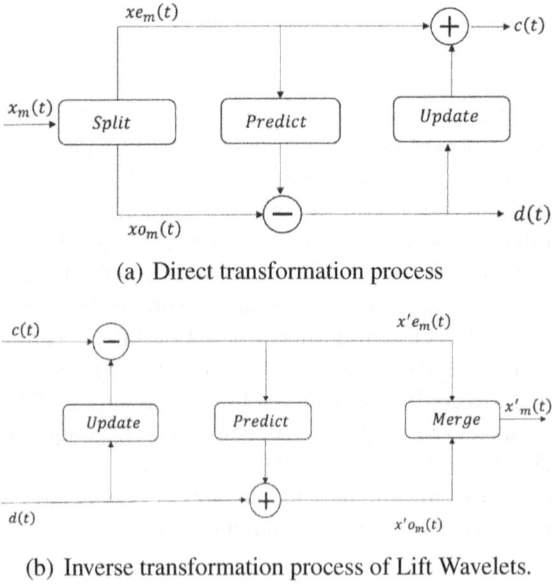

In the process of the row-lifting wavelet transform, the original data is eventually replaced by $c(t)$ and $d(t)$. Similarly, in the process of the column-lifting wavelet transform, the greater spatial correlation implies the smaller high frequency coefficient. When the first-level lifting wavelet transformation is finished on all rows and columns, the original data is converted into a low frequency part LL and three high frequency parts. For sake of readiness, only three quarters of the wavelet coefficients can be received at a sink node and a quarter of these coefficients is compressed.

Another extension to the wavelet approach is proposed in [11]. The extension uses the technique in [10] in order to jointly apply it with the routing tree.

4 Gossip Aggregation

A gossip protocol is a distributed communication paradigm inspired by the gossip phenomenon that can be observed in social networks. Initially born to efficiently disseminate information, as its human counterpart, it has been later used to solve several other problems, such as failure detection, data aggregation, distributed topology construction, and resource allocation, just to name a few. Gossip protocols tend to be used in contexts where both the scale and the dynamism of the underlying communication network makes the adoption of traditional communication protocols highly unpractical. The term epidemic protocol is used as a synonym for gossip protocol, because the way gossip spreads information can be modeled as the spread of a virus in a biological community. However, in [26], authors describe the term gossip as the probabilistic exchange of information between two nodes, and the epidemic as the information dissemination where a node randomly chooses another one. In this paper, no distinction is made between gossip and epidemic. In general, in a gossip communication, an initial node sends a message to a random subset of neighbours which repeat the propagation process. The model used in IoT and WSNs is the SI model [13], also called simple epidemics or anti-entropy. Three styles of anti-entropy protocols have been proposed:

- the *push* style: nodes periodically send (push) the content value to a node selected randomly;
- the *pull* style: nodes periodically ask (pull) new updates from randomly selected nodes;
- the *push-pull* style: push and pull are combined together.

In the next section, we propose an overview of the gossip protocols used for data aggregation in IoT and WSNs.

4.1 Gossip Approaches in IoT

The push-pull gossiping algorithm [22] can be used to perform an anti-entropy averaging aggregation process. In particular, this algorithm executes an epidemic protocol to perform a pairwise exchange of aggregated values between neighbour nodes. Periodically, each node randomly chooses a neighbour to send its current value, and waits for the response with the value of the neighbour. Each time a node receives a value from a neighbour, it sends back its current one and computes the new estimate (average), using both the received and sent values as parameters.

In [25] the push-sum protocol is proposed to compute aggregation functions, such as sum or average. Each node i, at a discrete time t, maintains and propagates information of a pair of values (s_{ti}, w_{ti}): s_{ti} represents the sum of the exchanged values, and w_{ti} denotes the weight associated to this sum at the given time t and node i. At each iteration, a neighbour is chosen uniformly at random, and half of the values are sent to the target node and the other half to the node itself. Upon receive, the local values are updated, adding each value from a received pair to its local component. The estimate of the aggregation function can be computed by all nodes, at each time t by s_{ti} and/or w_{ti}. The accuracy of the produced result will tend to increase progressively along each iteration, converging to the correct value. Furthermore, authors assume the existence of a fault detection mechanism, that allows nodes to detect when a message did not reach its destination.

GGAP (Gossip-based Generic Aggregation Protocol) [44] extends the push-synopses protocol presented in [25] to tolerate non contiguous faults. GGAP is a gossip protocol for continuous monitoring of aggregates, which is robust against failures that are discontinuous in the sense that neighbouring nodes do not fail within a short period. The protocol estimates the average of local state variables x_i of nodes i of a given network graph. The protocol implements two parallel passes. During the first pass, it computes the sum s_i, while the other pass computes the weight w_i. At the end of each iteration, s_i and w_i are distributed to nodes according to the matrix elements $\alpha_{i,j}$, whereby $\alpha_{i,j}$ is chosen in such a way that $\alpha_{i,j} \geq 0$ and $\sum_j \alpha_{i,j} = 1$. The elements $\alpha_{i,j}$ can be chosen differently from round to round, in which case matrix $A(t)$ is time-dependent. After each round, the local estimate α_i of the global average at node i can be obtained as $\alpha_i = s_i/w_i$.

In [8], Distributed Random Grouping (DRG) algorithm is simple and natural and uses probabilistic grouping to progressively converge to the aggregate value. DRG is local and randomized and is naturally robust against dynamic topology changes caused by link/node failures. DRG takes advantage of the broadcast nature of wireless transmissions: all nodes within the radio coverage can hear and receive a wireless transmission. The main idea is that, at each round of the algorithm, every node independently becomes a group leader with probability p_g and then invites its neighbours to join the group. Then, all the members in a group update their values with the locally derived aggregate (average, maximum, minimum, etc.) of the group. All values will progressively converge to the correct aggregate value.

SOFA [7] utilizes opportunistic anycast to drastically reduce the rendezvous times of asynchronous duty cycled nodes. Long rendezvous times are the key limitation of protocols operating under high densities conditions. SOFA combines the energy efficiency typical of low-power MAC protocols with the robustness and versatility of gossip-like communication. SOFA focuses on maximizing the messages exchanged locally among neighbours (1-hop), leaving the multi-hop dissemination and aggregation of information to the Gossip layer. During the data exchange phase, SOFA selects the first random neighbour that wakes up as the destination.

In [37] a local data management mechanism, implemented as an app on smart phones is proposed. The paper describes DIAS, the Dynamic Intelligent Aggregation Service for the computation of aggregation functions. DIAS can perform decentralized privacy-preserving data analytics, and it is applicable in the context of the Internet of Things because it can manage dynamic input data streams from sensors. DIAS can compute a wide spectrum of aggregation functions such as AVERAGE, SUM, MAX, MIN and STANDARD DEVIATION. Each node i of the DIAS network can contain a *disseminator* d_i and/or an *aggregator* a_i. The disseminator is the agent with which the data supplier is connected to, whereas the aggregator is the agent with which the data consumer is connected to. Disseminators discover aggregators in the network to which they send their local sensor data. Discovery is performed via a fully decentralized gossiping protocol, the peer sampling service [23] in which aggregators publish themselves and disseminators sample aggregators published. Disseminators classify and cache aggregators sampled from the peer sampling service in a pool of limited size and spread the local sensor data in the network periodically by pushing them to remote aggregators from the pool. Aggregators collect the input data for the computation of the aggregation functions. In Fig. 13 an overview of the component of DIAS is proposed.

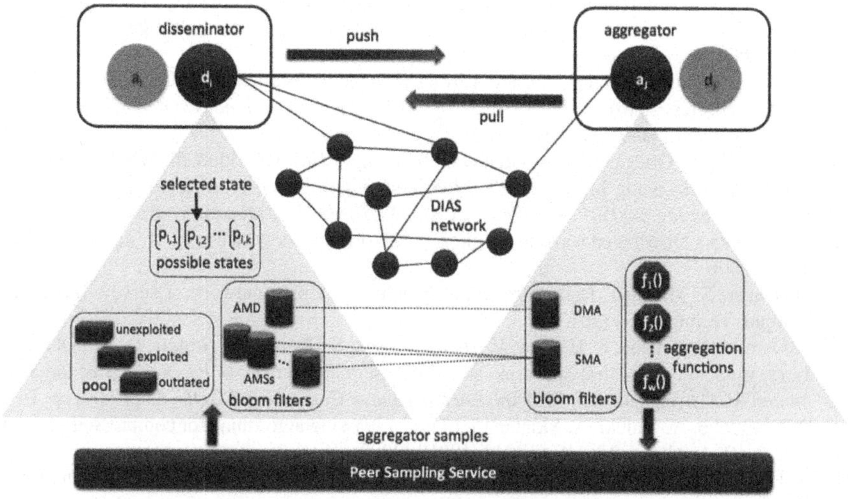

Fig. 13 Overview of the DIAS main components and interactions

5 Conclusion

In this paper we propose an overview of several data compression techniques used in IoT and in WSNs. In detail, we showed the following techniques: Space Filling Curves, the Quantile Digest, the Wavelet Transform, and the Compressive Sensing. Moreover, we proposed an overview of the gossip approaches proposed in WSNs for the data aggregation issue. We proposed, for each listed techniques, current proposals in which they are used, and we described in details how these techniques are used.

References

1. S. Abdollahzadeh, N.J. Navimipour, Deployment strategies in the wireless sensor network. Comput. Commun. **91**(C), 1–16 (2016)
2. A. Alghamdi, M. Alshamrani, A. Alqahtani, S.S.A. Al Ghamdi, R. Harrathi, Secure data aggregation scheme in wireless sensor networks for IoT, in *2016 International Symposium on Networks, Computers and Communications (ISNCC)* (IEEE, 2016), pp. 1–5
3. S.B. Baker, W. Xiang, I. Atkinson, Internet of things for smart healthcare: Technologies, challenges, and opportunities. IEEE Access **5**, 26521–26544 (2017)
4. W. Cai, M. Zhang, Data aggregation mechanism based on wavelet-entropy for wireless sensor networks, in *4th International Conference on Wireless Communications, Networking and Mobile Computing, 2008. WiCOM'08* (IEEE, 2008), pp. 1–4
5. E.J. Candès, M.B. Wakin, A sensing/sampling paradigm that goes against the common knowledge in data acquisition (2008)
6. E. Carlini, A. Lulli, L. Ricci, Dragon: multidimensional range queries on distributed aggregation trees. Future Gener. Comput. Syst. **55**, 101–115 (2016)
7. M. Cattani, M. Zuniga, M. Woehrle, K. Langendoen, SOFA: communication in extreme wireless sensor networks, in *European Conference on Wireless Sensor Networks* (2014), pp. 100–115
8. J.Y. Chen, G. Pandurangan, D. Xu, Robust computation of aggregates in wireless sensor networks: distributed randomized algorithms and analysis. IEEE Trans. Parallel Distrib. Syst. **17**(9), 987–1000 (2006)
9. L. Cheng, S. Guo, Y. Wang, Y. Yang, Lifting wavelet compression based data aggregation in big data wireless sensor networks, in *2016 IEEE 22nd International Conference on Parallel and Distributed Systems (ICPADS)* (2016), pp. 561–568
10. A. Ciancio, A. Ortega, A distributed wavelet compression algorithm for wireless multihop sensor networks using lifting, in *IEEE International Conference on Acoustics, Speech, and Signal Processing, 2005. Proceedings. (ICASSP'05)*, vol. 4 (2005), pp. iv–825
11. A.G. Ciancio, Distributed wavelet compression algorithms for wireless sensor networks. Ph.D. thesis (2006)
12. M. Dagar, S. Mahajan, Data aggregation in wireless sensor network: a survey. Int. J. Inf. Comput. Technol. **3**(3) (2013)
13. A. Demers, D. Greene, C. Hauser, W. Irish, J. Larson, S. Shenker, H. Sturgis, D. Swinehart, D. Terry, Epidemic algorithms for replicated database maintenance, in *Proceedings of the Sixth Annual ACM Symposium on Principles of Distributed Computing, PODC '87* (1987), pp. 1–12
14. D.L. Donoho, A. Maleki, A. Montanari, Message-passing algorithms for compressed sensing. Proc. Natl. Acad. Sci. **106**(45), 18914–18919 (2009)
15. W. Du, J. Gorce, T. Risset, M. Lauzier, A. Fraboulet, Compressive data aggregation on mobile wireless sensor networks for sensing in bike races, in *24th European Signal Processing Conference, EUSIPCO 2016*, Budapest, Hungary, 29 Aug–2 Sept 2016 (2016), pp. 51–55

16. Y.J. Fan, Y.H. Yin, L.D. Xu, Y. Zeng, F. Wu, IoT-based smart rehabilitation system. IEEE Trans. Ind. Inf. **10**(2), 1568–1577 (2014)
17. B. Guidi, A.D. Salve, L. Ricci, A data aggregation strategy based on wavelet for the internet of things, in *19th International Symposium on Symbolic and Numeric Algorithms for Scientific Computing, At Timisoara* (2017)
18. W.R. Heinzelman, A. Chandrakasan, H. Balakrishnan, Energy-efficient communication protocol for wireless microsensor networks, in *Proceedings of the 33rd Annual Hawaii International Conference on System Sciences, 2000* (2000), pp. 10
19. D. Hilbert, Über die stetige Abbildung einer Linie auf ein Flächenstück (Springer, Berlin, Heidelberg 1970), pp. 1–2
20. Z. Huang, L. Wang, K. Yi, Y. Liu, Sampling based algorithms for quantile computation in sensor networks, in *Proceedings of the 2011 ACM SIGMOD International Conference on Management of Data, SIGMOD '11* (2011), pp. 745–756
21. S.M.R. Islam, D. Kwak, H. Kabir, M. Hossain, K.S. Kwak, The internet of things for health care: a comprehensive survey. IEEE Access **3**, 678–708 (2015)
22. M. Jelasity, A. Montresor, Epidemic-style proactive aggregation in large overlay networks, in *Proceedings of the 24th International Conference on Distributed Computing Systems (ICDCS'04)* (2004), pp. 102–109
23. M. Jelasity, S. Voulgaris, R. Guerraoui, A.M. Kermarrec, M. van Steen, Gossip-based peer sampling. ACM Trans. Comput. Syst. **25**(3) (2007)
24. P. Jesus, C. Baquero, P.S. Almeida, A survey of distributed data aggregation algorithms. CoRR arXiv:1110.0725 (2011)
25. D. Kempe, A. Dobra, J. Gehrke, Gossip-based computation of aggregate information, in *44th Annual IEEE Symposium on Foundations of Computer Science, 2003. Proceedings* (IEEE, 2003), pp. 482–491
26. A.M. Kermarrec, M. van Steen, Gossiping in distributed systems. SIGOPS Oper. Syst. Rev. **41**(5), 2–7 (2007)
27. J. Lawder, The application of space-filling curves to the storage and retrieval of multi-dimensional data (2000)
28. A.B. Lee, B. Nadler, L. Wasserman, Treeletsan adaptive multi-scale basis for sparse unordered data. Ann. Appl. Stat. **2**(2), 435–471 (2008)
29. S. Madden, M.J. Franklin, J.M. Hellerstein, W. Hong, Tag: a tiny aggregation service for ad-hoc sensor networks. SIGOPS Oper. Syst. Rev. **36**(SI), 131–146 (2002)
30. Y. Matias, J.S. Vitter, M. Wang, Dynamic maintenance of wavelet-based histograms, in *Proceedings of the 26th International Conference on Very Large Data Bases, VLDB '00* (2000), pp. 101–110
31. G.M. Morton, A computer oriented geodetic data base and a new technique in file sequencing. Technical report (1966)
32. D. Needell, J.A. Tropp, CoSaMP: iterative signal recovery from incomplete and inaccurate samples. Commun. ACM **53**(12), 93–100 (2010)
33. S. Patil, S.R. Das, A. Nasipuri, Serial data fusion using space-filling curves in wireless sensor networks, in *2004 First Annual IEEE Communications Society Conference on Sensor and Ad Hoc Communications and Networks, 2004. IEEE SECON 2004* (IEEE, 2004), pp. 182–190
34. G. Peano, Sur une courbe, qui remplit toute une aire plane. Mathematische Annalen **36**(1), 157–160 (1890)
35. F. Piccialli, A. Chianese, J.J. Jung, Cultural heritage on internet of things (IoT) systems: trends and challenges. Concurr. Comput.: Pract. Exp. **29**(11), e4155–n/a (2017)
36. B. Pourghebleh, N.J. Navimipour, Data aggregation mechanisms in the internet of things: a systematic review of the literature and recommendations for future research. J. Netw. Comput. Appl. **97**(Supplement C), 23–34 (2017)
37. E. Pournaras, M. Warnier, F.M. Brazier, A generic and adaptive aggregation service for large-scale decentralized networks. Complex Adapt. Syst. Model. **1**(1), 19 (2013)
38. R. Rajagopalan, P.K. Varshney, Data-aggregation techniques in sensor networks: a survey. Commun. Surv. Tutor. **8**(4), 48–63 (2006)

39. H. Sagan, A three-dimensional Hilbert curve. Int. J. Math. Educ. Sci. Technol. **24**(4), 541–545 (1993)
40. N. Shrivastava, C. Buragohain, D. Agrawal, S. Suri, Medians and beyond: new aggregation techniques for sensor networks, in *SenSys '04* (2004), pp. 239–249
41. D. Singh, G. Tripathi, A.J. Jara, A survey of internet-of-things: future vision, architecture, challenges and services, in *2014 IEEE World Forum on Internet of things (WF-IoT)* (IEEE, 2014), pp. 287–292
42. J.S. Vitter, M. Wang, Approximate computation of multidimensional aggregates of sparse data using wavelets, in *Proceedings of the 1999 ACM SIGMOD International Conference on Management of Data, SIGMOD '99* (1999), pp. 193–204
43. J.S. Walker, *A Primer on Wavelets and Their Scientific Applications* (CRC press, 2008)
44. F. Wuhib, M. Dam, R. Stadler, A. Clem, Robust monitoring of network-wide aggregates through gossiping. IEEE Trans. Netw. Serv. Manag. **6**(2), 95–109 (2009)
45. X. Xu, R. Ansari, A. Khokhar, A.V. Vasilakos, Hierarchical data aggregation using compressive sensing (HDACS) in WSNs. ACM Trans. Sens. Netw. (TOSN) **11**(3), 45 (2015)
46. G. Yang, M. Xiao, S. Zhang, Data aggregation scheme based on compressed sensing in wireless sensor network. Inf. Comput. Appl. 556–561 (2012)
47. C. Zhao, W. Zhang, X. Yang, Y. Yang, Y.Q. Song, A novel compressive sensing based data aggregation scheme for wireless sensor networks, in *2014 IEEE International Conference on Communications (ICC)* (2014), pp. 18–23

Swarm Intelligence and IoT-Based Smart Cities: A Review

Ouarda Zedadra, Antonio Guerrieri, Nicolas Jouandeau, Giandomenico Spezzano, Hamid Seridi and Giancarlo Fortino

Abstract Smart cities are complex and large distributed systems characterized by their heterogeneity, security, and reliability challenges. In addition, they are required to take into account several scalability, efficiency, safety, real-time responses, and smartness issues. All of this means that building smart city applications is extremely complex. Swarm Intelligence is a very promising paradigm to deal with such complex and dynamic systems. It presents robust, scalable and self-organized behaviors to deal with dynamic and fast changing systems. The intelligence of cities can be modeled as a swarm of digital telecommunication networks (the nerves), ubiquitously embedded intelligence (the brains), sensors and tags (the sensory organs), and software (the knowledge and cognitive competence). In this chapter, swarm intelligence-based algorithms and existing swarm intelligence-based smart city solutions will be

O. Zedadra (✉) · H. Seridi
LabSTIC, 8 may 1945 University, P.O. Box 401, 24000 Guelma, Algeria
e-mail: zedadra_nawel1@yahoo.fr

H. Seridi
e-mail: seridihamid@yahoo.fr

A. Guerrieri · G. Spezzano · G. Fortino
CNR - National Research Council of Italy - Institute for High Performance Computing and Networking (ICAR), Via P. Bucci 7-11C, 87036 Rende, Cosenza, Italy
e-mail: antonio.guerrieri@icar.cnr.it

G. Spezzano
e-mail: giandomenico.spezzano@icar.cnr.it

G. Fortino
e-mail: giancarlo.fortino@unical.it

N. Jouandeau
LIASD, Paris 8 University, 93526 Saint Denis, France
e-mail: n@ai.univ-paris8.fr

G. Fortino
DIMES, Università della Calabria, Via P. Bucci, cubo 41c, 87036 Rende, Cosenza, Italy

F. Cicirelli et al. (eds.), *The Internet of Things for Smart Urban Ecosystems*, Internet of Things, https://doi.org/10.1007/978-3-319-96550-5_8

analyzed. Moreover, a swarm-based framework for smart cities will be presented. Then, a set of trends on how to use swarm intelligence in smart cities, in order to make them flexible and scalable, will be investigated.

Keywords Swarm intelligence · Swarm intelligence-based algorithms · IoT Smart cities

1 Introduction

Minds arise from the interaction with other minds and social behavior helps individual species members to adapt to their environment, especially by providing individuals with more information than their own senses can gather [1]. Swarm Intelligence (SI)-based computation focuses on the collective behavior of decentralized, self-organized systems [2] and it is inspired by the behavior of some social insects such as ants, termites and birds. SI is characterized by its emergent behaviors resulted from the local interactions between individuals which produce intelligent behaviors at group level [3]. It presents robust solutions to complex problems with dynamic and scalable properties.

Smart cities consider as primary goal the one of offering comfort, safety, and healthy living with an intelligent form of collaboration with the residents. Smart cities employ the prospects provided by pervasive and cooperative computing technologies to the benefits of human beings. In such a setting, digital environments are designed to perform a central role for handling the challenges in transportation systems, health care, supplies, learning and education, sensing city dynamics, computing with heterogeneous data sources, managing urban big data, and environmental protection issues including decreasing pollution and others [4, 5].

In this chapter, we present a SI-based framework for comprehensive smart cities. The proposed framework uses a decentralized control over its components in order to build scalable and flexible smart cities. In this context, we have already proposed an architecture to integrate and use SI-based algorithms in IoT-based systems [6]. Moreover, this chapter aims to disseminate recent research and development efforts in the area of smart cities, and investigates trends and challenges to make cities smarter. The chapter: (1) shows the progress in IoT-based smart cities, (2) can be used as an initial reading point to explore many of related IoT-based smart city applications, (3) presents trends on how the SI paradigm can be used to build an urban-scalable IoT system.

The chapter is organized as follows: First, we present typical SI-based algorithms and their scope of application in Sect. 2. Then, we present an overview of existing IoT-based smart cities in Sect. 3. After that, in Sect. 4, we present the proposed framework and some trends on how SI paradigm could be used to build flexible and scalable smart cities. Finally, we conclude the chapter in Sect. 5.

2 Swarm Intelligence-Based Algorithms: An Overview

In Artificial Intelligence, heuristics are methods giving the possibility to speed up processes. They could be developed for one problem and used for another one. In this Section, we present a set of typical SI-based algorithms with their scope of applications.

The main characteristics of swarm intelligence such as robustness, adaptability and self-organization, flexibility, scalability and decentralization, are getting a good solution to distributed problems. In recent years, SI-based algorithms have become more and more popular due to their capabilities to provide solutions to complex problems. Many of them have today reached a good level of maturity that allows their practical application to the realization of smart city applications and services. In particular, such algorithms are: Ant Colony Optimization (ACO) [7, 8], Artificial Bee Colony (ABC) [9], Honey Bee Mating Optimization (HBMO) [10], Firefly Optimization Algorithm (FFA) [11], Glow-worm Swarm Optimization Algorithm (GSO) [12], Bacterial Foraging Optimization Algorithm (BFOA) [13], Particle Swarm optimization (PSO) [14], Cuckoo Search Algorithm (CSA) [15], Bat Algorithm (BA) [16], Shuffled Frog Leaping Algorithm (SFLA) [17], Artificial Fish Swarm Algorithm (AFSA) [18], Wolf-inspired [19] and Cat Swarm Optimization (CSO) [20] algorithms. Table 1, presents typical SI-based algorithms described by five columns used to show: category of the algorithm (*Cat*), name of the algorithm (*Name*), the reference where the algorithm first appeared (*Appeared*), the source of inspiration (*Inspired*) and the main applications of the algorithm (*Applications*).

3 Smart Cities

Due to the urban population growth and rapid urbanization, the current city services and their governance will fail to deliver adequate added values to citizens. It is essential to improve means and systems that enhance the community's quality of life which requires the provisioning of such services in a dynamic and effective manner. The impressive advances in computing and wired/wireless communication technologies have brought with them the prospect of embedding different hierarchies of smartness and intelligence in the so called Smart Cities. In this section, we present

Table 1 Typical SI-based algorithms and their scope of application

Cat	Name	Appeared	Inspired	Applications
Insect-based	ACO	[7]	Pheromone trail laying and following behavior of real ants	Knapsack problem (KP), travel salesman problem (TSP), Job shop scheduling problem (JSSP), vehicle routing problem, packet-switched routing in internet-like networks, assignment problem, protein folding, set problem, device sizing problem in nanoelectronics physical design, image processing, data mining, predictive control for nonlinear processes
	ABC	[9]	Foraging behavior of honey bee swarm	Continuous optimization, dynamic clustering, train neural networks, medical pattern classification, TSP, network reconfiguration problem in a radial distribution system, task allocation, multi-level thresholding, routing problem
	HBMO	[10]	Marriage in real honey-bee	Clustering, TSP, process planning problem, vehicle routing problem, image processing, classification, global unconstrained optimization, process planning and scheduling problems, probabilistic optimal placement and sizing of distributed energy resources
	FFA	[11]	Flashing behavior of fireflies	Image processing, feature selection and fault detection, train neural network, semantic web composition, classification, clustering, routing in wireless sensor networks (WSN), heart disease prediction, JSSP, numerical optimization, load frequency control (LFC), synchronization
	GSO	[12]	Behavior of glowworms in emitting luciferin.	Image processing, KP, TSP, distributed resource management, data clustering, location optimization
Bacteria	BFOA	[13]	Social foraging behavior of Escherichia coli bacteria	Mathematical analysis, electrical engineering, neural network problems, pattern recognition, machine learning, image processing, WSN, JSSP, vehicle routing problem, dynamic economic dispatch problem, optimal location
Bird	PSO	[14]	Coordinate movement of fish schools and bird flocks	TSP, KP, flow shop scheduling problem, DNA sequence assembly, the quadratic assignment problem (QAP), JSSP, data clustering, data mining, image processing
	CSA	[15]	Breeding behavior of cuckoo birds	Data clustering, data gathering, chemical engineering, image segmentation, neural network training problem, scheduling problems, semantic web service composition problem, TSP

(continued)

Table 1 (continued)

Cat	Name	Appeared	Inspired	Applications
	BA	[16]	Echolocation behavior of micro-bats	Continuous optimization, combinatorial optimization and scheduling, inverse problems and parameter estimation, classification, clustering, data mining, image processing, discrete decision making
Amphibious	SFLA	[17]	Leaping and shuffling behavior of frogs to exchange information	Data mining, image processing, circuit design, robot control, manufacturing optimization, scheduling problems, TSP
	AFSA	[18]	Collective movement of the fish and their social behaviors	Artificial neural network training, function optimization, data mining, image processing, robot control optimization, routing problem, data clustering
Wild	Wolf	[19]	Predatory behaviors of the wolf colony	Global optimization problems, training multi-layer perceptrons, parameter estimation in surface waves, JSSP, TSP, control of a multi-area ST-Thermal power system, power dispatch problem, multi-criterion optimization, path planning for mobile robot
	CSO	[20]	Seeking and tracing modes of cats	Optimal linear phase FIR filter design, IIR system identification, binary discrete optimization, clustering, optimizing artificial neural networks, Solving multiobjective problems, workflow scheduling in cloud computing environment, power consumption in WSN, set covering problem, TSP

definitions of smart cities (Sect. 3.1), then we summarize smart cities related works (Sect. 3.2), and we finally present a qualitative comparison among them. All of this will be done with the aim to show the progress in using SI and other paradigms in smart cities modeling and development.

3.1 Definitions

There are multiple definitions regarding *smart cities* which vary with the context of use [21]. As defined by Harrison et al. [22], a smart city is: "*a city connecting the physical infrastructure, the IT infrastructure, the social infrastructure, and the business infrastructure to leverage the collective intelligence of the city*".

Toppeta [23] defines a smart city as a city "*combining ICT and Web 2.0 technology with other organizational, design and planning efforts to dematerialize and speed up bureaucratic processes and help to identify new, innovative solutions to city management complexity, in order to improve sustainability and livability*".

Technologically, smart cities are not only about automating routine functions, but also monitoring, understanding, analyzing, and planning the city and processes within it [24]. In this regard, Picon refers to the intelligence of cities where intelligence is defined in the sense of *the ability to learn, understand and reason using ICT technologies* [25].

3.2 Related Works

In this Section, we present works related to smart cities and compare them qualitatively. In particular, we categorize the following works in four categories: Smart Environment, Smart Government, Smart Mobility, and General Frameworks for smart cities and Security in smart cities.

3.2.1 Smart Environment

Waste Collection

Medvedev et al. propose [26] a waste collection system named Decision Support System (DSS), enhanced with IoT services which enable dynamic scheduling and routing in smart cities. The proposed system features an on-board surveillance system which makes the process of problem reporting and evidence collection more usable.

Gutierrez et al. [27] present a practical intelligent waste collection cyber physical system. Such system is based on an IoT sensing prototype which measures the waste level of trashcans and sends this data over the Internet to a server for storage and processing. Based on this data, an optimization process allows creating the most efficient collection routes, and forwards them to workers.

Anagnostopoulos et al. [28] present a system model which assumes two kinds of trucks for waste collection, the Low Capacity Trucks and the High Capacity Trucks, to propose a dynamic routing algorithm which is robust and copes when a truck is overloaded or damaged and needs replacement.

Anagnostopoulos et al. [29] consider the management of the trade-off between the immediate collection and its cost. The aim is to minimize the time required for serving high priority areas while keeping the average expected performance at high level. Authors propose four models for serving the immediate collection of high priority bins, and provide routing functionalities and routing adaptation for serving areas that are characterized as critical (high priority).

Medvedev and Zaslavsky [30] propose a conceptual architecture of a hybrid context storage and indexing mechanism that enables and supports the Context Spaces theory based representation of context for large-scale smart city applications. The proposed approach is illustrated using a solid waste management system scenario with adaptive garbage collection from IoT enabled garbage bins.

Aazam et al. [31] propose a Cloud-based Smart WAste Management (CloudSWAM), where waste bins are connected to the cloud and data is stored there in real time. There are separate bins for each category of waste (organic, plastic and bottles, and metal). Since the proposed smart waste management is associated with cloud, the statuses of all waste bins throughout the city or even country are accessible from the cloud. All the stakeholders, including recycling agencies, can take a note of that and plan accordingly. The waste collection is also done when it is required, helping the waste management to decide a cost-effective route while collecting the waste within a metropolis.

Castro et al. [24] propose a smart monitoring system for public trash cans. The system consists of a network of small and battery powered wireless sensors that are retrofitted to existing trash bins, a cloud based back-end that aggregates data from the sensors, and a front-end that visualizes the utilization levels of the bins. Three major requirements were identified: low cost sensor, simplicity, Open/transparent system. For system design, authors developed an addon sensor system, that is simple to install, inexpensive and based on open standards.

Lokuliyana et al. propose in [32] an IoT-based framework for waste management with the aim to automate the solid waste identification, localization and collection. Authors argue that higher efficiency and lower latency can be achieved by implementing the proposed framework.

Smart Grid

Los Angeles Smart Grid Project [33] was launched in 2010 and set to transform the Los Angeles municipal utility into a Smart Grid. It aims at: installing smart meters to thousands of customer premises, implementing and experimenting with demand response mechanisms, developing scalable machine learning algorithms trained over large amounts of data to forecast the demand at intervals of 15 min within a few hours or the next day.

Hurtado et al. propose in [34] a smart grid (SG) building energy management system (BEMS). The developed hierarchical agent structure allows lower level agents abstracting the information of their immediate environment into the form of single value information blocks for the higher level agents. A PSO optimizer is proposed to improve the MAS's capability in exploiting the building's flexibility for the SG.

Kane et al. propose in [35] a parallel computing and communication architecture called Reflex-Tree, inspired by the human nervous system, which uses several distinct hierarchical layers to process and react to millions of data streams of biological sensory information in real-time. The key element of the reflex-tree concept is the inclusion of automated "reflex" circuits in the sensing and distributed computing architecture. The architecture was tested on two case studies: city power supply network and gas pipeline management.

Smart Building

Asimakopoulou and Bessis [36] discuss the opportunity of how various participant users including critical infrastructures, cars, buildings, and humans could be connected via sensors and mobile APIs in order to capture data about their surrounding environment. Authors present a technical model for smart cities and crowd sourcing, in which the whole community contributes towards an integrated disaster management and at the same time receives personalized warnings, evacuation routes and other messages according to the current situation.

Molina et al. [37] present a model for an Area Sensor Network exploiting all the existing communication infrastructures to connect any type of sensor (ZigBee, Bluetooth, WirelessHART, Ultra Wide Band, also including Wifi, 3G and Internet) located in an area. The framework uses the Sensor Observation Service as a common sink of sensed data from Wireless Area Sensor Networks [38], from Body Sensor Networks [39] and from fixed networks in form of open data. The effectiveness of the approach has been shown through SmartLabBuilding, containing smart laboratories and workers wearing sensors.

3.2.2 Smart Government

E-Government

Cano et al. [40] present a model of conceptualization of the citizens as sensors in the design of digital architectures of smart cities. Authors consider the level of participatory governance as one of the main parameters to measure smart cities. The proposed model involves the citizens in decision making concerning e-governance by providing citizens and communities with access to information and high quality in responding to social needs. Authors present a model of citizen-sensor triangle which supports participation, urban digital democracy, interoperability and the openness of urban data.

City Monitoring

Zanella and Lorenzo propose in [41] a web-based approach for the design of IoT services, and the related protocols and technologies, discussing their suitability for smart city environments. The proposed framework is tested on the "Padua Smart City" project, which is a proof-of-concept deployment of an IoT island in the city of Padua (Italy) and interconnected with the data network of the city municipality.

Jamil et al. propose in [42] an air pollution monitoring system around metropolitan and industrial cities in Pakistan by using Zigbee based WSNs while deployed on public transport vehicles. The model has been developed for indoor and outdoor hazardous gases.

Some researches report that city areas can produce positive emotions in their citizens, as the contact with natural places. In this area, Roza and Postolache propose [43] a smartphone application that analyzes the citizen's emotions and shares it with others in order to improve the level of citizen's information about what kind of emotions the citizens feel when they visit a certain city area.

Shah and Mishra in [44] present an IoT-based system for environment monitoring to control temperature, humidity and CO_2. Apart from sensing temperature, humidity and CO_2, the sensor node is claimed to be low power. The system consists of a transmitter node that senses the data and transmits it to a receiver node through wireless communication. Then, the received data is transferred to a personal computer through a USB interface. The sensed data is graphically depicted and recorded in an excel sheet. This data is then transmitted to a database via internet and transferred to a smartphone.

Giordano et al. [45] propose a three-layer architecture called Rainbow, based on a Multi-Agent System (MAS). It is designed to bring the computation as close as possible to the physical part. The three layers are: (i) the *physical* one, enclosing sensors and actuators, which are directly immersed in the physical environment; (ii) the *Intermediate* one, where sensors and actuators from the physical layer are represented as Virtual Objects; and (iii) the *upper layer*, concerning the cloud part. Three smart city applications have been developed by using Rainbow: noise pollution, urban drainage networks, and smart street. In this latter, a set of sensor nodes have been deployed in the city (some nodes host noise, temperature, relative humidity and luminosity sensors, some other nodes host air quality sensors, measuring the concentration of CO, CO_2, NO and O_3). Information are collected by sensors, then filtered and aggregated. A web portal is available showing geo-referenced sensor measures as well as aggregated information indexes.

Kumar and Singh [46] propose a recognition system for pet animal (dogs) to recognize individual dogs using their facial images in smart cities. The recognition system does not need any extra hardware like RFID-tags, embedding microchips, hot iron, freeze-branding and sensors. They propose a Fisher Linear Projection and Preservation (FLPP) feature extraction and representation approach for unique recognition of pet animals. In the proposed recognition system, the surveillance camera performs the face detection of individual dogs and localizes the Region of Interest (ROI) of captured face images of dog. After detection, dog face images (e.g., different size of face image frames) are extracted from the captured video.

Cicirelli et al. propose iSapiens [47–49], an edge- and agent-based platform, as an effective IoT tool for the implementation of distributed smart city applications. A real case study, namely the Smart Street Cosenza has been implemented by using iSapiens. In this case, iSapiens allows to gather data from the Smart Street and to provide citizens with information about climatic and environmental wellness. Moreover, an alert system for the detection of anomalies that may occur in the covered city area has been developed. The design of the Smart Street has been realized through the Smart Environment Metamodel (SEM) [50, 51] framework which shows the modeling over two perspectives, namely functional and data requirements.

Alhussein proposes in [52] a cloud-based framework to monitor Parkinson's disease (PD) in smart cities. He implemented a complete framework which allows a client to upload a voice signal to the cloud. The cloud manager authenticates the client and receives the signal. The manager then sends the signal to the feature extraction and classification server, which extracts and selects some features from the signal, which is then classified as PD or healthy based on the selected features. Once the signal is classified, the cloud manager establishes a connection with an available registered doctor, sending the results and the voice samples to the doctor. The doctor then sends a prescription back to the cloud, and the cloud manager notifies the client with the decision and the prescription. Based on the prescription, the client obtains a treatment. Then, the city controls vehicles and traffic to provide the client with the service as smoothly and early as possible.

Calbimonte et al. [53] focus on allowing citizens to self-monitor their environment and assess the quality of the resources in their surroundings. Authors aim to construct an hybrid and heterogenous sensing network by interconnecting multiple participants: stationary stations, mobile sensing devices on public transportation and crowdsensing. Experiments were realized on air quality monitoring.

3.2.3 Smart Mobility

Traffic Regulation

Cosido et al. propose in [54] a method to automatically find a cycle route in the area of Santander. The model uses a combination of soft computing and geographic information system techniques. The resulting multi-objective NP-hard problem is solved with a population-based bio-inspired meta-heuristics (adaptation of the Ant Colony Optimization (ACO) and Multi Objective Ant Colony Optimization (MOACO) algorithms).

On the control of an underground railway intersection problem, Patrascu et al. propose [55] an agent-based scalable coordination architecture in the framework of smart cities concept and Intelligent Operation Centre (IOC) strategy.

Stolfi et al. [56, 57] propose an approach called *Red Swarm* which allows continuous distributed exchange of data between vehicles and several spots located at traffic lights (using V2I communications) in order to suggest a customized route to each vehicle. *Red Swarm* consists of: (1) Spots installed at the traffic lights over the city, which use a WiFi connection to suggest new routes to vehicles, (2) Rerouting Algorithm, which selects the route to be suggested, (3) The evolutionary algorithm which computes the configuration of the system, (4) User Terminal Units, such as mobile phones or tablets to communicate with the spots.

Authors in [58, 59] propose an open platform, composed of four layers: (1) layer 0, which integrates sensing/performance technologies, is characterized by an openness regarding the connection to sensor networks of different natures, (2) Low-level services layer, which provides services such as filtering of signals, normalization services, (3) Information fusion algorithms layer, which facilitates an intelligent management of the information obtained from the lower levels of the architecture. Here, intelligent agents specialized in Information Fusion are required (supervised learning and previous training, Hierarchical Ant Based Control H-ABC), (4) High-level services layer, which provides an innovative module that allows management and customization of services to end users.

Li et al. [60] propose an optimization-based framework for determining adaptive traffic signal time settings. Authors formulate a comprehensive bi-level optimization model. The upper-level determines the traffic signal settings to minimize the average travel time, while the lower-level aims at achieving the network equilibrium. Genetic Algorithms (GA) are used to find near-optimal solutions.

Liu et al. [61] propose a cloud-based taxi trajectory pattern mining and trajectory clustering framework for smart cities which includes: preprocessing raw Global Positioning System trace, proposing a distributed trajectory pattern mining algorithm, and proposing a distributed trajectory clustering algorithm.

Smart Parking

Horng proposes in [62] an innovative adaptive recommendation mechanism for smart parking. It helps drivers to quickly find a parking space and reduces traffic congestion by using cellular automata mechanism and cognitive radio network model. The author adopts the Artificial Fish Swarm Algorithm to build this two-parts parking recommendation mechanism.

Baroffio et al. propose in [63] a parking lot occupancy detection system based on a Visual Sensor Network (VSN) that can be operated using the analyze-then-compress paradigm. Camera nodes extract local features specifically designed for the task at hand and transmit them to a central controller where a classifier is used to retrieve the occupancy information.

Mohammad et al. propose in [64] a cloud based and IoT centered architecture for efficient reservation of parking named BlueParking which efficiently finds the best available parking slots near the driver's destination. The advantage is the procedure that helps the drivers to reserve the best location from source based on intelligent routing and time estimation of the trip. BlueParking has been simulated by using Vehicle Starter Kit. The scheme is developed and demonstrated on the BlueMix IoT platform.

Yavari et al. propose in [65] an IoT-based contextualization technique that considers the entire range of data that are being collected in smart cities and uses such data to provide hyper-personalized information to each user in the smart cities. Authors exemplify the proposed solution in a smart parking space recommender application/service.

Grodi et al. propose in [66] a prototype of smart parking system. The system makes use of wireless sensor networks and wireless networks for detecting parking spaces status and reporting it to a database in a real-time basis. Parking detection requires some sort of sensor that can check the occupancy status of a parking spot. Sensors are connected to a user notification system. The gathered information are put on the web allowing anyone with an internet connected device to visualize such information.

Aydin et al. propose in [67] a navigation and reservation based parking system. The aim is to find the minimum distance to the free parking slot. The route from the current position to the nearest free parking slot is determined by using a genetic algorithm.

3.2.4 General Frameworks for Smart Cities and Security in Smart Cities

Chourabi et al. [68] propose eight factors that are essential for understanding smart city initiatives: management and organization, technology, governance, policy, people and communities, economy, built infrastructure, and the natural environment. These factors have been brought together into a smart city framework which explains the relationships and influences between these factors and smart city initiatives. The factors in this framework are represented at two different levels of influence in order to mirror the differentiated levels of impact.

Bicocchi et al. [69] aim at realizing large socio-technical superorganisms to support collective urban awareness and activities. They propose a middleware architecture which enables superorganism features with connection to cloud-components. Authors give specific attention to the *awareness layer* which will be responsible of: filtering sensor data streams coming from heterogeneous sensors, classifying data using general or specific purpose algorithms, and fusing classification labels to obtain a coherent overall picture of the situation. Inspired by biologic mechanisms, it is able to autonomously self-reconfigure and to dynamically enable and disable specific data streams.

The SmartSantander project [70] targets the creation of an European experimental test facility for the research and experimentation of architectures, key enabling technologies, services and applications for the IoT in the context of smart cities. This facility aims to leverage key IoT-enabling technologies and to provide the research community with a unique in-the-world platform for large scale IoT experimentation and evaluation under real-world operational conditions.

Zambonelli proposes in [71] a reference architecture for an infrastructure to support the future vision of self-organizing urban super-organisms. The reference architecture and its configurable coordination laws can support coordination among heterogeneous agents, can express situation-awareness by integrating advanced classification techniques, and can support and control a variety of self-organizing coordination patterns.

Muvuna et al. [72] propose a standard and holistic model which integrates all the sub-systems within a smart city system. The proposed model considers the smart sub-systems as parts of smart cities in which it gathers data blocks from sub-systems and processes them into meaningful information easily understandable by users. The smart city platform operations pass through three stages: data production, data processing, and sharing of information. The model presents an integrated platform on which information is shared and made available to both citizens and decision makers.

Authors in [73] propose a service oriented middleware called SmartCityWare for effective integration and utilization of CoT and fog computing. SmartCityWare abstracts services and components involved in smart city applications as services accessible through the service-oriented model.

Kamienski et al. propose in [74] a contextaware management framework, where contexts are modeled as graphs and can be explicitly designed and their occurrences can be tracked down.

Van Zoonen [75] proposes a framework for exploring people's specific privacy concerns in smart cities. Smart city privacy challenges consider: personal data used for service purposes, personal data used for surveillance purposes, impersonal data used for surveillance purposes and impersonal data collected for service purposes. The framework was then used to explore how specific technologies (smart bin, smart parking), and data usage (predictive policing, social media monitoring) may produce variable privacy concerns.

Li et al. [76] present a mobile-cloud framework to solve the data over-collection problem, which immensely improves the security and saves storage space of smartphones. All the data is stored in the cloud, and smartphones only deal with some basic operations of applications. Every application that wants to use users data sends its request for accessing to the cloud, and the cloud access control service could provide detailed permissions for every application to every block of users data (Table 2).

3.3 Discussion

Table 2 shows a qualitative comparison of works presented in this Section. It consists of eleven columns, used to show: the sub-system (*SS*), the application (*App*), the research reference (*Reference*), which technologies and paradigms were used (*Cloud*, *Fog*), Swarm Intelligence (*SI*), Evolutionary Algorithms (*EA*), Agent-Based Computing (*ABC*) [77, 78], Self-Adaptive behaviors (*Self-A*)), whether experiments are realized (Real Implementation (*RI*)) or simulated (Computer Simulations (*CS*)).

Regarding the table rows, *S.Buil* and *E.G* stand for: Smart Building and E-Government respectively. General Frameworks for smart cities and Security in smart cities (GF-and-S) in Table 2 groups privacy concerns works and frameworks with no case study, while if frameworks are used with a case study, they are reviewed in the section addressing the case study itself. Signs: × and – are used to mark the paradigms as existing in current work or not respectively.

From Table 2, some conclusions can be inferred:

- Cloud provides complementary resources for storage and processing in IoT-based applications. Typically, all the presented works are based on cloud (about 75%).
- Fog computing gives a very important value in smart cities [79], since it allows real time processing and gathering of data. It is an emerging technology and about 18% of the discussed works rely on it.
- ABC is a paradigm that uses cognitive or reactive agents to solve complex problems. About 24% of the presented works consider the ABC paradigm. It has been used in the discussed works to allow: fog computing, adaptive behaviors and to provide objects with intelligence.

Table 2 Qualitative comparison of literature review

SS	App	Reference	Cloud	Fog	SI	EA	ABC	Self-A	RI	CS
Smart environment	Waste collection	Medvedev et al. [26]	×	–	–	–	–	×	–	×
		Gutierrez et al. [27]	×	–	–	×	–	×	–	×
		Anagnostopoulos et al. [28]	×	–	–	–	–	×	–	×
		Anagnostopoulos et al. [29]	×	–	–	–	–	–	–	×
		Medvedev et al. [30]	×	–	–	–	–	×	–	×
		Aazam et al. [31]	×	–	–	–	–	–	–	–
		Castro et al. [24]	×	×	–	–	–	–	×	–
		Lokuliyana et al. [32]	×	–	–	–	–	–	–	–
	S.Grid	Hurtado et al. [34]	×	–	×	–	×	×	–	×
		Simmhan et al. [33]	×	–	–	–	–	×	–	×
		Kane et al. [35]	×	×	–	–	–	–	–	×
	S.Buil	Molina et al. [37]	×	–	–	–	–	–	×	–
		Asimakopoulou and Bessis [36]	×	–	–	–	–	–	–	–
Smart government	E.G	Cano et al. [40]	–	–	–	–	–	–	–	–
	City monitoring	Cicirelli et al. [47–49, 51]	×	×	–	–	×	×	×	×
		Zanella et al. [41]	–	–	–	–	–	×	×	–
		Giordano et al. [45]	×	×	×	–	×	×	×	–
		Kumar and Singh [46]	–	–	–	–	–	–	–	×
		Alhussein [52]	×	–	–	–	–	–	–	×
		Jamil et al. [42]	×	–	–	–	–	–	–	–
		Roza and Postolache [43]	×	–	–	–	–	–	×	–
		Shah and Mishra [44]	×	–	–	–	–	–	×	–
		Calbimonte et al. [53]	×	–	–	–	–	–	×	–

(continued)

Table 2 (continued)

SS	App	Reference	Cloud	Fog	SI	EA	ABC	Self-A	RI	CS
Smart mobility	Traffic regulation	Cosido et al. [54]	–	–	×	–	–	×	–	×
		Patrascu et al. [55]	–	–	–	–	×	×	–	×
		Stolfi et al. [56]	–	–	–	×	–	×	×	–
		Stolfi and Alba [57]	–	–	–	×	–	–	–	×
		Chamoso et al. [58]	×	–	×	×	×	×	–	×
		Chamoso and Prieta Pintado [59]	×	–	×	×	×	×	×	–
		Li et al. [60]	–	–	–	×	–	×	–	×
		Liu et al. [61]	×	–	–	–	–	–	–	×
	S.Parking	Horng [62]	×	–	×	–	–	×	–	×
		Baroffio et al. [63]	–	–	–	–	–	–	×	–
		Mohammad et al. [64]	×	–	–	–	–	–	–	×
		Grodi et al. [66]	–	–	–	–	–	–	×	–
		Aydin et al. [67]	–	–	–	×	–	×	–	×
GF-and-S	GF-and-S	Sanchez et al. [70]	×	–	–	–	–	–	×	–
		Mohamed et al. [73]	×	×	–	–	×	–	×	–
		Kamienski et al. [74]	–	–	–	–	–	×	–	×
		Bicocchi et al. [69]	×	–	–	–	–	×	–	–
		Zambonelli [71]	×	–	–	–	×	×	–	–
		Li et al. [76]	×	–	–	–	–	–	×	–

- EA are bio-inspired technologies, used to provide self-adaptive and self-organized behaviors. 16% of the discussed works consider EA algorithms to allow self-configuration.
- A few works considered distributed collective behaviors (sensing, coordination or communication).
- SI paradigm has been considered in smart city sub-systems through SI-based algorithms in order to resolve optimization problems (13% of the presented works). However, SI paradigm as a modeling scheme has not been considered for smart cities.
- In order to adapt to dynamic changes and growing in smart cities, self-adaptive behaviors were widely considered in the presented works (49%).
- All the presented works (unless [72]), focus on smart city sub-systems and ignore the integration and the modeling of smart cities as a whole.

4 Swarm-Based Smart Cities

This Section proposes a SI-based framework for smart cities. Moreover, it presents some trends on how to use the SI paradigm to make smart cities flexible and scalable.

4.1 A SI-Based Framework for Smart Cities

We present in this Section, a comprehensive smart city architecture (see Fig. 1). In this architecture, we define a smart city as a whole system composed of several sub-systems. Each sub-system is composed by a set of applications and each application is a swarm of entities (sensors and actuators). As in nature law, smart city entities need or provide resources to other entities from the same or different families. Entities need to be complementary in order to survive and to achieve mutual benefits. We classify interactions in our model into two types: (1) *co-work interactions*, where sensors and actuators belonging to the same application exchange information in order to realize a global goal, (2) *social interactions*, where sensors and actuators belonging to different applications exchange information to realize specific goals.

According to the categorization presented in Sect. 3.2 and Table 2, the proposed architecture integrates four of the most studied trends: smart environment, smart mobility, smart government, general frameworks for smart cities and security in smart cities [72]. It could be extended to integrate the six pillar trends since there are many more applications in real world. To realize a robust, scalable and reliable smart city, the control in the system and over its components needs to be decentralized. We adopt an hierarchical structure, where in each level the system is decomposed into different *data blocks*. The lower level includes sensors and actuators. Actuators could be triggered subsequently by data captured using sensors from another application. As to formulate the system as a swarm, we suppose that system entities can exchange information between them directly or indirectly. Thus, data from each application is stored in *data blocks* at **Application Level** (e.g. *Data Block- Smart Building, Data Block- Smart Parking…*) and data from different applications belonging to the same trend is stored in *data blocks* at **Sub-System Level** (e.g. *Data Block- Smart Environment, Data Block- Smart Mobility…*) and data from all sub-systems is stored in *data blocks* at **System Level** (*Data Block- Smart City*). The communication is direct between entities belonging to the same application (e.g. traffic regulation and smart parking) and it is indirect through *data blocks* at **Application Level**, if entities belong to different applications but the same trend (e.g. smart building and security). While, if entities belong to different trends the communication is indirect through *data blocks* at **Sub-System Level**. Finally, *data blocks* at **System Level** are used to store the whole smart city data and to communicate it to other smart cities or external applications.

The hierarchical structure of the proposed SI-based architecture is detailed below, while the graphical representation is given by Fig. 1.

1. **System level**: It is the higher level which constitutes the whole system. It is composed of a collection of sub-systems. It conserves all the data gathered by all the sensors deployed in the city. Data is transferred by lower levels. Data is stored and then can be transmitted to smart city sub-systems, to government or even to other cities. This level data block could be deployed in the cloud.
2. **Sub-System level**: According to the categorization presented in Sect. 3.2, there are four sub-systems: smart environment, smart mobility, smart government and general frameworks and security in smart cities. Each sub-system is composed of a set of applications. Each sub-system stores the data derived from the *Application level*. Data is centralized in sub-systems data blocks, which can communicate in a decentralized manner to exchange information between them. Fog technology is required in this level to allow real time operations and high quality services.
3. **Application level**: In Fig. 1, we list a set of applications which include (but are not limited to): traffic regulation, smart parking, waste management, smart grid, smart building, security, E-Government and city monitoring. Each application is considered as a swarm of entities (sensors or actuators). Data gathered by sensors of each application is stored *Data Blocks* specific for this application. Data from this level is cloned to the appropriate sub-system level. Fog computing technology is also required in this level.
4. **Sensors/Actuators level**: This is the lowest level. It envelops sensors and actuators which could be machines, materials, computers or even humans. Sensors and actuators belonging to the same application can directly exchange information. While those belonging to different applications can communicate it from *data blocks* of the **Application Level**. ABC paradigm is used in this level (i.e. mobile software agents).

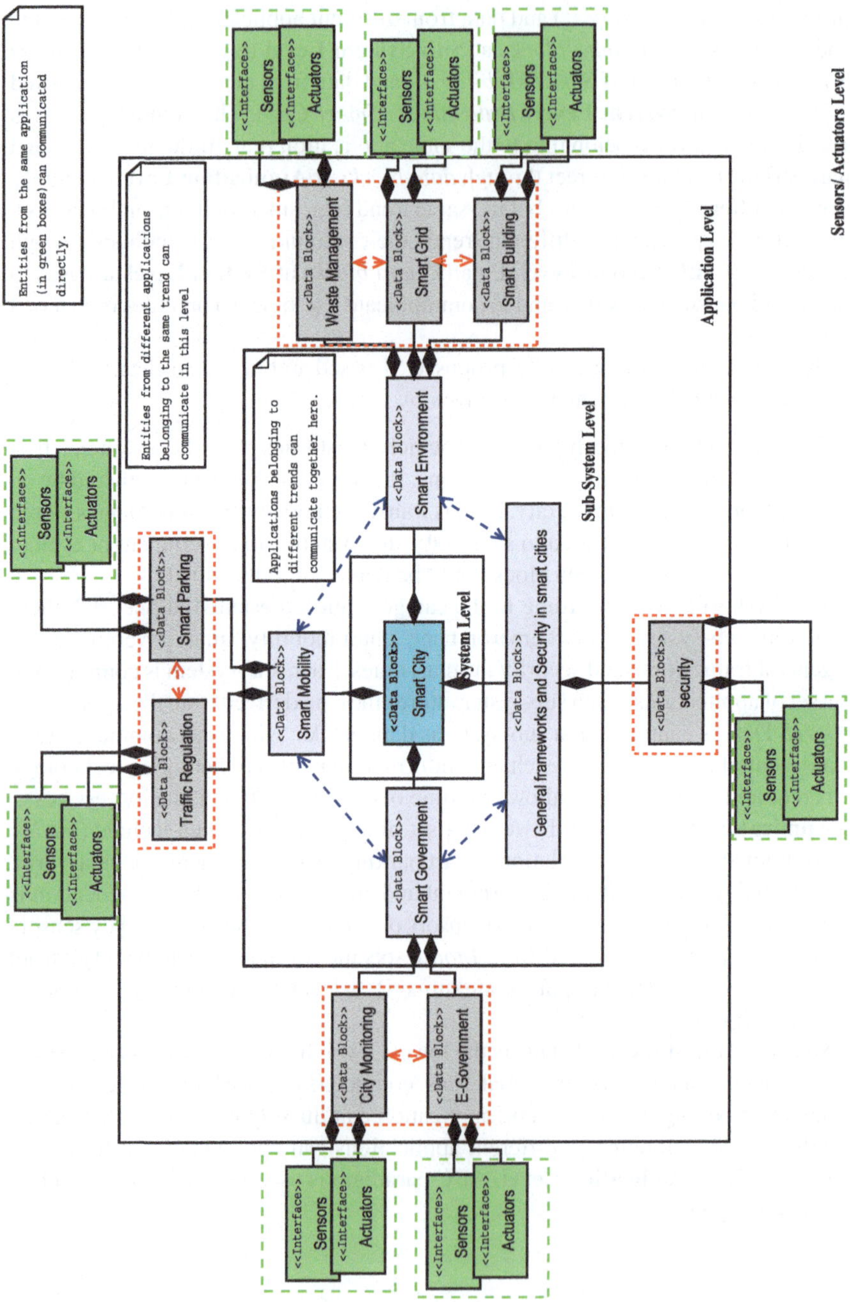

Fig. 1 A graphical representation of the proposed SI-based framework for smart cities

4.2 Swarm Intelligence in Smart Cities: Trends

To make scalable, robust and sustainable smart cities, in our point of view, many trends regarding SI need to be investigated:

- The social dimension of swarm individuals provide scalable and reliable solution to complex and dynamic problems. Thus, objects (devices) could be extended with social dimension: the complementary capabilities of devices need to be brought together, coordinated and organized through the rely on bio-inspired self-organizing approach, specifically the Swarm Intelligence one.
- Agent-Based Simulation (ABS) is a suitable tool to address new challenges for smart city solution development strategies, ranging from modeling social mechanisms to proposing new emergent phenomena and interactive systems design.
- Models and infrastructures should be created to make it possible to engineer large-scale systems. Such models and infrastructures should enable a variety of related self-organizing patterns to be put in place and working at the same time, and to concurrently express various forms of collective behaviors serving different purposes.
- Techniques of Swarm Intelligence Optimization (SIO) open the door for a promising future for geospatial technological developments, moreover they help in reducing its inherent fuzziness by application of suitable metaheuristics techniques.
- SI-based algorithms provide robust solutions to various combinatorial problems. Many of the challenges actually existing in smart cities were already considered in literature through SI-based algorithms. Hence, they could be adapted by including IoT infrastructures to meet smart city requirements and to solve smart city challenges. For example, waste management is a problem in smart cities which required solutions to different sub-problems such as scheduling, routing and optimization of trash locations. If we take a look at column *Application* in Table 1, we can see that the scheduling problem was resolved using several SI-based algorithms (CSO, FFA, ABC, HBMO, PSO, Wolf and CSA), also location optimization was resolved by GSO algorithms and vehicle routing was resolved by ACO, BFOA, AFSA and BA. So, these solutions could be used in smart city context by adapting them and integrating IoT technologies.
- MAS are suitable technologies to deal with uncertainty in dynamically changing environments. This technology plays a crucial role in the proposed model [80, 81].
- With the continuous growing of installed devices, IoT should be based on low-cost micro-controllers to build scalable complex systems.
- Decentralized control constitutes the best solution to build scalable systems but coordination issues need to be addressed.
- Swarm robotics could be a source of inspiration to solve coordination issues.

5 Conclusion

In this paper, the potentiality to use the Swarm Intelligence paradigm and algorithms in order to make Smart Cities much more smart and scalable has been analyzed. In the proposed vision, objects could be assimilated to animals with very low capabilities as individuals which produce high level and complex behaviors at group level. However, to reach this goal, many research challenges still need to be addressed, and suitable middleware infrastructures have to be developed. This chapter was divided into three fundamental parts: the first has presented the most common SI-based algorithms. The second has focused on smart city definitions and related works. Finally, in the third part, a SI-based framework has been proposed, and then guidelines on how to use SI within smart cities in order to make them flexible and scalable have been presented.

References

1. J. Kennedy, Swarm intelligence, in *Handbook of Nature-Inspired and Innovative Computing* (Kluwer Academic Publishers, Boston, 2001), pp. 187–219
2. R.S. Parpinelli, H.S. Lopes, New inspirations in swarm intelligence: a survey. Int. J. Bio-Inspired Comput. **3**(1), 1 (2011)
3. S. Garnier, J. Gautrais, G. Theraulaz, The biological principles of swarm intelligence. Swarm Intell. **1**(1), 3–31 (2007)
4. V. Angelakis, E. Tragos, H.C. Pöhls, A. Kapovits, *Designing, Developing, and Facilitating Smart Cities* (Springer International Publishing, Cham, 2017)
5. G. Garofalo, A. Giordano, P. Piro, G. Spezzano, A. Vinci, A distributed real-time approach for mitigating CSO and flooding in urban drainage systems. J. Netw. Comput. Appl. **78**, 30–42 (2017)
6. O. Zedadra, C. Savaglio, N. Jouandeua, A. Guerrieri, H. Seridi, G. Fortino, Towards a reference architecture for swarm intelligence-based internet of things, in *the Proc. of the 10th International Conference on Internet and Distributed Computing Systems (IDCS 2017)* (Springer, December, Fiji, 2017), pp. 11–13
7. M. Dorigo, M. Birattari, T. Stutzle, *Ant colony optimization* IEEE Computational Intelligence Magazine **1**(4), 28–39 (2006)
8. F. Cicirelli, A. Forestiero, A. Giordano, C. Mastroianni, Transparent and efficient parallelization of swarm algorithms. ACM Trans. Auton. Adapt. Syst. **11**(2), 14:1–14:26 (2016)
9. D. Karaboga, An idea based on honey bee swarm for numerical optimization. Technical Report-tr06, Erciyes University, Engineering Faculty, Computer Engineering Department (2005)
10. H.A. Abbass, MBO: marriage in honey bees optimization-a Haplometrosis polygynous swarming approach, in *Proceedings of the 2001 Congress on Evolutionary Computation (IEEE Cat. No. 01TH8546)*, vol. 1 (IEEE, 2000), pp. 207–214
11. X.S. Yang, X. He, Firefly algorithm: recent advances and applications. Int. J. Swarm Intell. **1**(1), 36–50 (2013)
12. K.N. Krishnanand, D. Ghose, Detection of multiple source locations using a glowworm metaphor with applications to collective robotics, in *Proceedings 2005 IEEE Swarm Intelligence Symposium, 2005 (SIS 2005)*, vol. 2005 (IEEE, 2005), pp. 84–91
13. K.M. Passino, Biomimicry of bacterial foraging for distributed optimization and control. IEEE Control Syst. Mag. **22**(3), 52–67 (2002)

14. R.C. Eberhart, J. Kennedy, Particle swarm optimization, in *Proceedings of the International Conference on Neural Networks, Piscataway*, vol. 1000, pp. 1942–1948 (1995)
15. X.S. Yang, S. Deb, Engineering optimisation by cuckoo search. Int. J. Math. Model. Numer. Optim. **1**(4), 330 (2010)
16. X.-S. Yang, A new metaheuristic bat-inspired algorithm. Stud. Comput. Intell. **284**, 65–74 (2010)
17. M. Eusuff, K. Lansey, F. Pasha, Shuffled frog-leaping algorithm: a memetic meta-heuristic for discrete optimization. Eng. Optim. **38**(2), 129–154 (2006)
18. X. Li, J.-x. Qian, Studies on artificial fish swarm optimization algorithm based on decomposition and coordination techniques. J. Circuit Syst. **1**, 1–6 (2003)
19. C. Liu, X. Yan, C. Liu, W. Hua, The wolf colony algorithm and its application. Chin. J. Electron. **20**(2), 212–216 (2011)
20. S.C. Chu, P.W. Tsai, Computational intelligence based on the behavior of cats. Int. J. Innov. Comput. Inf. Control **3**(1), 163–173 (2007)
21. L.G. Anthopoulos, Understanding the smart city domain: a literature review, in *Transforming City Governments for Successful Smart Cities* (Springer, 2015), pp. 9–21
22. C. Harrison, B. Eckman, R. Hamilton, P. Hartswick, J. Kalagnanam, J. Paraszczak, P. Williams, Foundations for smarter cities. IBM J. Res. Dev. **54**(4), 1–16 (2010)
23. D. Toppeta, The smart city vision: how innovation and ICT can build smart, "livable", sustainable cities. Innov. Knowl. Found. **5**, 1–9 (2010)
24. A.C. Lundin, A.G. Ozkil, J. Schuldt-Jensen, Smart cities: a case study in waste monitoring and management. in *Proceedings of the 50th Hawaii International Conference on System Sciences* (2017)
25. A. Picon, *Smart Cities: A Spatialised Intelligence* (Wiley, 2015)
26. A. Medvedev, P. Fedchenkov, A. Zaslavsky, T. Anagnostopoulos, S. Khoruzhnikov, Waste management as an IoT-enabled service in smart cities, in *Conference on Smart Spaces* (Springer, 2015), pp. 104–115
27. J.M. Gutierrez, M. Jensen, M. Henius, T. Riaz, Smart waste collection system based on location intelligence. Procedia Comput. Sci. **61**, 120–127 (2015)
28. T. Anagnostopoulos, A. Zaslavsky, A. Medvedev, Robust waste collection exploiting cost efficiency of loT potentiality in smart cities, in *2015 International Conference on Recent Advances in Internet of Things (RIoT)*, pp. 7–9, Apr 2015
29. T. Anagnostopoulos, K. Kolomvatsos, C. Anagnostopoulos, A. Zaslavsky, S. Hadjiefthymiades, Assessing dynamic models for high priority waste collection in smart cities. J. Syst. Softw. **110**, 178–192 (2015)
30. A. Medvedev, A. Zaslavsky, M. Indrawan-Santiago, P.D. Haghighi, A. Hassani, Storing and indexing IoT context for smart city applications, in *Internet of Things, Smart Spaces, and Next Generation Networks and Systems* (Springer, 2016), pp. 115–128
31. M. Aazam, M. St-Hilaire, C.-H. Lung, I. Lambadaris, Cloud-based smart waste management for smart cities, in *2016 IEEE 21st International Workshop on Computer Aided Modelling and Design of Communication Links and Networks (CAMAD)* (IEEE, 2016), pp. 188–193
32. S. Lokuliyana, J.A.D.C. Anuradha Jayakody, L. Rupasinghe, S. Kandawala, IGOE IoT framework for waste collection optimization, in *National Conference on Technology and Management (NCTM)* (IEEE, 2017), pp. 12–16
33. Y. Simmhan, S. Aman, A. Kumbhare, R. Liu, S. Stevens, Q. Zhou, V. Prasanna, Cloud-based software platform for big data analytics in smart grids. Comput. Sci. Eng. **15**(4), 38–47 (2013)
34. L.A. Hurtado, P.H. Nguyen, W.L. Kling, Smart grid and smart building inter-operation using agent-based particle swarm optimization. Sustain. Energy Grids Netw. **2**, 32–40 (2015)
35. J. Kane, B. Tang, Z. Chen, J. Yan, T. Wei, H. He, Q. Yang, Reflex-tree: a biologically inspired parallel architecture for future smart cities, in *2015 44th International Conference on Parallel Processing* (IEEE, 2015), pp. 360–369

36. E. Asimakopoulou, N. Bessis, Buildings and crowds: forming smart cities for more effective disaster management, in *2011 Fifth International Conference on Innovative Mobile and Internet Services in Ubiquitous Computing* (IEEE, 2011), pp. 229–234
37. B. Molina, C.E. Palau, G. Fortino, A. Guerrieri, C. Savaglio, Empowering smart cities through interoperable sensor network enablers, in *2014 IEEE International Conference on Systems, Man, and Cybernetics (SMC)*, vol. 2014-Janua. (IEEE, 2014), pp. 7–12
38. A. Guerrieri, G. Fortino, A. Ruzzelli, G.M.P. O'Hare, A WSN-based building management framework to support energy-saving applications in buildings, in *Advancements in Distributed Computing and Internet Technologies: Trends and Issues* (IGI Global, 2012), pp. 258–273
39. G. Fortino, R. Gravina, A. Guerrieri, G. Di Fatta, Engineering large-scale body area networks applications, in *Proceedings of the 8th International Conference on Body Area Networks* (ICST (Institute for Computer Sciences, Social-Informatics and Telecommunications Engineering), 2013), pp. 363–369
40. J. Cano, C.E. Jimenez, S. Zoughbi, A smart city model based on citizen-sensors, in *2015 IEEE First International Smart Cities Conference (ISC2)*, vol. 81 (IEEE, 2015), pp. 1–2
41. A. Zanella, N. Bui, A. Castellani, L. Vangelista, M. Zorzi, Internet of Things for smart cities. IEEE Internet of Things J. **1**(1), 22–32 (2014)
42. M.S. Jamil, M.A. Jamil, A. Mazhar, A. Ikram, A. Ahmed, U. Munawar, Smart environment monitoring system by employing wireless sensor networks on vehicles for pollution free smart cities. Procedia Eng. **107**, 480–484 (2015)
43. V.C.C. Roza, O.A. Postolache, Citizen emotion analysis in smart city, in *2016 7th International Conference on Information, Intelligence, Systems & Applications (IISA)* (IEEE, 2016), pp. 1–6
44. J. Shah, B. Mishra, IoT enabled environmental monitoring system for smart cities, in *2016 International Conference on Internet of Things and Applications (IOTA)* (IEEE, 2016), pp. 383–388
45. A. Giordano, G. Spezzano, A. Vinci, Smart agents and fog computing for smart city applications, in *International Conference on Smart Cities* (2016, June), pp. 137-146
46. S. Kumar, S.K. Singh, Monitoring of pet animal in smart cities using animal biometrics. *Future Gener. Comput. Syst.* (2016)
47. F. Cicirelli, A. Guerrieri, G. Spezzano, A. Vinci, An edge-based platform for dynamic smart city applications. Future Gener. Comput. Syst. **76**, 106–118 (2017)
48. F. Cicirelli, A. Guerrieri, G. Spezzano, A. Vinci, O. Briante, G. Ruggeri, iSapiens: a platform for social and pervasive smart environments, in *2016 IEEE 3rd World Forum on Internet of Things (WF-IoT)* (IEEE, 2016), pp. 365–370
49. F. Cicirelli, A. Guerrieri, G. Spezzano, A. Vinci, O. Briante, A. Iera, G. Ruggeri, Edge computing and social internet of things for large-scale smart environments development. IEEE Internet of Things J. **4662**(c), 1–1 (2017)
50. F. Cicirelli, G. Fortino, A. Guerrieri, G. Spezzano, A. Vinci, Metamodeling of smart environments: from design to implementation. Adv. Eng. Inform. **33**, 274–284 (2017)
51. F. Cicirelli, G. Fortino, A. Mercuri, A. Guerrieri, G. Spezzano, A. Vinci, Exploiting the sem framework for modeling smart cities, in *International Conference on Internet and Distributed Computing Systems* (Springer, 2017)
52. M. Alhussein, Monitoring Parkinson's disease in smart cities. IEEE Access **5**(c), 19835–19841 (2017)
53. J.-P. Calbimonte, J. Eberle, K. Aberer, Toward self-monitoring smart cities: the OpenSense2 approach. Informatik-Spektrum **40**(1), 75–87 (2017)
54. O. Cosido, C. Loucera, A. Iglesias, Automatic calculation of bicycle routes by combining meta-heuristics and GIS techniques within the framework of smart cities, in *2013 International Conference on New Concepts in Smart Cities: Fostering Public and Private Alliances (SmartMILE)* (IEEE, 2013), pp. 1–6
55. M. Patrascu, M. Dragoicea, A. Ion, Emergent intelligence in agents: a scalable architecture for smart cities, in *2014 18th International Conference on System Theory, Control and Computing (ICSTCC)* (IEEE, 2014), pp. 181–186

56. D.H. Stolfi, E. Alba, Red Swarm: reducing travel times in smart cities by using bio-inspired algorithms. Appl. Soft Comput. **24**, 181–195 (2014)
57. D.H. Stolfi, E. Alba, Eco-friendly reduction of travel times in European smart cities, in *Proceedings of the 2014 Conference on Genetic and Evolutionary Computation (GECCO'14)* (ACM Press, New York, USA, 2014), pp. 1207–1214
58. P. Chamoso, F. De Prieta, F. De Paz, J.M. Corchado, Swarm agent-based architecture suitable for internet of things and smartcities, in *Advances in Intelligent Systems and Computing*, vol. 373 (Springer International Publishing, Cham, 2015)
59. P. Chamoso, F. De La Prieta, Swarm-based smart city platform: a traffic application. ADCAIJ: Adv. Distrib. Comput. Artif. Intell. J. **4**(2), 89 (2015)
60. Z. Li, M. Shahidehpour, S. Bahramirad, A. Khodaei, Optimizing traffic signal settings in smart cities. IEEE Trans. Smart Grid **8**(5), 2382–2393 (2017)
61. J. Liu, X. Yu, Z. Xu, K.-K.R. Choo, L. Hong, X. Cui, A cloud-based taxi trace mining framework for smart city. Softw. Pract. Exp. **47**(8), 1081–1094 (2017)
62. G.-J. Horng, The adaptive recommendation mechanism for distributed parking service in smart city. Wirel. Pers. Commun. **80**(1), 395–413 (2015)
63. L. Baroffio, L. Bondi, M. Cesana, A.E. Redondi, M. Tagliasacchi, A visual sensor network for parking lot occupancy detection in smart cities, in *2015 IEEE 2nd World Forum on Internet of Things (WF-IoT)* (IEEE, 2015), pp. 745–750
64. M. Amin, R. Kawaguchi, N. Shirmohammad, M. Sato, BlueParking, in *Proceedings of the Second International Conference on IoT in Urban Space - Urb-IoT '16*, vol. 24–25-May (ACM Press, New York, New York, USA, 2016), pp. 86–88
65. A. Yavari, P.P. Jayaraman, D. Georgakopoulos, Contextualised service delivery in the internet of things: parking recommender for smart cities, in *2016 IEEE 3rd World Forum on Internet of Things (WF-IoT)* (IEEE, 2016), pp. 454–459
66. R. Grodi, D.B. Rawat, F. Rios-Gutierrez, Smart parking: parking occupancy monitoring and visualization system for smart cities, in *SoutheastCon 2016* (IEEE, 2016), pp. 1–5
67. I. Aydin, M. Karakose, E. Karakose, A navigation and reservation based smart parking platform using genetic optimization for smart cities, in *2017 5th International Istanbul Smart Grid and Cities Congress and Fair (ICSG)* (IEEE, 2017), pp. 120–124
68. H. Chourabi, T. Nam, S. Walker, J.R. Gil-Garcia, S. Mellouli, K. Nahon, T.A. Pardo, H.J. Scholl, Understanding smart cities: an integrative framework, in *2012 45th Hawaii International Conference on System Sciences* (IEEE, 2012), pp. 2289–2297
69. N. Bicocchi, A. Cecaj, D. Fontana, M. Mamei, A. Sassi, F. Zambonelli, Collective awareness for Human-ICT collaboration in smart cities, in *22nd IEEE International WETICE Conference* (WETICE, 2013, June), pp. 3-8, IEEE
70. L. Sanchez, L. Muñoz, J.A. Galache, P. Sotres, J.R. Santana, V. Gutierrez, R. Ramdhany, A. Gluhak, S. Krco, E. Theodoridis, D. Pfisterer, SmartSantander: IoT experimentation over a smart city testbed. Comput. Netw. **61**, 217–238 (2014)
71. F. Zambonelli, Engineering self-organizing urban superorganisms. Eng. Appl. Artif. Intell. **41**, 325–332 (2015)
72. J. Muvuna, T. Boutaleb, S.B. Mickovski, K.J. Baker, Systems engineering approach to design and modelling of smart cities, in *International Conference for Students on Applied Engineering (ICSAE)* (IEEE, 2016), 437–440
73. N. Mohamed, J. Al-Jaroodi, I. Jawhar, S. Lazarova-Molnar, S. Mahmoud, SmartCityWare: a service-oriented middleware for cloud and fog enabled smart city services. IEEE Access, **5**(c), 17576–17588 (2017)
74. C.A. Kamienski, F.F. Borelli, G.O. Biondi, I. Pinheiro, I.D. Zyrianoff, M. Jentsch, Context design and tracking for IoT-based energy management in smart cities. IEEE Internet of Things J. **5**(2), 687–695 (2018)
75. L. van Zoonen, Privacy concerns in smart cities. Gov. Inf. Q. **33**(3), 472–480 (2016)
76. Y. Li, W. Dai, Z. Ming, M. Qiu, Privacy protection for preventing data over-collection in smart city. IEEE Trans. Comput. **65**(5), 1339–1350 (2016)

77. G. Fortino, D. Grimaldi, L. Nigro, Multicast control of mobile measurement systems. IEEE Trans. Instrum. Meas. **47**(5), 1149–1154 (1998)
78. G. Fortino, W. Russo, Using P2P, GRID and agent technologies for the development of content distribution networks. Future Gener. Comput. Syst. **24**(3), 180–190 (2008)
79. G. Aloi, G. Caliciuri, G. Fortino, R. Gravina, P. Pace, W. Russo, C. Savaglio, Enabling IoT interoperability through opportunistic smartphone-based mobile gateways. J. Netw. Comput. Appl. **81**, 74–84 (2017)
80. G. Fortino, A. Garro, W. Russo, An integrated approach for the development and validation of multi-agent systems. Int. J. Comput. Syst. Sci. Eng. **20**(4), 259–271 (2005)
81. F. Aiello, G. Fortino, A. Guerrieri, R. Gravina, Maps: a mobile agent platform for WSNs based on java sun spots, in *Proceedings of ATSM* (2009)

Cost Saving and Ancillary Service Provisioning in Green Mobile Networks

Muhammad Ali, Michela Meo and Daniela Renga

Abstract Mobile Network Operators (MNOs) are facing huge operational costs, due to the staggering increase of mobile traffic and to substantial bandwidth reliability requirements needed to enable the services of Smart Urban Ecosystems. With the purpose of reducing the cost due to power supply, dynamic load adaptation techniques are often implemented in Mobile Networks, in order to save energy when the traffic demand is low. Moreover, renewable energy (RE) sources are commonly introduced to power base stations, further contributing to decrease the operational expenditures. Finally, in a Demand Response context, the Smart Grid (SG) may actively ask its customers to dynamically adapt their consumption, by means of monetary incentives. The MNO is interested in improving the interaction with the SG, since mutual benefits can be obtained: cost reduction for the MNO and ancillary service provisioning from the SG side. We investigate via simulation a mobile access network where WiFi offloading techniques are combined with a properly designed energy management strategy, in order to reduce the load and better satisfy the SG requests. The impact of WiFi offloading is analyzed in different scenarios, including those envisioning the use of RE to power base stations (BSs) and/or the application of Resource on Demand (RoD) strategies, that activate or deactivate BSs based on traffic demand. Real data about traffic, RE production and SG requests are adopted. WiFi offloading results effective both in improving the probability of providing ancillary services and in reducing operational costs in any scenario, even when no RE is available. Furthermore, its impact is even more significant than the application of RoD strategies. Positive revenues are also possible for the MNO when RE are used, even when photovoltaic panels with relatively small capacity are installed.

M. Ali · M. Meo · D. Renga (✉)
Dipartimento di Elettronica e Telecomunicazioni, Politecnico di Torino, Turin, Italy
e-mail: daniela.renga@polito.it

M. Ali
e-mail: muhammad.ali@studenti.polito.it

M. Meo
e-mail: michela.meo@polito.it

F. Cicirelli et al. (eds.), *The Internet of Things for Smart Urban Ecosystems*, Internet of Things, https://doi.org/10.1007/978-3-319-96550-5_9

1 Introduction

The staggering increase of the mobile traffic is currently leading Mobile Operators to deploy denser mobile access networks, especially in urban environments, thus entailing a considerable rise in their operational costs. Furthermore, with the extensive penetration of the Smart City paradigm, cellular networks are going to play a fundamental role in urban environments. Indeed, mobile networks will be essential in enabling several of the countless services deployed in Smart Urban Ecosystems, including smart mobility, safety and environmental protection, smart building, and enforcement of energy efficiency and awareness. In this context, mobile networks should provide high bandwidth capacity, fast speed Internet access for the services that are becoming the more and more demanding, and high reliability in terms of service continuity. Considering that mobile traffic is growing twice as fast as fixed IP and that 27 billion of networked devices and connections are expected by 2021, leading to an annual mobile traffic exceeding half a zettabyte, the energy consumed by mobile access networks is bound to incredibly rise, and so is the energy bill for Mobile Operators [1]. At the same time, within the Smart City ecosystem, the Smart Grid represents a new paradigm replacing the traditional electric grid, by envisioning several distributed energy producers, possibly from renewable energy sources, rather than a single energy supplier, and enabling bidirectional flows both of information and energy. One of the challenges of new electricity generation and distribution networks is coping with the mismatches between electricity demand and supply, mismatches that are more difficult to predict due to the presence of renewable sources. However, by making the grid "smart" and capable of interacting with the customers, it is possible to introduce new paradigms of electricity supply and consumption in which the customers take actively part in the process. With the electric power grid becoming a *Smart Grid* (SG), the Demand Response (DR) policies applied by the SG operator (SGO) may highly affect the energy costs, by introducing huge variability in the energy prices over time. The DR approaches aim at maintaining the demand-supply balance by adapting demand to the production curve. Production excesses are coped with by setting low energy prices that incentivize end users to increase their demand; conversely, users are induced to reduce their consumption when the production is low, by letting energy prices rise. By dynamically modulating their power consumption in accordance with the SG requests, users can contribute to provide the so-called *ancillary services* (ASs).

In this context, big consumers can have a central role. While they can significantly contribute to DR objectives by adapting their (huge) demand, their convenience consists in the possibility to obtain important electricity cost reductions. Telecommunication mobile operators are an example of such big consumers. Today, operators are experiencing huge costs to power their devices and this is expected to get even worse in the future. At the same time, they have some possibility to act on their network to adapt consumption to the SG requests.

We consider in this study the use of WiFi offloading (WO) technique. WiFi offloading consists in transferring a portion of the mobile traffic from the BSs to some nearby

WiFi Access Points (APs). This technique is commonly adopted to relieve mobile access networks from a part of their traffic load during peak hours in heterogeneous networks. In our case, WiFi offloading is introduced as a way to decrease the traffic load whenever the SG requests a reduction of the consumption. WiFi offloading alone may not lead to remarkable reduction in the consumption, due to the limited load proportionality of energy consumption. However, when applied in a scenario where the base stations (BSs) can also be switched on and off, WiFi offloading allows to further decrease the number of active BSs, hence to further reduce the system consumption. Hence, we investigate the effect of WiFi offloading in improving the interaction of the mobile network with the SG, both in the case of the normal network operation as well as in the case in which Resource on Demand (RoD) strategies are used to reduce the overall consumption of the network. RoD techniques reduce the consumption by dynamically switching on and off the base stations based on the traffic load. Furthermore, we investigate whether the WiFi offloading approach may further benefit from the introduction of a local renewable energy (RE) generator in the mobile access network. Several examples of mobile networks powered by RE can already be found in the literature [2]. The irregular and unpredictable nature of RE production is addressed by the installation of an energy storage device, where extra amounts of produced RE can be harvested for future usage, for example during the night. The expected impact of local RE generation results from two reasons. On the one hand, the RE production allows the BSs to become more independent from the grid, hence decreasing the demand from the SG, with consequent lowering of the cost. On the other hand, the presence of batteries allows also to store energy, when it is convenient, for future usage.

RoD approaches implying BS sleep modes are commonly adopted to make mobile networks more energy efficient and the use of RE to power the BSs is emerging as a technique to make mobile networks more independent from the grid [2–4]. Studies start to be available related to the application of BS sleep mode based RoD strategies to RE powered BSs [5–8]. Finally, WO represents a widely adopted technique allowing to migrate traffic from cellular networks to WiFi networks to tackle the recent explosion of mobile data traffic [9, 10]. However, the interaction between a renewable powered mobile access network and the SG in a DR framework and with usage of BS sleep mode RoD techniques is not well investigated in literature and only few recent studies focus on this issue [11, 12]. In addition, the impact of WO on improving the energy efficiency and reducing operational costs in a renewable powered mobile access network interacting with the SG, with possible application of BS sleep mode RoD strategies, is an aspect that still remains to be investigated.

The study presented hereafter investigates a portion of a RE powered mobile access network interacting with the Smart Grid in a Demand Response framework. WiFi offloading techniques are introduced to enhance the response of the cellular network to the SG requests. The role of WiFi offloading in reducing the energy bill is analyzed via simulation and its impact on improving the capability of providing ancillary services is examined as well. Results show that the introduction of WiFi offloading provides mutual benefits for the Mobile Network Operator (MNO) and the SG. Indeed, the application of WiFi offloading techniques combined with a properly designed energy management strategy allows to save up to 100% of the energy bill in our scenario, when WiFi offloading is applied jointly with RE powering. Furthermore, the introduction of WiFi offloading is effective in increasing the capability of providing ancillary services, in particular by raising the probability of satisfying the SG requests of decreasing the grid consumption by up to almost 75% in the scenario investigated in our study. Finally, under the application of WiFi offloading, an increase in the RE generation system size contributes to significantly increase the network performance in terms of both raising the probability of providing ancillary services and reducing operational costs, even achieving positive revenues.

2 Mobile networks in a Demand Response framework

The increasing penetration of distributed renewable energy generators in the SG, integrated alongside traditional synchronous centralized generators, has been responsible of the huge diffusion of the Demand Response paradigm in recent years. On the one hand, the presence of several additional distributed power generators may more frequently lead to overrun the absorptive capacity of the traditional power grid, causing system instability or failure [13]. On the other hand, the erratic nature of renewable energy production increases the risk of requiring the activation of additional high inertia generators, in case of peak power demand and low renewable energy production, and this is neither desirable nor effective in providing the needed energy supply in a short time. In particular, *Price Based Programs* are often adopted as effective methods to flatten the pattern of energy demand, by adjusting the energy prices over time, depending on the varying energy load. Users are incentivized to shift their demand from high peak to low peak periods, by receiving a monetary reward or penalty, depending on whether they are able or not to accomplish the request from the SG of increasing or decreasing their consumption [13]. By satisfying the SG requests, users therefore contribute to provide ancillary services.

The mobile operator (MO) may play a relevant role in the DR framework, by making the mobile access network actively interact with the SG and dynamically modulate its energy consumption in accordance with SGO requests. The DR policy adopted in our work envisions that every day the MO forecasts its energy consumption for the following day depending on the expected traffic, which is estimated based on historical data. The information is hence provided to the SGO, with a granularity of half an hour. During the next day, every half an hour the SG announces to the end users its current request, that may be of type *increase* (U), *decrease* (D) or *nothing-to-do* (N). In case of U request, the SG is asking its users to increase the energy consumption from the grid with respect to the forecast energy; in case of D, the request is for decreasing the consumption with respect to the reference amount that had been foreseen the day ahead; in case of N, no specific behavior is required. Whenever the MO is able to satisfy the SG request of U or D, it receives a reward, proportional to the increase or reduction in the consumption, respectively; in case it does not accomplish and even opposes the request, a penalty is computed, proportional to the

Table 1 Energy prices, rewards and penalties in (€/MWh)

Price, reward or penalty	p_G	r_U	r_D	p_U	p_D
Value (€/MWh)	37	24	60	12	30

gap with respect to the expected demand. Table 1 reports the values of energy prices, rewards and penalties adopted in this study [14]. p_G is the cost for unit of energy bought from the grid. r_U and r_D are the rewards per unit of energy traded to provide ancillary services in case of U and D, respectively. p_U and p_D are the penalties paid for each unit of energy opposing the SG request with respect to the predicted consumption, in case of U and D, respectively. Let us denote with C_f the forecast network consumption in a given timeslot, and with C_a the actual consumption in the same timeslot. When a request of type U is issued, if $C_a > C_f$ a reward is provided to the Mobile Operator, corresponding to $r_U \cdot (C_a - C_f)$. Conversely, if $C_a < C_f$ a penalty is due, resulting in an additional cost of $p_U \cdot |C_a - C_f|$. Similarly, when a request of type D occurs, a reward is provided to the MO if $C_a < C_f$, corresponding to $r_D \cdot |C_a - C_f|$, whereas if $C_a > C_f$ an additional cost of $p_D \cdot (C_a - C_f)$ must be paid by the MO.

In this work, we investigate the potential of improving the interaction with the Smart Grid of a mobile access network, by designing energy management strategies aiming to better react to the requests from the SG. These strategies improve the interaction with the Smart Grid by exploiting RE sources to power the BSs and the implementation of RoD and WO techniques, with the purpose of reducing the electric cost for the MO and enhancing the ancillary service provisioning. The considered scenario is depicted in Fig. 1. It consists of a green mobile access network where a macro BS and multiple micro BSs provide coverage over a given area. As it will be further detailed later on, the cluster of BSs can be powered by the traditional electric grid, by the renewable energy that is locally produced by PV panels, and by the energy previously harvested in some local storage. A central controller takes care of energy management and it is in charge of applying the RoD and WO techniques to the network system.

3 Mobile network response to the SG

To respond to the SG requests, the mobile operator uses WiFi Offloading. Currently, WiFi APs are increasingly deployed by more and more operators and private users, making the WiFi technology become available almost everywhere. Therefore, WO looks a very promising solution to increase the capacity of cellular networks [9]. In our case, WO is exploited to increase the chance of satisfying the requests from the grid of decreasing the consumption, by reducing the traffic load right when the

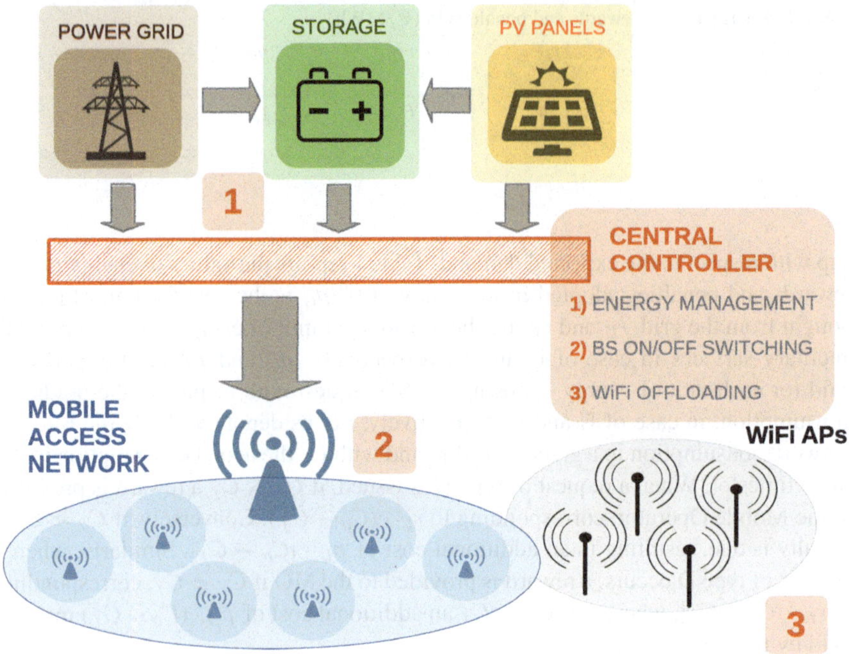

Fig. 1 The green mobile access network scenario

reward for providing ancillary service is higher. This technique can also be combined with RoD to further increase its effectiveness. Indeed, the impact of WiFi offloading in improving the interaction with the SG is evaluated in two different scenarios. In the first scenario, all the BSs are always kept active, even during off-peak periods. In the latter, a RoD strategy is applied in order to adapt over time the network power consumption to the actual traffic demand when the SG requires to reduce the energy amount drawn from the grid, hence saving energy and receiving rewards. The proposed strategy exploits the possibility of putting the BSs into deep sleep mode, in which negligible consumption can be assumed. We consider a cluster composed of one macro BS and a few micro BS that can increase the capacity of the cluster. When a request of type D is issued, the RoD scheme operates by switching off one or more micro BSs, depending on the current traffic that has to be handled by each of them. If the traffic load decreases below a given threshold, denoted as ρ_{min}, the micro BS can be switched off and its traffic moved to the macro BS, as long as there is still enough capacity on the macro BS to handle the additional load [5].

4 RE to improve the interaction with the SG

The application of WiFi offloading techniques in mobile access networks can help to better answer to the requests form the SG, in particular in those periods in which the SG asks for a reduction of the consumption. However, their impact may be constrained on the one hand by the minimum capacity to be guaranteed even during off-peak periods for an acceptable QoS and, on the other hand, by the maximum amount of traffic load that can be offloaded to the WiFi APs. Furthermore, this approach does not provide huge margine to satisfy the requests to increase consumption from the grid, since the cluster load cannot be forced to grow.

The introduction of photovoltaic (PV) panels to power the BSs cluster makes it more energy self-sufficient and further increases the probability of satisfying the SG when a reduction of the consumption is required. In addition, batteries are needed to address the typical intermittent and unpredictable nature of renewable energy production, since extra amounts of produced RE can be harvested into storage for future usage, when the RE is not available. At the same time, storage is useful not only for powering the BSs when the request from the SG is to decrease the consumption, but also for storing some extra energy drawn from the grid when the request is to increase consumption.

5 Energy management policy

The energy management strategy applied to respond to the SG requests is now presented. A general description is provided assuming the presence of a RE generator, with a set of PV panels whose capacity is S_{PV} and a set of batteries whose size is expressed in terms of the number of storage units, denoted as B. The same operation holds in case no RE generator is envisioned. In this case, S_{PV} and B are simply equal to 0. C_f denotes the network consumption for each half an hour forecast the day ahead, while C_a is the actual network consumption. In our simulations we assume that the predicted consumption C_f is computed assuming that no RoD strategy and no WiFi offloading are applied and all the energy required for the network operation is drawn from the grid. Every half an hour, different actions can be taken, based on the type of grid request and the amount of produced RE. The BSs need C_a may be derived from the produced RE, from the storage (always respecting the maximum discharging rate) or from the grid. RE may be used to power BSs, harvested into battery or wasted in some cases. The RE management is based on a *first use—then harvest* principle, meaning that the produced RE can be directly used to power the BSs and only the extra amounts are possibly harvested afterwards. In order to better answer U requests, an additional amount of energy, denoted as E_b, can be drawn from the grid and harvested into battery for future usage. E_b is the maximum amount that can be injected into the battery in half an hour, constrained by the maximum charging rate and the maximum available battery capacity, denoted as C_B. The RE produced every half an hour is denoted as R_p, while E_g represents the energy drawn from the grid for each half an hour. E_r denotes the residual energy need of the access network after R_p has been used to power the BSs. Every half an hour, energy management decisions are taken according to Algorithm 1. Note that, when the WO is active and a D request occurs, part of the traffic load is transferred from some micro BSs to neighboring APs. Afterwards, in case RoD is active and if a D request is issued by the SG, the energy management system checks the load: if it is sufficiently low and there is enough capacity in the macro BS, some of the micro BSs can be switched off and their traffic is moved to the macro BS, as long as there is still enough residual capacity. Afterwards, C_a is updated accordingly. WO and RoD are hence exploited to timely respond to the SG requests when needed, i.e. when the SG asks to decrease the consumption from the grid. The behavior of this energy management algorithm in terms of RoD application is different with respect to the algorithm proposed in [15]. In that paper, when the RoD strategy is active, it was always applied, if possible, regardless the type of SG requests, with the main purpose of decreasing the overall energy consumption. Hence, when the traffic is low, some micro BSs are turned off even in case of U or N requests. Hence, in case of RoD strategy application, the network demand forecast the day ahead, C_f, was computed assuming the application of RoD.

In case of U request, a reward is obtained if the total energy taken from the grid, E_g, is larger than the predicted consumption C_f. The reward is proportional to the extra amount of energy taken from the grid. By converse, in case it is lower, a penalty is paid proportional to the energy gap with respect to the reference level C_f predicted the day ahead. Similarly, in case of D, a reward is obtained if $E_g < C_f$, and vice versa for the penalty.

6 Simulating the mobile network-Smart Grid interaction

As depicted in Fig. 1, we consider a portion of a mobile access network is studied, consisting of a macro BS, providing coverage over a given area, and 6 micro BSs, providing additional capacity for peak periods. Real traffic traces from a residential area, provided by an Italian mobile operator [5], are used to simulate the traffic variation over time in the system. The macro and micro BS consumptions are derived by applying the EARTH model [16], which takes into account the partial load proportionality. The BS cluster draws its energy demand from the traditional electric grid, but it may also be powered by RE. The BS cluster interacts with the SG and answers to its requests according to the energy management policy described above. Unlike our previous work [15], where SG requests (U, D, N) were randomly generated according to a uniform distribution, so that each type of request had the same occurrence probability, in our simulations realistic traces of SG requests are considered. These traces are obtained by considering real energy prices, derived from the Australian energy market [17], that dynamically vary over time, either within a single day, from day to day or over seasons. The pricing data from the Australian

Algorithm 1 Energy management algorithm

1: $E_r = 0$; $E_g = 0$;
2: **switch** *grid request* **do**
3:   **case** U:
4:     **if** $(R_p < C_a)$ **then**
5:       power BSs from the grid;
6:       harvest extra energy, E_b, from grid into the battery;
7:       waste R_p;
8:       $E_g = C_a + E_b$;
9:     **else**
10:       use R_p to power BSs;
11:       store $(R_p - C_a)$ into the battery, waste the extra amount;
12:     **end if**
13:     **if** $(E_g > C_f)$ **then**: a reward is received;
14:     **else if** $(E_g < C_f)$ **then**: a penalty is paid;
15:     **end if**
16:   **end case**
17:   **case** D:
18:     **if** WO is active **then**: offload some traffic;
19:     **end if**
20:     **if** RoD is active **then**: switch off some micro BSs;
21:     **end if**
22:     use R_p to power BSs;
23:     compute residual need $E_r = max(0, C_a - R_p)$;
24:     **if** $(E_r = 0)$ **then**
25:       store $(R_p - C_a)$ into the battery, waste the extra amount;
26:     **else**
27:       draw E_r from the battery and, if needed, from grid;
28:       $E_g = max(0, E_r - Battery\ Level)$;
29:     **end if**
30:     **if** $(E_g > C_f)$ **then**: a penalty is paid;
31:     **else if** $(E_g < C_f)$ **then**: a reward is received;
32:     **end if**
33:   **end case**
34:   **case** N:
35:     code from line 21 to 28;
36:   **end case**
37: **end switch**

energy market have been considered instead of the Italian energy prices in order to study a scenario where the SG request dynamics are highly affected by the extensive penetration of RE sources. The energy prices adopted for this study are those registered in the Australian State of New South Wales during the year 2016. These prices are set by the SG operator and they are provided with a timestep granularity of 30 minutes. Their values are mapped into three price ranges, corresponding to the three types of SG requests. Low range values are mapped as requests of type U, whereas high prices are mapped as requests of type D. Intermediate prices correspond to N requests. Thresholds to define the three ranges are set so that the U and D requests are equiprobable (40% each), whereas N represents 20% of the total SG istances.

By mapping the actual sequence of energy prices into SG requests, a more realistic performance analysis can hence be conducted with respect to [15]. The energy prices defined by the Smart Grid Operator may be highly affected, among several factors, by the dynamic variations of the local RE production. Hence, the study should consider the typical solar radiation profiles observed in a city located in this state. In particular, the city of Sydney has been selected to derive, by means of PV Watts, real profiles of solar energy production, as it will be described later on.

Neighbor WiFi Access Points (APs) may be exploited to transfer and handle some of the traffic from the micro BSs, in order to reduce the load of the mobile access network and better answering the requests from the SG. Typically, up to 20% of the maximum mobile traffic load can be trasferred to neighbor APs by means of WO [18]. Two different scenarios are considered. In the first one, the macro and micro BSs are always active, providing full capacity even during off-peak periods. In the latter, the RoD strategy described in Sect. 2 is introduced to accomplish requests of type D, by adapting the consumption to the actual traffic load; the threshold ρ_{min} has been set equal to 0.37, since this value optimizes the energy saving [5]. A centralized controller is in charge of energy management (which is performed according to the proposed policy), decisions about when each micro BS should be switched off (based on the RoD strategy) and traffic transferred to the APs, and about when the WiFi offloading approach is applied.

The system performance in both scenarios has been tested by simulation over an entire year in baseline conditions, i.e., without any RE generator. The performance has been evaluated in terms of yearly energy consumption, probability of providing ancillary services, and costs, including cost for energy, rewards for providing ancillary services and penalties for opposing the SG requests. In addition, the WiFi offloading has been applied in both scenarios to study its impact. When the WiFi offloading is introduced, the energy management policy envisions that, in case of a D request, some of the traffic can be moved to neighboring APs, hence reducing the overall load and allowing, in the RoD scenario, the switching off of additional BSs. We assume that the maximum capacity available for WiFi offloading corresponds to 10% of the peak traffic load handled by the 6 micro BSs. This allows to achieve an average offloading efficiency approximately between 10 and 20%, depending on the specific case, envisioning a scenario in which the number of available neighboring APs is limited [18–20].

Furthermore, a RE generator and a storage system are considered to power the cluster of BSs, in order to investigate the effect of local solar energy production and the presence of batteries on the interaction with the SG. The RE generator consists in a set of various PV panels, built up by standard modules of crystalline Silicone, whose efficiency is 15% [21]. The nominal capacity of the modules, denoted as S_{PV}, is the maximum DC output power that can be produced under standardized environmental conditions from the sun radiation and it is measured in peak W (Wp). The range of S_{PV} considered in this work is [5, 25] kWp. Real data about RE production are derived on a half an hour basis from the tool PVWatts [21] in the city of Sydney, over the Typical Meteorological Year. In [15], the RE powered network interacting with the SG was investigated in the city of Turin, in Italy. Here, since the energy prices available are derived from the Australian Energy Market Operator (AEMO) [17], the city of Sydney has been selected as location to conduct the performance analysis. Both locations are placed in comparable latitude bands (45.07°N for Turin, 33.87°S for Sydney) in terms of solar radiation, and the seasonal patterns of RE production follow a quite similar average behavior. A DC-to-AC inverter efficiency of 96% and typical performance losses of 14% observed during actual operation are assumed for the PV system.

The battery set is composed by 5–25 lead-acid batteries, each with capacity 200 Ah and voltage 12 V, which represent a common storage technology adopted in RE system [22]. The number of batteries is denoted as B. The total nominal capacity of the entire battery bank may sum up to a value between 12 and 60 kWh, depending on the value of B. A maximum Depth of Discharge (DoD) of 70% is assumed, in order to prevent charging efficiency impairment [23], battery aging [24] and capacity loss [22]. A lifetime duration of 500–600 cycles can be assumed under DOD = 70% [25]. Therefore, assuming 5–25 battery units, the maximum storage capacity that is actually available is limited by the DOD, resulting between 8, 4 kWh ($B = 5$) and 42 kWh ($B = 25$). Charging and discharging losses, amounting to 25%, are considered as well in our work [26]. Finally, high charging and discharging rates may reduce battery lifetime and impair its efficiency, especially in case of frequent charging and discharging events occurring in RE systems. Constraints on the maximum charging and discharging rate are therefore taken into account, by assuming a conservative value for the maximum charge and discharge current of C/10 A, where C is the numerical value of the battery capacity in Ah [27]. This means that, for example, the energy drawn in an from a battery of 200 Ah will be at most 240 Wh, from which efficiency losses must then be curtailed. The same holds in case of battery charging.

7 Performance analysis

The system performance has been investigated in terms of energy yearly taken from the grid, probability of providing ancillary services, effect of system dimensioning and cost analysis, considering the possible application of WO, either alone or combined with RoD, and examining the impact of the introduction of a PV system. The results derived in our work are extensively detailed hereafter. Table 2 summarizes the main performance results, along with the experimental settings under which our simulations have been conducted.

7.1 Energy demand from the grid

Figure 2 reports the amount of energy taken from the grid over one year, E_G, for the cases without RE production (orange bars) and under RE supply (green bars), under various possible configuration settings:

Table 2 Performance results

Performance metrics	RE generation	PV panel size (kWp)	Number of battery units	Values of performance metrics			
				No strategy	WO	RoD	WO+RoD
E_G (MWh)	none	–	–	12.15	11.47	11.77	11.46
	✓	10	15	7.86	7.53	6.84	6.68
p_U	none	–	–	0	0	0	0
	✓	5	5	0.87	0.87	0.82	0.82
	✓	10	15	0.35	0.34	0.33	0.33
p_D	none	–	–	0.00	0.75	0.37	0.75
	✓	5	5	0.71	0.87	0.78	0.88
	✓	10	15	0.95	0.98	0.97	0.98
E_{U_+} (MWh)	none	–	–	0	0	0	0
	✓	10	15	4.01	3.84	3.59	3.51
E_{D_+} (MWh)	none	–	–	0	0.68	0.38	0.69
	✓	10	15	4.65	4.77	4.74	4.78
Cost saving (%)	none	–	–	0	115.34	201.87	87.61
	✓	5	5	14.57	118.7	214.28	92.63
	✓	10	15	8.12	121.38	217.59	101.11
	✓	10	15	14.91[a]	122.82[a]	222.82[a]	102.92[a]

[a] In this case extra amounts of produced RE are sold back to the SG

1. no strategy is applied (baseline scenario);
2. only WO is applied;
3. only RoD is applied;
4. both WO and RoD strategies are applied.

In the baseline scenario, without RE production, $E_G = 12.15$ MWh; this value is considered as the reference value, E_G*, with respect to which the energy saving is computed. Under RoD, 11.77 MWh are consumed and drawn from the grid, with a 3.1% saving with respect to E_G*. The application of WO allows to reduce E_G by 5.6% without RoD and by 5.7% under RoD, again with respect to E_G*. Interestingly, the impact of WO in reducing the energy taken from the grid is higher than RoD. Furthermore, the additional reduction in E_G provided by RoD when WO is running is almost negligible. Theoretically, under RoD strategy higher energy savings would be expected, since the energy saved by completely switching off a micro BS is higher than the energy amount that can be saved by transferring the traffic load to neighboring APs. Nevertheless, in our case it is more advantageous to apply WO rather than RoD, since at every timeslot in which a D request is issued, it is more likely that the network is able to offload traffic rather than to switch off micro BSs. This is due to the fact that whereas both WO and RoD can be applied only in case of D requests, two additional constraints further limit the application of RoD: a micro BS can be switched off only if its traffic load is below a given threshold and if the

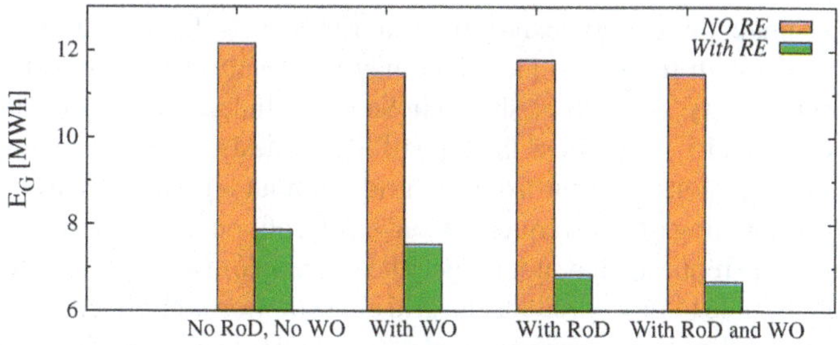

Fig. 2 Grid demand, E_G, without and under RE production

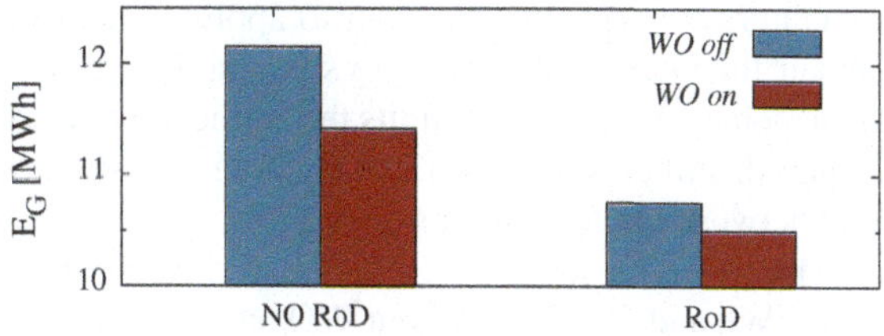

Fig. 3 Grid demand, E_G, without RE production, under the energy management strategy proposed in [15]

macro BS still has enough residual capacity to carry the traffic moved from the micro BS.

This trend is different with respect to results obtained in [15] under the same configurations, that are shown in Fig. 3. This figure reports E_G in two scenarios, without and under RoD, comparing the case without WiFi offloading (blue bars) and the case in which WiFi offloading is applied (red bars). The application of RoD shows a larger impact than WiFi offloading only, reducing E_G by 11.5%, almost twice the 6% reduction obtained under WO. In addition, by applying RoD on the top of WO a more significant reduction of E_G is achieved with respect to results reported in Fig. 2, with a saving more than doubled if compared against the WO-only policy (from 6% to 13.6%). These differences are due to a twofold reason. First, the energy management algorithm proposed in [15] is slightly different with respect to that adopted in this work. In [15], when the RoD strategy is active, it is run at any time, disregarding the type of SG requests. Only WO is applied in case of a request of type D is issued. Conversely, according to the energy management algorithm proposed in this study, when either WO or RoD strategies are active, they both operate just in case of request of type D. Second, as already mentioned, the impact of RoD in reducing the energy drawn from the grid is further limited by additional constraints with respect to the case of WO. Indeed, under WO some of the traffic can always be transferred to neighboring APs without any constraint. Conversely, under RoD strategy, micro BSs can only be switched off if their traffic load is below the threshold ρ_{min} and as long as the macro BS has still enough capacity to carry the additional traffic moved from the micro BSs. Hence, RoD may not be applied at any timeslot in which a request of type D is issued.

Although under this energy management alorithm the RoD shows a smaller impact in reducing E_G with respect to [15], this may not result in relevant differences in terms of energy costs, as it will be shown in Sect. 7.3. Indeed, the application of RoD at any timeslot in [15], regardless the type of SG requests, may contribute to reduce the overall energy consumption from the grid, without significantly improving the capability of matching the SG requests. Conversely, although the application of RoD only in case of D requests may not reduce the overall E_G to the same extent, it may nevertheless be able to timely reduce the grid consumption when needed, i.e. when D requests are issued and rewards are provided for accomplishing the SG requests, thus achieving a better demand-supply match and still providing significant cost saving, without impairing the bandwidth availability if not strictly needed. Note that Mobile Operators are typically reluctant to apply on their networks any strategy that may impair the Quality of Service to some (although limited) extent. Hence, an energy management strategy that limits the application of RoD to periods when it is strictly required, still capable to effectively reduce cost, may represent a good tradeoff for real network implementations.

The case with RE production is also shown in Fig. 2 (green bars), assuming S_{PV} = 10 kWp and B = 15. Without WO and RoD running, a reduction of 35.2% in E_G can be achieved. The introduction of WO allows to reduce E_G by 38%, whereas RoD, unlike the case when no local RE is produced, results to be more effective than WO in decreasing the energy drawn from the grid, showing a trend coparable to that observed in [15]. Indeed, the provided E_G reduction is 43.7% for RoD alone and 45% when RoD is applied on the top of WO. This means that, when a local RE generator is installed, in order to further reduce E_G, it is more convenient to apply a RoD strategy rather than WO alone, although this may not necessarily reflect a relevant difference in terms of cost reduction, as it will be shown.

7.2 Providing Ancillary Services

Here we investigate the capability of WO to enhance the interaction of the mobile network with the SG and improve the ancillary service provisioning, either alone or when combined with RoD. Furthermore, we analyse the effect of local renewable energy production on the system performance. Finally, some results are presented to show the impact of the RE system dimensioning on the capability of the cellular network to accomplish the requests from the SG under various configuration setting. The capability of providing ancillary services is evaluated both in terms of probability of responding to U and D requests from the grid and in terms of the amounts of energy traded with the SG to provide ancillary services, for which proportional rewards are envisioned.

Probabilities of providing ancillary services are defined as follows:

- P_{U_+}: probability of satisfying a grid request of type U (the SG asks to increase the consumption from the grid and the network is able to do that);

- P_{D_+}: probability of satisfying a grid request of type D (the SG asks to decrease the consumption from the grid and the network is able to do that).

The energy amounts traded with the SG are defined as follows:

- E_{U_+} represents the portion of the yearly cumulative amount of energy drawn from the grid that exceeds the forecast consumption, in case of U requests;
- E_{D_+} corresponds to the absolute value of the yearly cumulative amount of energy that represents the negative energy gap with respect to the predicted consumption from the grid in case of D requests.

7.2.1 Role of WiFi Offloading

Fig. 5 shows the results in case no RE is locally produced. U requests can never be satisfied as it can be seen from $P_{U_+} = 0$ under any scenario (WO on or off, without and under RoD). This is due to two reasons: first, the traffic patterns are assumed to be exactly known in advance so that the forecast traffic load results perfectly accurate; second, the absence of any energy storage does not allow to draw and harvest extra energy from the grid to answer its U request. WO allows to achieve a P_{D+} of 74.7%, twice the value provided in case of RoD. applied, due to the same assumption of exact predictability of traffic. Indeed, in that case, RoD is always applied under any type of SG request and the grid consumption is forecast assuming the application of RoD, hence no reduction can be achieved in E_G by simply applying RoD and without RE production. Conversely, in this study, the application of RoD just in case of D requests allows to achieve a $P_{D+} = 37.2\%$ (against $P_{D+} = 0$ in [15]). This is a relevant result. Note that the reward for providing ancillary services in case of D is 2.5 times higher than for answering U requests. Furthermore, combining WO and RoD is not useful at all to increase the capability of providing ancillary services, since P_{D_+} is exactly the same as the case of WO only application. This behavior is consistent with results obtained in [15] without the use of RE. In that case, not only the combined application of RoD on the top of WO is not helpful, but it even reduces P_{D_+} (hence the probability of obtaining the corresponding rewards) by more than 30%. The probabilities of not answering U or D requests clearly result complementary to P_{U_+} and P_{D_+}. However, no penalties must be paid under any scenario, since no opposition to the SG requests is ever registered (Fig. 4).

7.2.2 Effect of local renewable energy production and system dimensioning

In Fig. 6 the probabilities of providing ancillary services are reported for the case in which RE is produced, comparing two different system dimensioning. Fig. 6a shows the values of P_{U_+} and P_{D_+} under different scenarios, in the case with an intermediate system syzing, i.e. a PV panel with capacity 5 kWp and a storage consisting of 5 battery units. The probability of satisfying either U or D requests is always above 70%

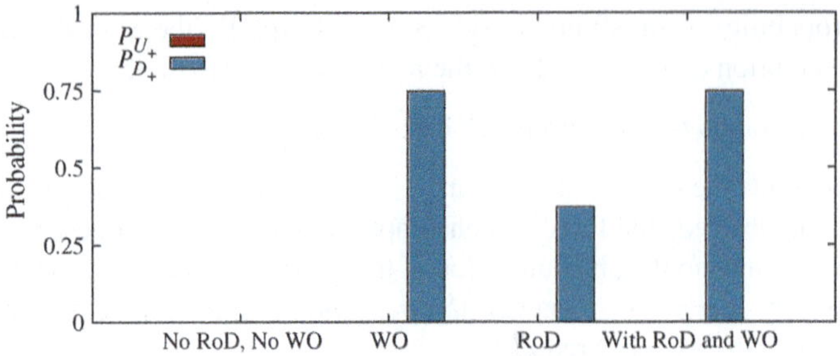

Fig. 4 Probability of providing ASs under no RE production

under any strategy configuration. WO allows to increase P_{D_+} by up to 22.9%, at the price of a negligible reduction in P_{D_+}. RoD strategy provides a smaller contribution in increasing P_{D_+}, of up to 13.9% at most, at the price of lower values of P_{U_+}. Fig. 6b depicts the case with a larger RE system sizing, i.e. $S_{PV} = 10\,kWp$ and $B = 15$. Whereas P_{U_+} results to be more than halved, the probability of answering D requests, P_{D_+}, is almost equal to 1 ($P_{D_+} = 0.95$–0.98). Unlike the case with intermediate sizes for PV panels and storage, under a larger RE system dimension no remarkable differences can be observed between the various combination of strategies, although WO can provide slightly higher increase in P_{D_+}. This means that the effect of higher RE production is dominant in this case. P_{U_+} is decreased when using a larger system sizing, due to a higher average level of battery charge determined by the higher levels of RE generation. Although this leads to a lower probabilility of receiving rewards for satisfying UP requests, a higher average level of battery allows to better answer D requests. Hence, P_{D_+} results significantly increased, leading to higher probability of receiving a reward that is twice the reward envisioned in case of UP, thus providing remarkable gains.

In order to better show the impact of RE system dimensioning, Figs. 7 and 8 report the probabilities of providing ancillary services for increasing values of S_{PV} and B, respectively, in the case in which both WO and RoD are applied.

According to Fig. 7, showing P_{U_+} and P_{D_+} for increasing panel size, the probability of accomplishing D requests is not significantly affected by the PV panel dimension. Very high values of P_{D_+} (> 0.93) are guaranteed even under the smallest PV panels, whereas only slight increases in P_{D_+} can be observed as S_{PV} grows larger. Conversely, the impact of the PV panel size on the probability of providing ancillary services in case of U requests results more relevant, especially for small values of PV panel capacity. Indeed, from $P_{U_+} = 0.52$ for 1 kWp PV panel capacity, it gradually decreases as S_{PV} becomes higher. However, after P_{U_+} has been reduced by 40% at $S_{PV} = 10$ kWp, no significant additional reduction can be detected for further increasing the PV panel size.

Figure 8 highlights a similar behavior for the impact of the battery size on the capability of providing ancillary services. Indeed, by increasing the number of battery units, the value of P_{D_+}, which is as high as 0.92 under the smallest storage size,

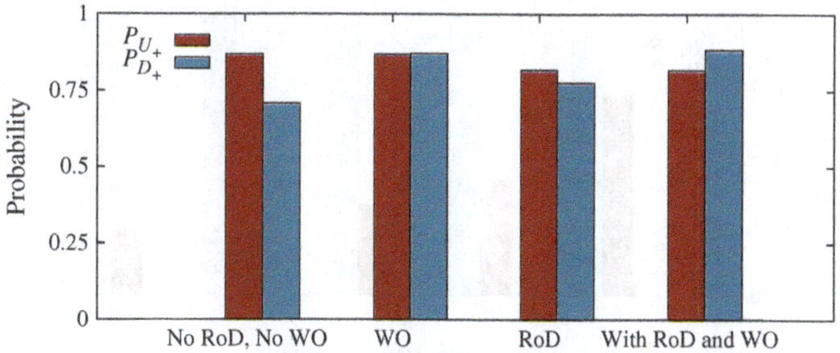

(a) $S_{PV} = 5\ kWp, B = 5$

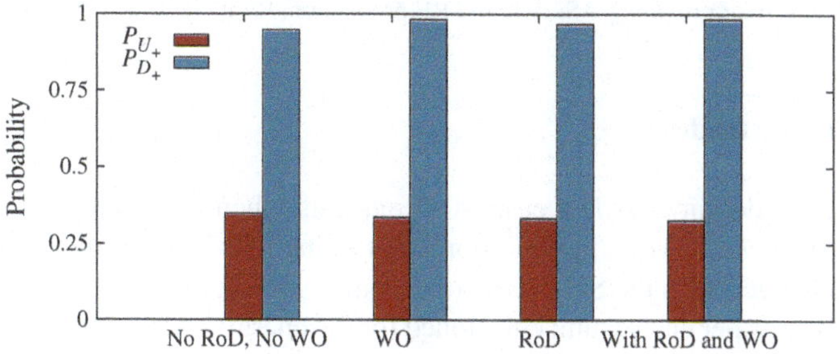

(b) $S_{PV} = 10\ kWp, B = 15$

Fig. 5 Probability of providing ASs, comparing cases under different sizes of RE generation system

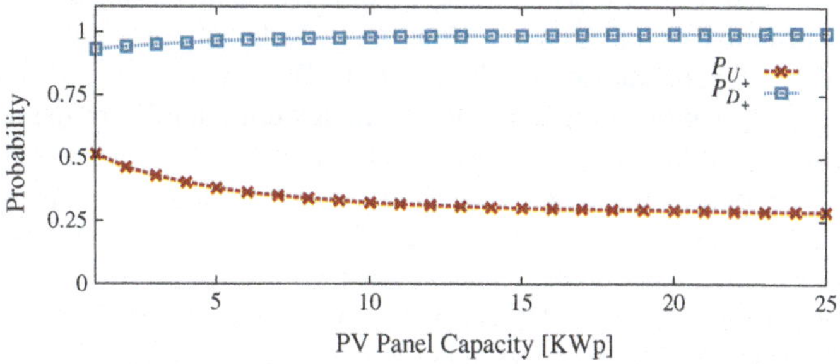

Fig. 6 Probability of providing ASs, without RE production, for increasing PV panel size

slightly raise to values close to 1 for the largest battery set. On the contrary, the increase in the battery dimension remarkably impairs the capability of accomplishing U requests. A sharp decrease of P_{U_+}, by 44%, is observed moving from 5 to 10 battery units, whereas the P_{U_+} reduction becomes less relevant for further growing of the storage dimension. This behavior sounds rather counterintuitive, since a better response to requests of type U would be expected using a larger battery, capable of storing more energy drawn from the grid in case of UP requests.

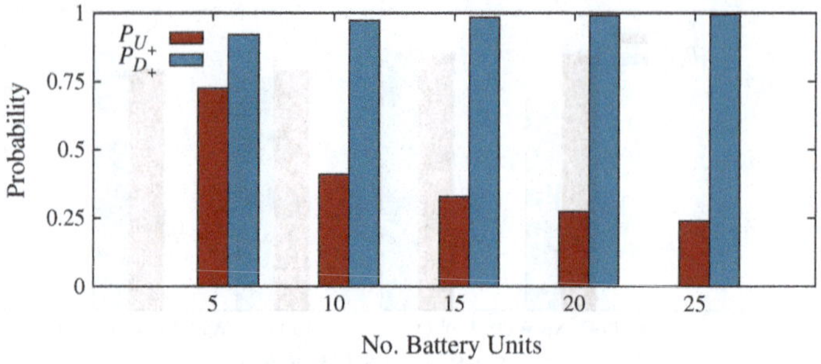

Fig. 7 Probability of providing ASs, without RE production, for increasing battery size

7.2.3 Energy trades

The energy trades involved in case of D requests when they can be satisfied are represented by E_{D_+}. They provide dominant gains with respect to the amount of energy exchanged with the SG in case of accomplishment of U requests, i.e. E_{U_+}. In fact, the rewards per energy unit envisioned for the energy trades represented by E_{D_+} are 2.5 fold higher than those obtained for E_{U_+}. Our investigation hence focuses on the analysis of E_{D_+} in various scenarios. The values of E_{D_+} are shown in Fig. 9 in the cases without local RE generation and under RE production. When no local RE supply is provided (Fig. 9a), WO results very effective in increasing the level of E_{D_+} with respect to the baseline scenario. Conversely, RoD only provides slightly more than 50% the E_{D_+} gain that can be achieved under WO application. Furthermore, the joint application of WO on the top of RoD strategy does not determine remarkable gain in terms of further E_{D_+} increase. This is consistent with the trend already observed in the probability of accomplishing DOWN requests, P_{D_+}, as reported in Fig. 5. In this case, since P_{U_+} is always null, the values of E_{U_+} are equal to zero in all cases (see Table 2), due to the reasons explained in Sect. 7.2.1.

Figure 9b shows E_{D_+} when a RE generator and a storage are present, with capacity $S_{PV} = 10$ kWp and B=15, respectively. In this case, the raise in E_{D_+} can be as high as 7 fold the maximum increase that can be obtained without the presence of local RE supply. Nevertheless, no substantial difference in the levels of E_{D_+} can be observed under the application of WO and/or RoD strategy, confirming that the effect of RE generation on the E_{D_+} gain prevails over the application of WO rather than RoD strategies. Under RE production, the values of E_{U_+} are generally lower than E_{D_+}, resulting up to about 4 MWh at most (see Table 2).

7.3 Cost analysis

The impact of WO on the mobile network operational cost is now evaluated, in order to investigate to what extent operational expenditures (OPEX) can be reduced

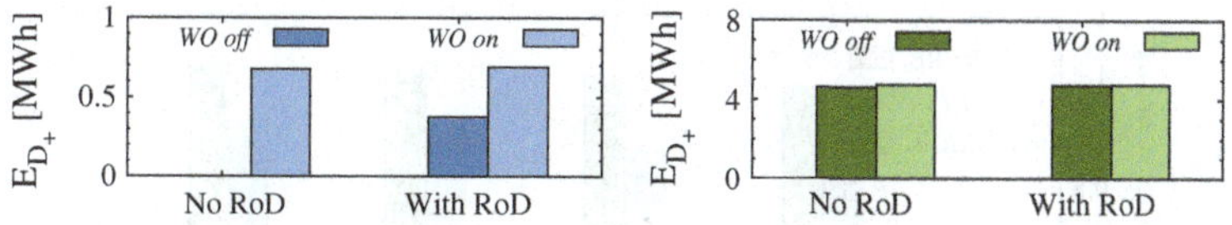

(a) No RE (b) Under RE production

Fig. 8 Energy for providing ancillary services (E_{U_+}, E_{D_+}), without and under RE production

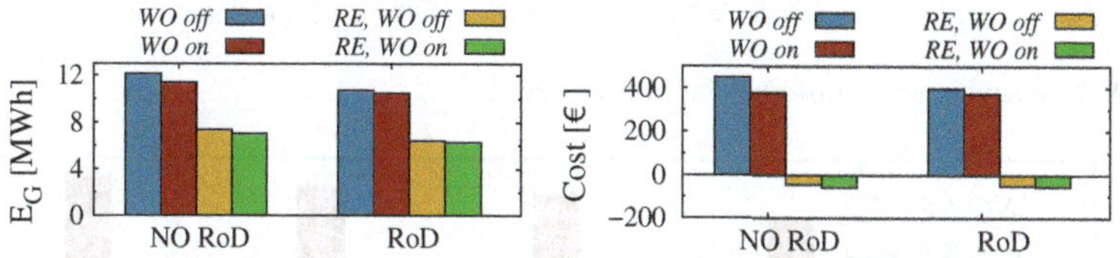

(a) E_G (b) OPEX

Fig. 9 Energy drawn from the grid, E_G, and OPEX cost when no RE is produced and under RE production ($S_{PV} = 10\ kW_p$, $B = 10$)

by applying this strategy, either alone or in combination with RoD, and in case of the presence of a local RE generator. It is worth to be noted that, in a Demand Response context, the cost reduction does not depend strictly on the overall decrease in E_G, that is the amount of energy drawn from the grid over a year, but rather on the capability of raising and decreasing the consumption from the grid at the right moment. Indeed, a good energy management strategy aims at dynamically adjust the energy demand from the grid in order to accomplish the current requests from the grid at each timestep. Therefore, remarkable cost savings may be achieved even in case the total E_G is not significantly decreased, since it is more important to timely increase or decrease the grid consumption when required rather that reducing the overall grid consumption without properly responding to the SG requests, hence missing monetary rewards and receiving penalties to be paid. This can be evinced from Fig. 10, that reports the values of E_G and the operational cost per year. These preliminary results were derived according to the algorhitm proposed in [15]. They refer to the city of Turin and the grid requests are assumed to be issued according to a uniform probability distribution, like in [15]. Costs have been investigated in the cases without WO or with active WO, assuming that RoD is either active or deactivated, and considering the scenarios with or without a RE generator ($S_{PV} = 10$ kWp, $B = 10$). Fig. 10a shows the energy bought from the grid, E_G, in the various cases, while Fig. 10b reports the operational costs (including energy cost, rewards and penalties) for the corresponding cases. In the baseline scenario, when no RE is produced and no strategy is applied, $E_G = E_G* = 12.15$ MWh and the corresponding reference cost, denoted as c_*, is 450 €. WO alone allows to reduce c by about 16% with respect to c_* and by about 17% when used in a RoD scenario, against a maximum reduction in E_G of 13.6%. Furthermore, when the RE production is introduced, the impact on cost is even more remarkable, with a cost reduction by up to 113% when

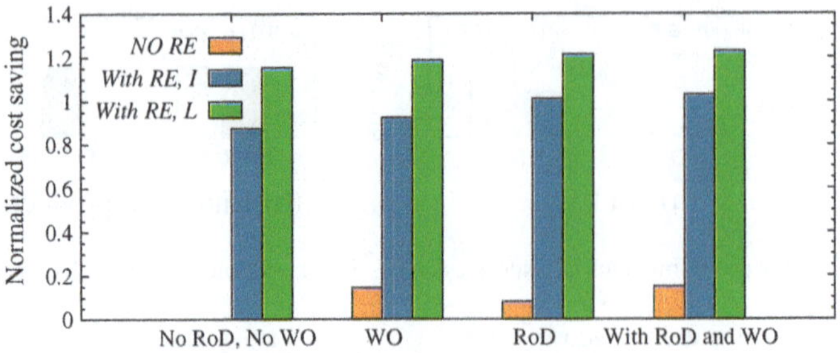

Fig. 10 Cost saving under different RE system size

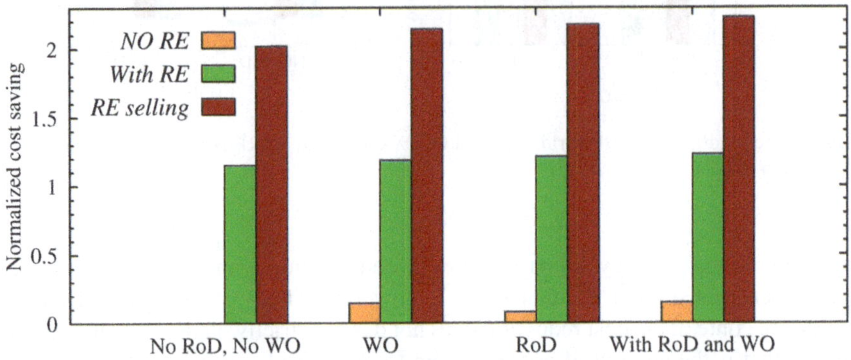

Fig. 11 Cost saving under three scenarios: (1) NO RE—no RE is produced; (2) With RE—under RE production ($S_{PV} = 10\ kW_p$, $B = 15$); (3) RE selling—assuming that the extra amounts of produced RE are sold back to the grid

WO is running. This means that, despite a reduction in E_G by at most 48.3% when RE is locally produced, the operational cost is not only nullified but even becomes negative, thanks to the joint action of the WO strategy and the RE generation, along with storage, in improving the interaction with the SG, providing huge rewards that compensate for energy cost and penalties.

Cost savings have been further investigated in the RE powered mobile access network described in Sect. 6, assuming the city of Sydney as a location, realistic patterns for the SG request occurrences and the energy management algorithm proposed in Sect. 5. Fig. 11 reports the cost savings obtained when no RE is locally produced and in two cases in which a RE generator is present, either with intermediate (denoted as I: $S_{PV} = 5$ kWp, $B = 5$) or large (denoted as L: $S_{PV} = 10$ kWp, $B = 15$) size, and under the possible application of WO and RoD. When no RE is available (orange bars), the application of WO, either alone or running with RoD, allows to achieve up to 14.9% cost saving, almost double with respect to those obtained under RoD strategy only. Indeed, the RoD strategy is less effective, with only about 8.1% cost reduction. The additional cost reduction provided by RoD when applied jointly with WO is quite limited, resulting only 0.3% points higher with respect to the case in which WO is applied alone. When some local RE supply is available, the cost savings are significantly higher (between 7 and 15 times the value of c_*), allowing to completely nullify the energy bill even with intermediate RE system sizing (I, blue bars). Furthermore, results show that the RE generation system dimensioning has a relevant impact on the possible cost savings. Indeed, the installation of PV panels and storage with larger capacity (L, green bars) allows to obtain savings up to 120% of the energy cost, corresponding to positive revenues provided by the SG to the Mobile Network operator. Finally, in presence of RE, the benefits provided by the application of WO are lower with respect to the application of RoD alone, unlike the case when no RE generator is present. A RE generation system with intermediate size provides 88% cost reduction, and the application of RoD permits to nullify the energy cost. With a large RE system, positive revenues are always guaranteed. Focusing only on positive revenues paid by the SG to the mobile network operator, the impact of RoD is more evident, since the revenues are increased by 27% under WO only, by 40% under RoD alone, and by 53% when WO and RoD are combined.

Clearly, in case of RE production, additional capital expenditures (CAPEX) should be considered, due to the installation of the PV panels and the set of batteries. However, although the initial expenditure might be considerable, based on our results derived for several combinations of S_{PV} and B, it would be possible to select the proper dimensioning allowing to minimize the CAPEX+OPEX, taking into account the PV panel and battery lifetime duration. Furthermore, according to [28], after 8 years of RE powered mobile network operation the break-even point for CAPEX cost can be achieved. In addition, the introduction of a new energy management strategy could be envisioned, allowing to sell back to the SGO extra amounts of RE that are not immediately used and, in case they cannot be harvested in the storage, are usually wasted. The cost saving that could be obtained under the application of a similar energy management strategy are shown in Fig. 12, where they are compared against the case without any RE production and the case with a RE generator with intermediate size, assuming the application of the standard energy management strategy. The price per each energy unit that is sold to the grid is assumed to be half the price due for each energy unit bought by the mobile network operator from the grid, i.e. p_G. Cost savings result remarkably high when extra produced RE can be sold back to the grid, being more than twice the energy bill registered in baseline conditions in case no RE is present and no WO or RoD strategies are applied, c_*. WO alone allows to increase the cost saving by 12 additional percentage point with respect to the case in which no WO is applied. Up to 223% cost saving can be achieved by applying WO in conjunction with RoD. However, in relative terms, WO and RoD are equally effective in raising the cost saving, since they both provide savings that are about 1.8 fold higher those assured with the standard energy management strategy. These revenues may play a relevant role in compensating the higher CAPEX faced for the initial installation of a RE generation system.

Table 3 summarizes the effects on the performance metrics of the various combinations of resource management strategies, either without any local RE generator or under RE production. The main motivations at the basis of the different observed effects in terms of performance are highlighted in the last column.

Table 3 Effects of strategies on performance indicators

RE	RoD	WO	E_G	Cost	p_U	p_D	Motivations
–	–	✓	$\downarrow$	$\downarrow$	$\uparrow\uparrow$	–	a
–	✓	–	$\downarrow$	$\downarrow$	$\uparrow$	–	b
–	✓	✓	$\downarrow$	$\downarrow$	$\uparrow\uparrow$	–	a, b, c
✓	–	–	$\downarrow\downarrow$	$\downarrow\downarrow$	$\uparrow\uparrow$	$\uparrow\uparrow\uparrow$	d
✓	–	✓	$\downarrow\downarrow$	$\downarrow\downarrow$	$\uparrow\uparrow\uparrow$	$\uparrow\uparrow\uparrow$	a, d
✓	✓	–	$\downarrow\downarrow\downarrow$	$\downarrow\downarrow\downarrow$	$\uparrow\uparrow$	$\uparrow\uparrow$	b, d
✓	✓	✓	$\downarrow\downarrow\downarrow$	$\downarrow\downarrow\downarrow$	$\uparrow\uparrow\uparrow$	$\uparrow\uparrow$	a, b, c, d

Motivations a—Reduced traffic in case of D request; b—Reduced number of active BSs (hence E_G) in case of D request; c—Increased probability of switching off additional BSs; d—Improved capability of timely reacting to U and D requests

8 Conclusions

This work investigates the impact of WiFi Offloading, possibly combined with a RoD strategy, on improving the interaction of a mobile access network with the SG in a Demand Response context, considering also the case in which a RE source is locally available to power the BSs.

Results suggest that WiFi offloading can be very effective in enhancing the capability of mobile networks to provide ancillary services to the SG operator. Under WO, up to almost 75% of SG requests of consumption decrease can be accomplished in our scenario even when no RE production is locally available. The introduction of a small local RE generation system combined with some storage remarkably improves the probability of positively reacting to the SG requests, that becomes close to 1 under larger RE generators. Furthermore, WO results beneficial even in reducing operational cost, with cost savings that are twofold those obtained under RoD only. When RE is locally produced, substantially higher cost saving can be obtained, that are affected by the PV panel and battery sizing more than the probability of ancillary service provisioning is. Even with relatively small PV panel and storage dimension, the energy bill can be completely nullified, whereas under larger RE system size positive revenues can be achieved. Finally, a reduction of less than 50% in the energy drawn from the grid may correspond to cost saving higher than 100%, resulting in positive revenues, and confirming that a good energy management strategy does not operate by reducing the total grid consumption, but by timely increasing or decreasing the grid consumption exactly when required by the SG. Additional revenues obtained by selling back to the SG the extra produced RE may contribute to compensate for the CAPEX faced by Mobile Network Operators for the installation of RE generation systems. To this aim, optimization algorithms could be deployed as future work with the purpose of properly dimensioning the RE generation system, in order to minimize both CAPEX and OPEX cost for the MNO.

References

1. Cisco, in *Cisco Visual Networking Index: Global Mobile Data Traffic Forecast Update 2016–2021 White Paper*, Feb 2017
2. H. Al Haj Hassan, L. Nuaymi, A. Pelov, Renewable energy in cellular networks: A survey, in *2013 IEEE Online Conference on Green Communications (GreenCom)*, Oct 2013, pp. 1–7
3. S. Buzzi et al., A survey of energy-efficient techniques for 5G networks and challenges ahead. IEEE J. Sel. Areas Commun. **34**(4), 697–709 (2016)
4. J. Wu et al., Energy-efficient base-stations sleep-mode techniques in green cellular networks: a survey. IEEE Commun. Surv. Tutor. **17**(2), 803–826 (Secondquarter 2015)
5. M. Dalmasso, M. Meo, D. Renga, Radio resource management for improving energy self-sufficiency of green mobile networks. Perform. Eval. Rev. **44**(2), 82–87 (2016)
6. M. Deruyck, Reducing the impact of solar energy shortages on the wireless access network powered by a PV panel system and the power grid, in *IEEE 27th Annual IEEE International Symposium on Personal, Indoor and Mobile Radio Communications—(PIMRC): Mobile and Wireless Networks, Valencia, Spain*, Sept 2016, p. 2016
7. H. Ghazzai et al., Green networking in cellular hetnets: a unified radio resource management framework with base station ON/OFF switching. IEEE Trans. Veh. Technol. **PP**(99), 1–1 (2016)
8. S. Zhou, J. Gong, Z. Niu, Sleep control for base stations powered by heterogeneous energy sources, in *2013 International Conference on ICT Convergence (ICTC)* Oct 2013, pp. 666–670
9. Y. He et al., OnWiFi offloading in heterogeneous networks: various incentives and trade-off strategies, in *IEEE Communications Surveys Tutorials 18.4 (Fourthquarter 2016)*, pp. 2345–2385
10. F. Rebecchi et al., Data offloading techniques in cellular networks: a survey. IEEE Commun. Surv. Tutor. **17**(2), 580–603 (Secondquarter 2015)
11. H.A.H. Hassan et al., Renewable energy usage in the context of energy- efficient mobile network, in *2015 IEEE 81st Vehicular Technology Conference (VTC Spring)*, May 2015, pp. 1–7
12. D. Renga et al., Improving the interaction of a green mobile network with the smart grid, in *2017 IEEE International Conference on Communications (ICC)*, May 2017, pp. 1–7
13. A. Malik, J. Ravishankar, A review of demand response techniques in smart grids, in *2016 IEEE Electrical Power and Energy Conference (EPEC)*, Oct 2016, pp. 1–6
14. RTE-France, (Reseau de transport d'electricite), www.rtefrance.com (2015)
15. M. Ali, M. Meo, D. Renga. WiFi offloading for enhanced interaction with the Smart Grid in green mobile networks, in *2017 IEEE 14th International Conference on Networking, Sensing and Control (ICNSC)*, May 2017, pp. 233–238
16. M.A. Imran et al., *Energy efficiency analysis of the reference systems, areas of improvements and target breakdown* (Technical Report, ICT-EARTH deliverable, 2011)
17. Australian Energy Market Operator (AEMO), http://www.aemo.com.au/Electricity/National-Electricity-Market-NEM/Data-dashboard (2016)
18. K. Lee et al., Mobile data offloading: how much can WiFi deliver? IEEE/ACM Trans. Netw. **21**(2), 536–550 (2013)
19. E. Bulut, B.K. Szymanski, WiFi Access Point Deployment for Efficient Mobile Data Offloading, in *Proceedings of the First ACM International Workshop on Practical Issues and Applications in Next Generation Wireless Networks. PINGEN '12* (ACM, Istanbul, Turkey, 2012), pp. 45–50. ISBN: 978-1-4503-1531-9
20. E.M.R. Oliveira, A. Carneiro, Routine-based network deployment, in, 2014 IEEE Conference on Computer Communications Workshops (INFOCOM WKSHPS). Apr 2014, pp. 183–184
21. A. P. Dobos. PVWatts Version 5 Manual, Sept 2014
22. Shounan Hua et al., Application of valve-regulated lead-acid batteries for storage of solar electricity in stand-alone photovoltaic systems in the northwest areas of China. J. Power Sour. **158**(2), 1178–1185 (2006)
23. J. W. Stevens, G.P. Corey, A study of lead-acid battery efficiency near top-of-charge and the impact on PV system design, in *1996 Conference Record of the Twenty Fifth IEEE Photovoltaic Specialists Conference*, May 1996, pp. 1485–1488

24. H. Al-Sheikh, N. Moubayed, Health status and diagnosis of batteries in renewable energy systems: an overview, in *2012 International Conference and Exposition on Electrical and Power Engineering (EPE)*, Oct 2012, pp. 922–927
25. M. Jafari et al., Technical issues of sizing Lead-Acid batteries for application in residential renewable energy systems, in, *2015 4th International Conference on Electric Power and Energy Conversion Systems (EPECS)*, Nov 2015, pp. 1–6
26. H. Gharavi, R. Ghafurian, IEEE recommended practice for sizing lead-acid batteries for stand-alone photovoltaic (PV) systems IEEE Std 1013–2007. Proc. IEEE. **99**(6), 917–921 (2011)
27. D.P. Jenkins, J. Fletcher, D. Kane, Lifetime prediction and sizing of lead-acid batteries for microgeneration storage applications. IET Renew. Power Gener. **2**(3), 191–200 (2008)
28. M. Meo et al., Dimensioning the power supply of a LTE macro BS connected to a PV panel and the power grid, in 2015 *IEEE International Conference on Communications (ICC)*, June 2015, pp. 178–184

Structural Health Monitoring (SHM)

Raffaele Zinno, Serena Artese, Gabriele Clausi, Floriana Magarò, Sebastiano Meduri, Angela Miceli and Assunta Venneri

Abstract This work tries to fit structural health monitoring into the Internet of Things (IoT), the main topic of the research carried out in the context of the PON-DOMUS project [1]. The structural analysis has always used electrical and electronic methods for defining the deformation state of a structure. Examples are the displacement transducers, the strain gauges, the accelerometers. Here we have tried to coordinate and to connect the activities and the information of these sensors through the controls and the information transfer that allow the capabilities of IoT. The problems faced and solved by the other groups operating in the research project have also been transferred to the structural engineering part, thus allowing an on-line and real-time assessment of the structural health of the building and, therefore, an interaction with the subjects in charge maintenance or for the facilitation of the emergency management phases [2, 3].

R. Zinno (✉) · S. Artese · G. Clausi · F. Magarò · S. Meduri · A. Miceli · A. Venneri
University of Calabria, DIMES, Ponte P. Bucci, Arcavacata di Rende, CS, Rende, Italy
e-mail: raffaele.zinno@unical.it

S. Artese
e-mail: serena.artese@unical.it

G. Clausi
e-mail: gabriele.clausi@unical.it

F. Magarò
e-mail: floriana.magaro@unical.it

S. Meduri
e-mail: sebastiano.meduri@gmail.com

A. Miceli
e-mail: angela.miceli.smartlab@gmail.com

A. Venneri
e-mail: assuntavenneri80@gmail.com

F. Cicirelli et al. (eds.), *The Internet of Things for Smart Urban Ecosystems*, Internet of Things, https://doi.org/10.1007/978-3-319-96550-5_10

1 Introduction

Rarely the building's users have cognition about its structural health, both for not having technical competences, so disregarding warning signals, and for hidden or slowly damaging processes, with fast and unexpected failure, sometimes effects of external causes. Furthermore, in emergency scenarios, the user is often unprepared to handle the situation, especially if panic attacks are triggered, complicated by incorrect information, which confuse the situation even more.

Moreover, the high seismicity of most Italian territory requires the adoption of appropriate and effective measures to limit the effects of earthquakes on structures. The preservation of national building heritage against seismic risk can benefit by the in-depth knowledge of the dynamic behavior of buildings. In this context, Structural Health Monitoring (SHM) has been growing in the last few years as a set of time-scale measures at significant points in the structure, for diagnosis and control of constructions. It is, therefore, a non-destructive in situ technique of analysis and structural features that underlines the real-time control and monitoring of a phenomenon by measuring physical-mechanical parameters describing the interaction between the environment and the state variables of the structure under consideration. A monitoring system usually consists of sensors of various nature to detect environmental magnitudes and the structural response to stresses. The purpose of such a system is not only the assessment of the resistance to a building's earthquake: the performance control of the structure under operating conditions is also of great importance for both the protection against vibration risks and the reinforcement of critical structures or of strategic interest. The typical architecture of the monitoring systems involves the use of peripheral sensors directly connected by cables to a central data acquisition system. However, a large number of sensors involves a large number of measurements and the acquisition system may be overloaded, especially if it has to process the data for damage identification: its algorithms would be particularly burdensome from a computational point of view.

The present research work was intended to make the building not only able to communicate its structural health, activating automatic procedures or, if necessary, asking human intervention, but also capable of preparing for probable future events. The modification of the principal vibration frequencies and vibration modes over time, obtained from the sensors by using seismic noise, when the variations between one and the other overpass the alert threshold, make possible alerts to be used for maintenance or to manage emergency situations.

A prototype was realized on a school building (Industrial Technical Institute "A. Monaco"—Cosenza), demonstrating the validity of the proposal.

The work is organized following the various phases, even temporal ones, that have been developed during the research. First of all, the prototype building had to be geometrically detected, also in order to verify the reliability of the available design data found in the archives of the school and/or of the managing offices. Geometric data confirmed the initial informations. Subsequently, non-destructive and/or semi-destructive investigations were carried out to confirm also the mechanical data of the materials constituting the load-bearing structure (concrete and steel reinforcement). The values obtained have deviated from those expected within the limits related to the age of the building. The electronic devices used for the evaluation of the dynamic behavior of the prototype structure are then explained. The previous phases, less usual in the IoT, are only confirmatory and useful for a better structural response evaluation. In fact, it will be easy to understand from the method explanation, that it is sufficient to compare the dynamic response obtained in two phases (pre- and post-event) to understand if a damage has occurred and if the safety limits have been exceeded. Finally, it is described how the comparison between the theoretical dynamic data and/or between two experimental surveys occurs, and how this information is processed locally and transferred via the network.

2 Geomatics Techniques for Structural Monitoring

The geomatics techniques for the survey and representation of territory and buildings had an exceptional development in recent years. The applications are innumerable and range from land planning to geophysics, from mitigation of hydrogeological risk to monitoring of buildings, from cultural heritage to medicine. With regard to the structures, we can distinguish the applications based on the dimensions and location of the objects to be detected. Depending on dimensions we can have applications on a local scale, necessary to reference the object of study in its environmental context, and on a detailed scale (loading test in the laboratory, identification and control of cracks). Technologies used can be divided into three categories: (a) techniques based on the image acquisition and processing, (b) techniques based on the measurement of angles and distances and (c) combinations of the previous techniques. The approach to existing structures requires the knowledge, as precise as possible, of their geometry to obtain the as built. This 3D model will be used either for comparison with the project—for testing and accounting the works actually realized—as well as for documentation purposes [4].

In the case of old structures, very often it is not possible to have the working drawings, so the survey activity also serves to reconstruct the stages of the building process and to disaggregate the structural organism into the elements that had been considered and sized in the design phase [5, 6]. It is also possible to obtain relevant information about the eventual historical-constructive evolution, the stratigraphic composition of the structural elements, the state and the performances of the materials, and their compatibility with the new materials, used for the conservation and restoration purposes. The use of geomatics techniques, such as photogrammetry [7, 8] and laser scanner [9, 10], allow knowing the behaviour of the materials, as well as the complete deformation state of the object and the extent of deformations in real time. The geometric models obtained from the surveys are used by the structural engineers to obtain the finite element modelling and to deduce the structural behaviour under static loads, mobile loads, and seismic actions. The case study is the Auditorium of the Industrial Technical Institute (I.T.I.) "A. Monaco" in Cosenza

Fig. 1 The Auditorium of the I.T.I. “A. Monaco”

(Fig. 1), which is one of the independent building blocks that make up the whole Institute. The building, with reinforced concrete bearing structure, occupies an area of about 400 m2 and consists of two distinct areas: zone A and zone B, both with rectangular shape and height of 10.20 m. Zone A has dimensions of 13.65 m × 9.85 m and has 3 levels. The area B, with dimensions of 24.00 m × 11.15 m, has two levels. The second level of zone B coincides with the third level of zone A and can only be reached through the stairs present in this area. At the second level of zone B, there is a footbridge of about 7.50 m in length, which is the covered passage, on pilotis, between the auditorium and the school complex. The first operations carried out were the acquisition of the 1976 working project and the visual analysis of the structure, in order to identify in situ a cracking map and the possible degradation phenomena. Concerning the survey of the plants (Fig. 2), the Leica DISTO™ D8 was used. It is an Electronic Distance Meter (EDM) with a100 m range and 2 mm accuracy.

The comparison between the dimensions of the original project and those obtained from the survey showed a substantial correspondence, taking into account the tolerances of reinforced concrete works (a few cm) and the thickness of the coating. For this reason, for the area of the attic (not accessible), the foundations and the roof, the project drawings were considered valid. The Leica 1201 + robotic total station and the Nikon D3200 Digital Camera were used for the elevations survey. The total station has a range of 1 km, without a prism, with a precision of 1 mm + 1 ppm; the angular precision is 1 second. The Nikon D3200 Digital Camera has a 24.2-megapixel APS-C (1.5x) sensor. The lens used, 55 mm Nikkor, was calibrated using the Toolbox Camera Calibration for Matlab®. The knowledge of the calibration parameters of the coupling lens-camera body allows resampling the images eliminating the distortions. The reconstruction of the elevations was carried out by photographic rectifying (Fig. 3), with an analytical procedure. Using at least four coplanar points of known coordinates, measured by total station, orthogonalized images were obtained. Once obtained the rectified frames, for each zone of the facade, we started the mosaicking. This technique allowed us to generate a single image of the facade, mosaic of all

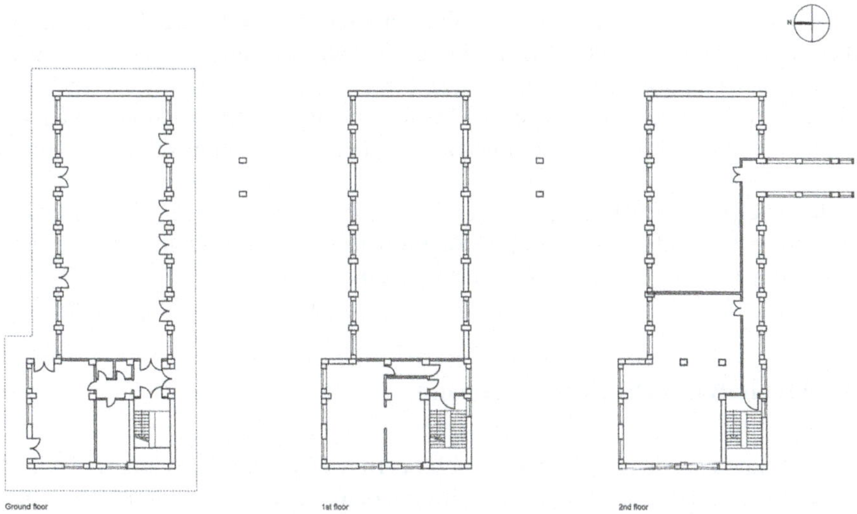

Fig. 2 Plans of the various floors of the building

Fig. 3 Reconstruction of an elevation from orthogonalized images

photos reproducing the various zones. The rectified image of the facade was subsequently used as a texture of the auditorium 3D model, in order to make it more realistic. The position of the openings was consistent with that obtained through the measurements inside the building.

Once all the elevations have been reconstructed and, therefore, all the relevant orthogonalized images have been obtained, the degradation state has been analyzed. The analysis showed that: (a) on the facades there are no cracks or traces of settlements; (b) the painting run-off on the whole cornice and in correspondence of the drainpipes is visible; (c) the marble slabs are missing at the corners of the building; (d) there are some stains and there is vegetation.

In order to be able to assess the presence of any instability due to movements of the foundation, the verticality of the edges of the building was verified. The measurements were performed with a Leica 1201 + total station. The instrument has been positioned, on different station points, in such a way as to be able to collimate the various edges optimally. Points were scanned on each corner, lying on a vertical plane passing through the station axis (z). For each scan, the x and y average were obtained. It is observed that the edges are all vertical. The deviations are contained within the limits of manufacture tolerances and reach maximum values of about 1 cm.

In conclusion, a 3D model of the building has been realized, with particular attention to the reinforced concrete bearing structure, to be used for the finite element modelling, in order to evaluate its structural vulnerability.

3 Mechanical Properties Testing

The accurate knowledge of mechanical properties of structural materials, obtained from experimental tests, allows having some parameters that can optimize and make numerical analysis more reliable, such as finite element modeling. Estimation of the in situ concrete strength has a fundamental importance in seismic assessments of existing reinforced concrete buildings, according to the most recent regulations. This evaluation must to be based on Destructive Testing (DT), Non-Destructive Testing (NDT) and, eventually, on Semi-Destructive Testing (SDT) in order to obtain representative property values for the whole structure. The only non-destructive tests are not reliable to estimate concrete compressive strength, if not calibrated with core test results. The first step is the execution of an experimental campaign of NDT to evaluate the variability of concrete and to delimit homogeneous zones or portions of the structure.

The most widely used methods for the characterization of structural concrete are those based on: sclerometer test results (NDT); ultrasonic tests results (NDT); pull-out test results (SDT); extraction of core samples (DT). These methods are described below, except for ultrasonic tests [11] that have not been used in the experimental activity. Sclerometric tests [12] have been carried out on some structural elements (beams and columns) identified inside the building.

All tests were preceded by: (1) Carbonation tests in order to evaluate the degradation process in concrete; (2) Pacometric investigations for the verification/identification of the position of steel bars immersed in concrete.

Sclerometric test (Fig. 4) consists in causing the impact of a conventional mass against the material surface and in measuring the rebound height of the mass; the measure is expressed in terms of percentage of the rebound height with respect to the distance made by the mass between the moment it is released and when it hits the concrete surface. This percentage is known as rebound index IR. The use of the rebound index involves many uncertainties and the probable accuracy in estimating the in situ compressive strength of concrete, which is about $\pm 25\%$. For this reason, this kind of test is used to evaluate the uniformity of in situ concrete and to identify and delimit areas characterized by poor quality or deteriorated concrete. Sclerometric tests performed on the structural elements of the Auditorium showed a good degree

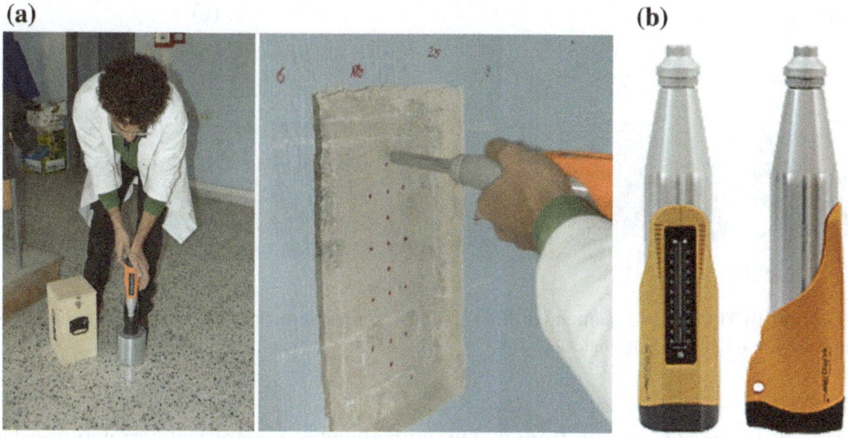

Fig. 4 **a** Calibration check e sclerometric test, **b** Ectha1000 DRC Sclerometer

Fig. 5 **a** Pull-out test equipment, **b** BOVIAR *Thoro* dowel, **c** Surface rupture

of homogeneity of concrete, providing an average value of the rebound index I_R equal to 41. To this index value, corresponds a compressive strength of about 30 N/mm^2. By cautiously fixing the RcK value equal to 70% of the compressive strength, and using the relationship E=5700 sqrt(RcK), a value for the Young's module of about 25,000 N/mm^2 is obtained.

Pull-out test is based on the correspondence between the maximum compressive load of concrete and the force necessary to extract a metal bar anchored to the structural element through a metal expansion plug inserted in a special housing in the hardened concrete. A pull-out test [13, 14] is performed by a manual pressurisation unit equipped with a pressure gauge (with a peak indicator) that activates a hydraulic jack connected to the plug inserted into the material to be tested (Fig. 5a). Because of its shape, the metal dowel is pulled out with a cone of concrete whose surface slope is approximately 45° to the vertical (Fig. 5b, c), highlighting that the collapse condition of the material is due to the overcoming of the tensile strength on the lateral surface of the cone.

The ultimate force obtained from the test is then used to define the strength parameters of concrete through calibration curves and/or by relating the test with other non-destructive testing methods. Therefore, this test consists of a direct measure-

Fig. 6 **a** HILTI DD150-U diamond drilling systems, **b** Measurement of hardened concrete carbonation depth, **c** Hole sealing material

ment of the tensile strength from which, with essentially empirical correlations, the compressive strength of concrete is deduced. The compressive strength is therefore deduced with empirical correlations but it shows a smaller margin of error if compared to the sclerometric test. Test results were obtained considering as extraction force:

$$F[kN] = 0,3047 * P[bar] \tag{1}$$

(where P is the pressure value read on the digital pressure gauge) and the following relation between the pull-out force and the equivalent concrete cube strength:

$$1[kN] = 0,94\left[N/mm^2\right] \tag{2}$$

The pull-out tests carried out gave compressive strength values varying between 22 and 27 MPa. The Core test [15] consists in the extraction of cylindrical specimens from the structural element through perforation with a hollow steel tube called diamond core drill (Fig. 6a). A kind of sled is fixed on the structure, on which the core drilling machine is mounted; the adoption of a water cooling system makes the coring operation rather invasive if carried out on buildings in operating conditions. A carbonation test is generally performed on each extracted cored specimen: the phenolphthalein indicator solution is applied on the carrot's surface causing a concrete color change from transparent white, in the carbonated part, to violet red, in the non-carbonated part (Fig. 6b). All holes resulting from the extraction of the carrots must be sealed adequately with thixotropic, premixed, expansive or non-shrink cementitious grout (Fig. 6c).

Identification of the areas of structure in which to perform drilling is a delicate operation. It is necessary to choose those parts of the structure where the concrete is less stressed (for example at half height in columns).

Laboratory direct compression tests were performed on each cored specimen to determine the compressive strength of concrete. Concrete core drilling is the best method to determine the compressive strength of concrete because it consists in a direct measurement of the original material of the structure. The limit to the number of cores resides, as well as operational and economic issues, in the damage to the structure that must always be reduced. Correlation between the sample resistance and the cylindrical or cubic strength of concrete is defined by well-known standardized procedures. Carbonation tests on the cylindrical concrete specimens provided values of carbonation depth ranging between 3 and 5 cm, while an average compressive strength equal to 29 MPa was found from compression tests.

4 The HVSR Analysis for the Valuation of the Dynamic Behavior

The object of this study is to detect the dynamic behavior of a strategic structure, subject to shaking induced by seismic noise or environmental microtremor; the expected phenomenon is the transfer of energy from the ground to the structure through the foundations. The purpose is the assessment of the damage status and the seismic response of the building. It is clear that the most suitable analysis in the diagnosis of structural damage passes through the comparison between a numerical model and the real structure, searching for any discrepancies. About that, a seismic station survey was conducted using Experimental Tromograph. The instrumental geophysical method (HVSR), points to identify the presence of behavior of a structure considered critical in terms of seismic response. The evaluation of the presence of the effect of double resonance between site and building is very important, although it is not explicitly included in the technical standards because it can cause an increase in the seismic action on the structure. This possible synchronization is evaluated simply by comparing the natural frequency of the site with that of the building. The geophysical measurements of passive seismic structure also allow identifying torsional effects due to damage; their presence may increase the stress on some structural elements, thus increasing the vulnerability of the building. For engineering studies, understanding these factors can be of fundamental importance, especially to quantify the seismic risk of buildings. By exploiting the H/V spectral ratio techniques, it is possible to determine the amplification of the spectral ordinates of the horizontal motion of a site with respect to a reference one through the Standard Spectral Ratio (SSR) [16], or it is possible to return the transfer function through the ratio of the horizontal component of the motion with respect to the vertical one through the Horizontal to Vertical Spectral Ratio (HVSR) [17]. These techniques require a statistically significant number of data, with a good Signal/Noise ratio, in order to represent the average properties of the propagation medium. The investigations for the determination of the resonance frequency of the structure were conducted with experimental instrumentation (Fig. 7a, b); the acquisitions took place according to the indications of the SESAME project (European project, Site Effects Assessment Using Ambient Excitations) and the post-processing took place using Geopsy® software. The single-station microtremor measurements are carried out by means of

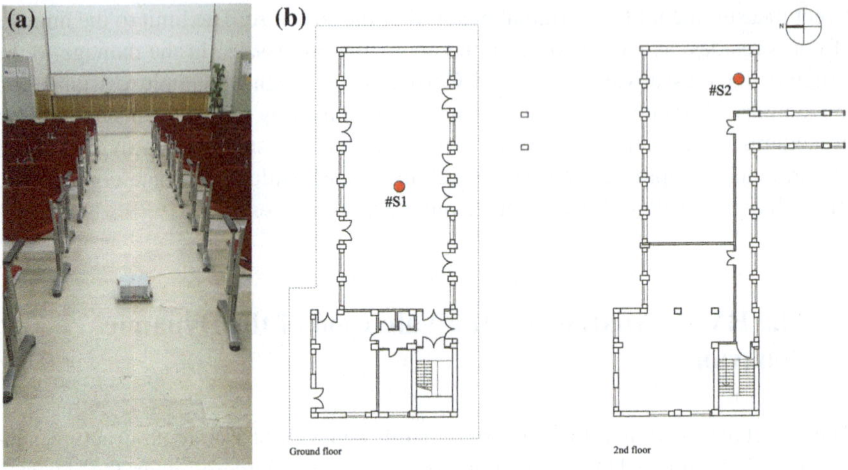

Fig. 7 **a** Experimental Tromograph in acquisition, **b** Position of the Tromograph inside the building

sufficiently sensitive three-part seismometers in the frequency range of engineering interest 0.1–20 Hz, corresponding to the frequencies of the vibrating modes of most structures. In order to be sensitive enough, an instrument must be able to detect signal even in the most silent points of the earth's surface. This technique is an experimental evaluation of the spectral amplitude ratios between the horizontal components (H) and the vertical component (V) of the environmental vibrations on the surface of the ground measured by specific three-component seismometers. For this reason, the test is commonly called HVSR (Horizontal to Vertical Spectral Ratio) or Nakamura test. The frequencies at which the H/V curve shows maximum peaks are linked to the resonant frequencies of the terrain considered below the measurement point or of the building in question.

The main purpose of the HVSR test is to bring to evidence the presence of seismic resonance, for example between the investigated land and the structure, in order to allow an estimate of the frequencies at which the motion of the ground or of the structure can be amplified due to these phenomena. In general, the estimation of the resonant frequency f is more precise as the greater the seismic impedance contrast responsible for this phenomenon is, that is, where the potentially dangerous effects are greater. The measurements collected in the context of this work were carried out by means of a digital seismometer of the Experimental Tromograph type (Fig. 8), an instrument specially designed for microtremor measurements in the frequency range of 0.1–500 Hz.

This tool was developed at the SmartLab laboratory of the Dimes department of the University of Calabria. The instrument is portable, light (2 kg) and small (18 × 14 × 8 cm) with low energy consumption. It employs very small masses and a non-resonant rigid structure built with non-diffractive materials. The sensors consist of a set of critically damped velocimeters that transmit the signal to a digital acquisition system,

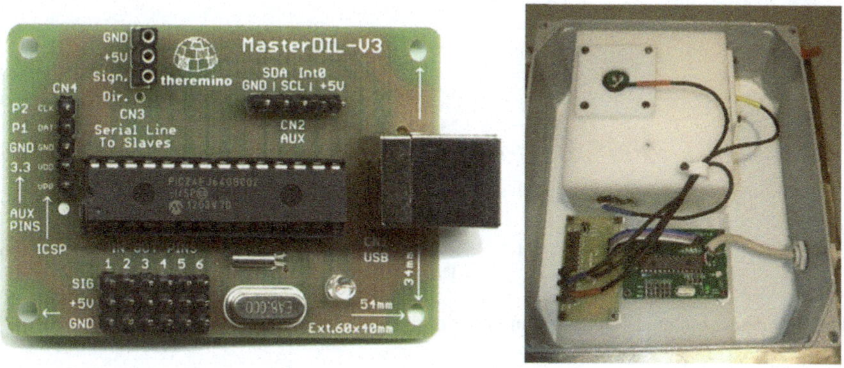

Fig. 8 Experimental tromograph

via the Theremino Datalogger card. The Theremino Datalogger (freeware hardware) is an economic an open source system, data acquisition system (ADConverter) but at the same time having great performance and being ease of use. Theremino is an open source system, completely free and open, designed to interface Windows, Linux, OSX and Android with the outside world. The modules of the system can communicate with each other, even via the Web. It is an Input/Output device, directly connectable via USB to various types of PC, without needing for special drivers, suitable for experimentation with sensors of various kinds and very simple in the mode of operation. The device is managed using the open source software HAL (Hardware Abstraction Layer), which is easy to understand, easy to manage and well documented. The dynamic investigation was carried out by acquiring the recordings of seismic noise at the various interfloors and subsequently elaborating the traces of the horizontal components, depurated of the subsoil effect. The registrations were made at Level 0 (level of walking surface) (Fig. 9) and in succession on the other interplans, up to Level 3 (Fig. 10), keeping the acquisition parameters with 500 Hz sampling frequency constant for each station, for a recording time of 20 min and with activation of only high-gain velocimetric channels for recording environmental micro-tremors. Subsequently, the recorded traces were processed using the Geopsy software, thus obtaining the H/V spectrum so as to immediately verify possible soil-structure resonance effects and quickly identify the dominant frequency of the building.

The characteristic frequency of the site in question is 2.3 Hz, most likely linked to the lithological change due to the 20 m of depth where there is the seismic bedrock corresponding to the passage to sand-gravel. Another seismic discontinuity, at 13 Hz, corresponds to the transition to poorly consolidated gravels. The fundamental frequency of the building corresponds to 5.1 Hz. In case of an earthquake, the building does not resonate with the ground since the fundamental frequency of the ground is 2.3 Hz. Even if the ground on which the building is founded is formed of alluvial sediments of the Quaternary, seismic amplification phenomena should not occur.

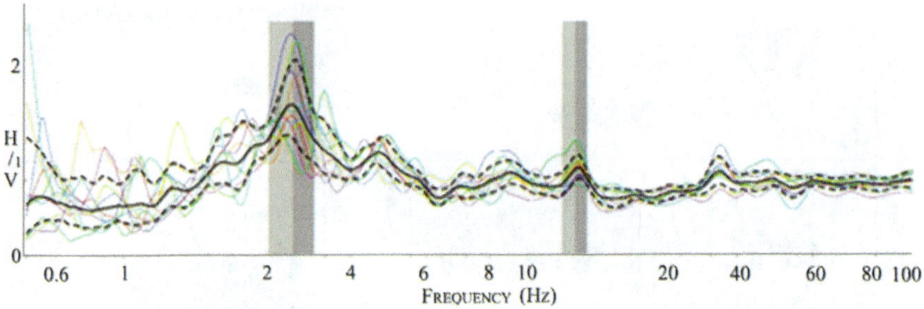

Fig. 9 Acquisition made on the ground floor of the building

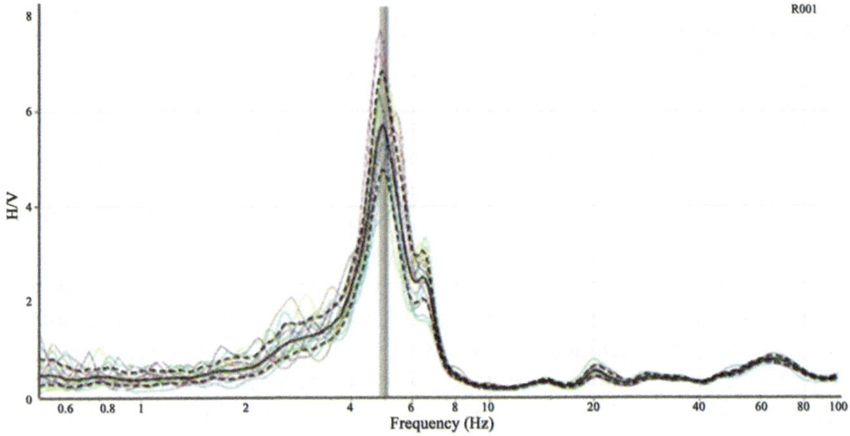

Fig. 10 Acquisition on the third floor of the building

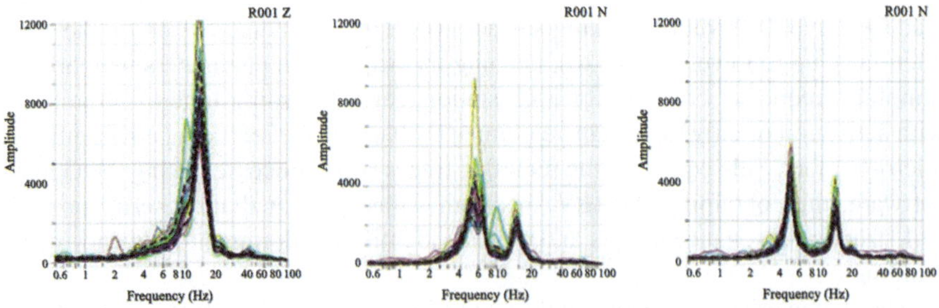

Fig. 11 Spectrum Analysis of Horizontal (E-N) and Verticals (Z) components

As shown in the Analysis of the Spectra of the Horizontal and Vertical components (Fig. 11), the North component of the signal is equivalent to the East one, while the vertical component, given the development in height of the building, tends to increase.

From these studies, performed with the Experimental Tromograph on the structure in question, two objectives have been reached: (a) verify whether the measurements in terms of frequency performed with the Tromograph are in agreement with those obtained from the accelerometric transducers; (b) provide a first measurement of the building frequencies, which will be measured again in the future to determine if the building conditions have been subject to change.

5 Vibration Measurement and Accelerometric Network

A vibrating body is an oscillating motion around a reference point whose oscillation parameters are not generally constant over time. Vibrations can be generated by deterministic or random motions; they may have periodic and non-periodic or irregular characteristics and may be harmonic. Deterministic vibrations usually overlap with random vibrations, which can be considered disturbances (or noise). Vibration measurement can be performed in different ways and for different purposes, such as to investigate the level of vibration transmitted to the human body by a mechanical system in order to evaluate the induced mechanical stresses; to evaluate the ability of a force to vibrate a system; to perform the experimental identification of the frequency response function of a single body or a system of bodies and estimate its own frequencies and its own modes of vibrating (**E***xperimental* **M***odal* **A***nalysis*). Furthermore, vibrations can be divided into two categories based on what generates them: free vibrations and forced vibrations. Free vibrations occur when a mechanical system oscillates without being subjected to any force, but because its initial conditions at the time considered are different from zero or in a certain instant the system is in a different condition from that of equilibrium. The most general free motion is the superposition of all the proper modes. Each mode participates in a specific part, depending on the initial conditions. In the DOMUS project, the analysis of the dynamic behavior of the structure selected as a demonstrator of the project was carried out following two approaches: analytical approach and experimental approach. The analytical approach determines the modal parameters of the system (natural frequencies, damping factors, and modal forms) starting from the knowledge of the structure's geometry, the boundary conditions and the characteristics of the materials, the mass distribution and the stiffness and damping of the structure. The experimental approach, instead, determines the dynamic parameters of the structure starting from the measurement of the structural response stressed by mechanical devices or environmental excitation, which can be considered a white noise: in this case the input on the structure is not known (nor can it be measured), but it is possible to evaluate dynamic parameters of the structure using only the output data. A sensor network consists of a series of devices distributed in space and which use their sensors to cooperate in monitoring events or environmental situations. Typically, the original signal is translated into a voltage or current quantity in such a way as to allow its acquisition and possible processing in digital format. The sensor networks consist of a set of nodes. Each node of the network consists of one or more

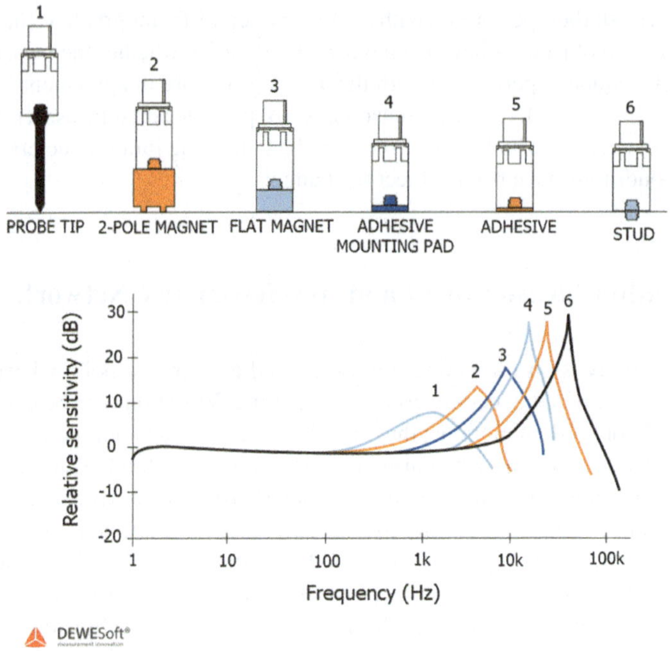

Fig. 12 Frequency response of a piezoelectric accelerometer

sensors, a data transmission system, a microcontroller for information processing and an energy source. The sensor networks can be wired or wireless; it is obvious that a structure of wired networks has a higher cost compared to a wireless network, but at the same time the use of a wireless network involves the implementation of more sophisticated techniques to ensure the consistency of data. For the modal identification of the I.T.I. "Monaco" Auditorium, a wired network composed of eight high-sensitivity monoaxial piezoelectric accelerometers was created. In general, the choice of accelerometers to be used considered the acceleration levels to be measured so that they fall within the measurement range and the sensitivity of the sensor, that is, the potential difference produced on the element for a given level of acceleration. This last parameter is imposed when the accelerometer is manufactured and cannot be changed, it is measured in mV/g (i.e., in tension per unit of acceleration). The minimum and maximum measurable values are obtained experimentally during calibration. Below, Fig. 12 is a typical trend of the frequency response of a piezoelectric accelerometer (Fig. 13).

As far as the electrical aspect is concerned, when an accelerometer is mounted on a conducting surface there is the possibility that the sensor will be disturbed by signals coming from other electrical devices or even from the power supply devices of the system itself, forming a ground loop (Fig. 14), they concatenate with the acquired signals. It is therefore convenient to isolate the accelerometers from the structure on which they are mounted. Accelerometers must be connected to the acquisition system

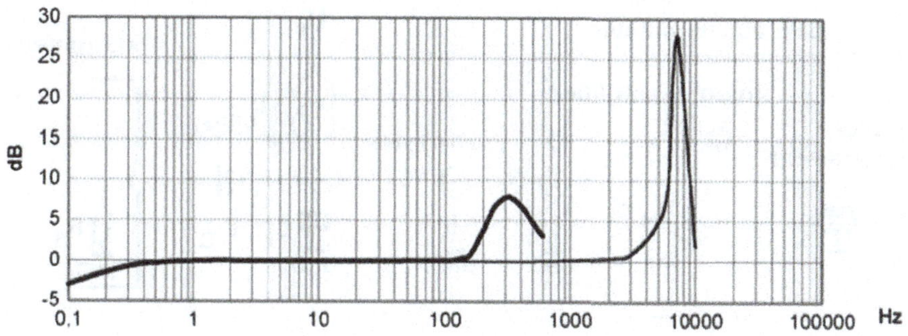

Fig. 13 Frequency response related to the type of assembly (*Source* DEWESoft)

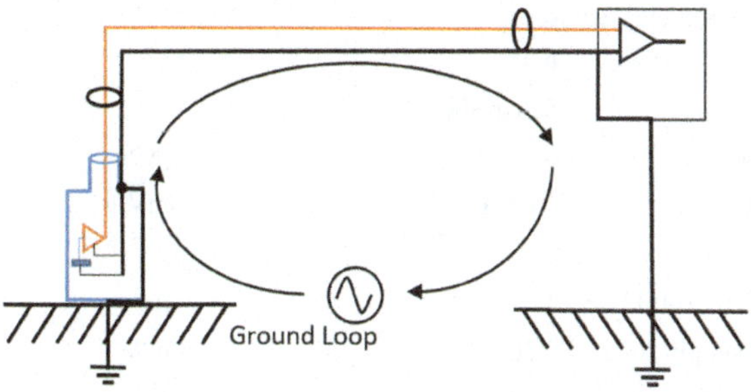

Fig. 14 Ground Loop

through strong connections to ensure continuous transmission without intermittences that would cause data loss. The output signal from a piezoelectric sensor is low, which means that a very sensitive signal conditioning circuit is required. ICP type accelerometers, which incorporate a micro-amplifier which converts the signal from high to low impedance, were used for this application.

The transducer needs a constant current supply which is supplied by specific acquisition systems and, in this case, it has been possible to use standard cables, even of considerable length, without the signal degrading significantly. As part of the DOMUS project, experimentation was carried out by creating an accelerometric network whose nodes were formed by high-quality piezoelectric accelerometers and the output of the ICP transducer permitted the use of a RG59 coaxial cable (Fig. 15).

The accelerometer used is the KS48 model, and Table 1 shows its main features.

6 Structural Analysis and Prototype Applications

The **Auditorium of I.T.I. "Monaco"** was modeled in the context of FEM (Finite Element Method) in order to know its mechanical behavior [18]. The spatial model

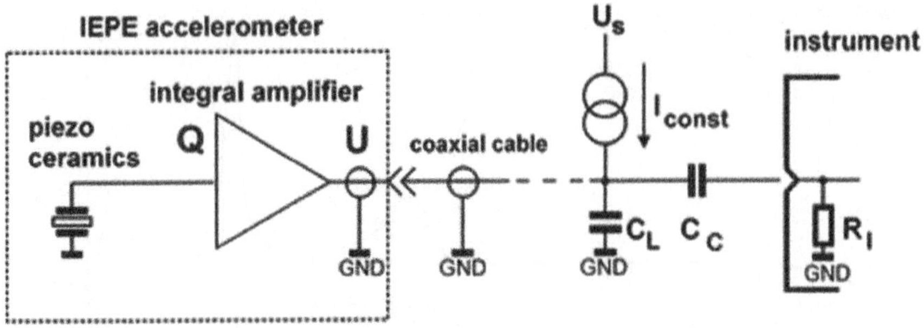

Fig. 15 Equivalent circuit of an ICP transducer and acquisition system

Table 1 The accelerometer features

OUTPUT	ICP®	
Piezo design	Scher/Shear	
Charge sensitivity	–	pC/g
Voltage sensitivity	1000 ± 20%	mV/g
Range	6	g
Residual noise (20.50 000 Hz)	4	μg
Constant current supply	2.20	mA
Output bias voltage	8.12	V
Resonant frequency	>6,5	kHz
Transverse sensitivity	<5	%
Capacitance without cable	–	nF
Operating temperature range	−20/90	°C
Temp. coefficient of sensitivity	−0,16	%/K

of the experimentation structure was realized with the help of the finite element software SAP2000 (Fig. 16).

Considering that there are no signs of deterioration of the material from the visual surveys, it has been considered a concrete that has the same characteristics as the original one. The concrete used for the construction of the building, $\gamma_{cls} = 2500$ kg/m^3, was characterized by an elastic modulus of E = 25000 N/mm^2 and a Poisson coefficient, ν, equal to 0.27. Beams and columns have been schematized as beam elements, while for the floors and stairs has been made to the shell area elements. The results that can be obtained from a beam element are the torques, forces, displacements and rotations at the nodes. The *thick shell,* used in the modeling of the floor of the Great Hall, represent a part of the casting shell elements category which, compared to the *thin shell* elements, also consider the deformability shear. It is flats *latero-cementitious*, 25 m thickness in the zone A and 20 cm thickness in the zone B. Unlike the beam elements, for which the shape functions are well studied and it is possible to arrive at an exact solution also considering elements at two nodes, for

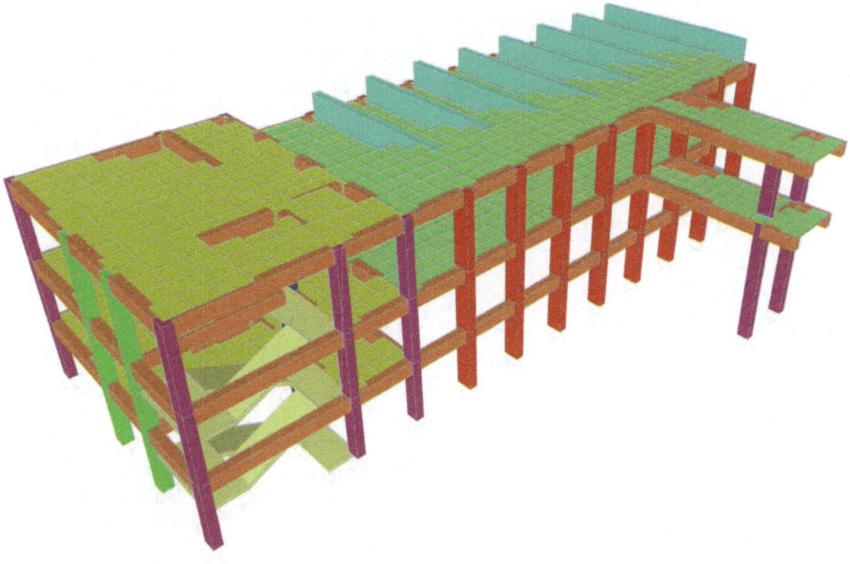

Fig. 16 Spatial model of the building studied

shell elements it is necessary to proceed with the *meshatura*, which consists in subdividing them into sub-elements having a ratio next to one between the sides. The stairs were modeled as *shell-thin* elements of a thickness of 15 cm bound to the beams and perimeter columns by interlocking. As regards non-load-bearing structural elements, infill panels and partitions, it was decided to represent them as loads acting on the beams. The analysis of construction details showed that the roof is simply resting on brick walls along the perimeter and in the central part and does not have a rigid behavior on its floor. Therefore, since its contribution to the overall stiffness is small and difficult to assess, the roof has not been geometrically modeled. In any case, its own weight and the loads acting on it (snow load) were taken into consideration and applied directly to the relative members. Given the lack of documentation concerning the foundation structures and the type of soil, the soil-structure interaction has been imposed on the ground floor by binding the ends of the pillars with perfect interlocking constraints. Once the finite element model of the structure was completed, the modal analysis was performed [19], which provided results in terms of frequencies and periods of vibration [20]. The contribution of the generic mode of vibration to the structural response, considered as a translational or rotational contribution, is established on the basis of the percentage of participant mass. While, the number of modes to consider is fixed, according to the current legislation [21], by the percentage of total participant mass that must be higher than 85%. In the present case, Table 2, the structural response is defined by the first eight vibration modes.

By examining the deformations of the first three modes of vibrating of the structure (Fig. 17), we have that: the first mode vibrates at a frequency of 4.9 HZ and is

Table 2 Translational and rotational contributions of the vibration modes

		Translaton X		Translaton Y		Rotaton Z	
Mode	Frequency (Hz)	Mass (%)	Sum (%)	Mass (%)	Sum (%)	Mass (%)	Sum (%)
1	4,9001	1,58	1,58	50,00	50,00	87,00	87,00
2	6,5951	59,00	60,58	3,20	53,20	0,53	87,53
3	9,1927	0,24	60,81	19,00	72,20	1,06	88,59
4	11,519	0,43	61,24	5,13	77,33	0,10	88,69
5	14,759	21,00	82,24	0,03	77,36	1,52	90,22
6	15,335	0,00	82,24	7,00	84,35	0,46	90,68
7	15,747	0,87	83,11	0,39	84,74	2,55	93,23
8	17,003	2,35	85,45	2,65	87,39	1,03	94,26

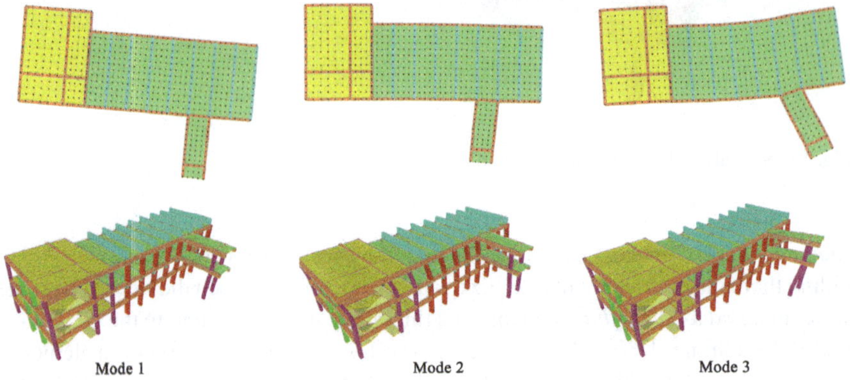

Fig. 17 Deformed shapes of the first three vibration modes

translational in the y direction with non-symmetrical deformation since the distribution of the stiffness is irregular. In fact, in zone A, the presence of the stairs and the greater thickness of the floor greatly increase the rigidity with respect to zone B. The second mode, which corresponds to a frequency of about 6.6 HZ, is translational in the direction x. The third mode is torsional and vibrates with a frequency of 9.19 HZ.

From the modal shapes, we can identify the areas that undergo greater deformations for the planning of the positioning of the sensorial instrumentation (Fig. 18). It is precisely at the structural parts affected by excessive displacements that the transducers are positioned. In the case studied, there were eight accelerometers available. Following the analysis carried out with the SAP2000 software, which allowed us to get an idea of the structure's own modes, we chose to place them at the points indicated in the figure, which also show the share of each accelerometer and the direction reading, chosen so as to obtain information also on the twisting component of the motion [22]. To avoid aliasing problems, it is important that the sampling rate is greater than or equal to twice the highest frequency contained in the signal.

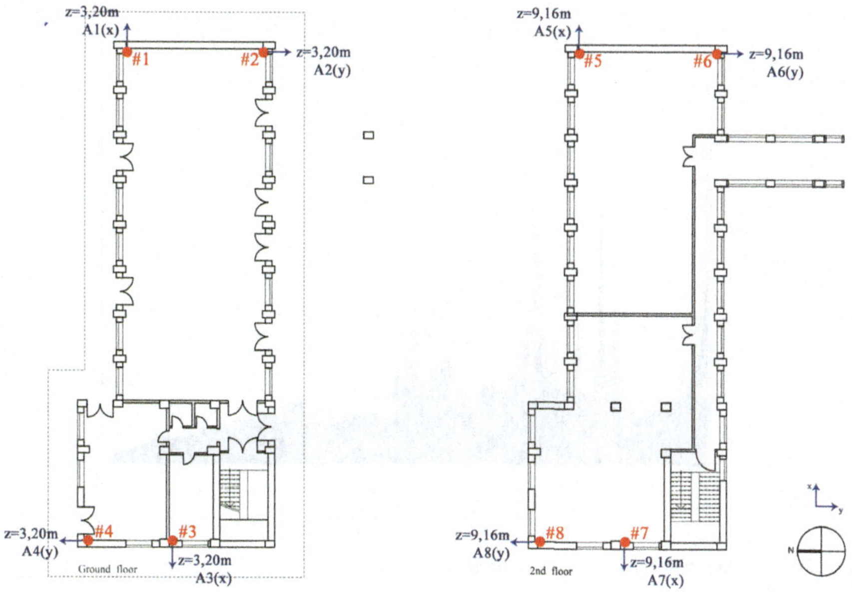

Fig. 18 Position, elevation and reading direction of the accelerometers at the ground floor and at the 2nd floor

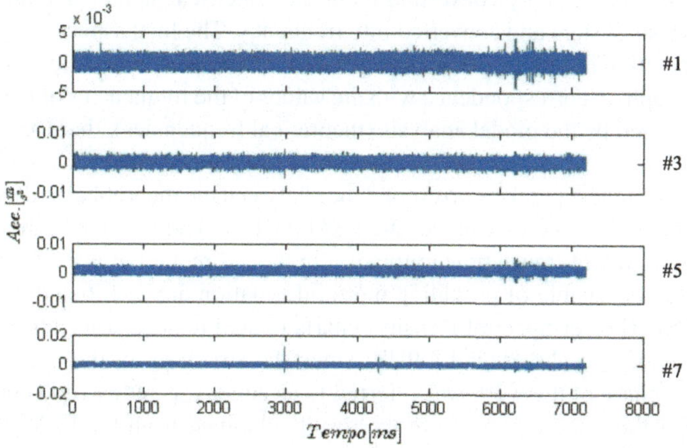

Fig. 19 Acquisitions of accelerometers no. 1-3-5-7

Considering that the range of frequencies to be explored is between 3 and 30 HZ, a band-pass filter has been applied, that is a device that allows the passage of the frequencies present within the range of interest and attenuates the frequencies outside its. The acquisitions were all made within 6 h. As an example, the accelerations recorded by accelerometers 1, 3, 5 and 7 are reported (Fig. 19).

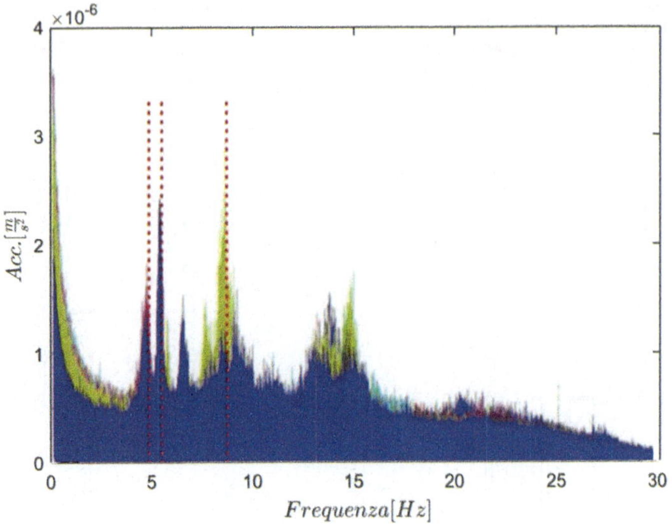

Fig. 20 Overlap of spectra in terms of frequency (PSD)

These sensors measure the amount of force applied, the displacement and acceleration of the structure by converting them into electrical signals that are recorded by the analysis system and converted into frequency. The final aspect is a graph with the superposition of power spectra relative to the various recorded signals (Fig. 20). From the graph, in correspondence with the values of the frequencies of the first three modes provided by the modal analysis (theoretical frequencies), there are increases, peaks, in terms of experimental frequency (frequency measured by the sensors). This identifies a sufficient precise correspondence between the theoretical and experimental frequencies. For the first mode, the FEM analysis had provided a frequency of 4.9 HZ. As a way of translational vibration in the y direction, it is expected that the accelerometers capable of providing more information are 2, 4, 6 and 8. Although this, the PSD (Power Spectral Density) charts related to accelerometers 4 and 8 do not have evident peaks around 4.9HZ, since the structure, in the part that in the description phase of the work was referred to as zone A, is rather rigid compared to zone B. For the same reason, from the results obtained from the FEM model it is possible to notice the non-symmetry of the deformed one justifies the fact that there are also high peaks in the PSD graphs related to accelerometers 1 and 5. Finally, the data taken into account for the determination of the first frequency are those related to accelerometers 1, 3, 5 and 7, i.e. those placed at the corners at the end of the zone B, that is, at the part of the structure most affected by the translational motion in the direction y. By way of example, the graph of the PSD relative to the accelerometer number 6 is shown with the indication of the experimental frequency associated with the first mode of vibrating (Fig. 21).

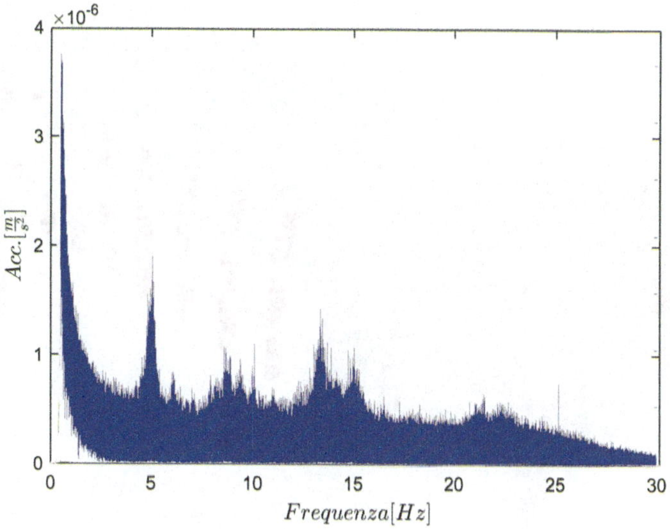

Fig. 21 Indication of the first frequency (accelerometer no.6)

From the average of the frequencies provided by the graphs of the power spectral density relative to the accelerometers of interest, it was possible to determine an experimental frequency for the first mode equal to 4,852 HZ. For the second vibration mode, which corresponded to a theoretical frequency of 6.59 HZ, the FEM analysis of the model showed how it was a translational mode in the *x* direction. Therefore, it is expected that the greatest information is provided by accelerometers 1, 3, 5 and 7. Actually, the greatest peak is present in the PSD graph related to accelerometer 7. This is due to the fact that, in zone A, passing from the intermediate level to the last level, there is a considerable reduction in rigidity due to the absence of the stairs. This aspect is highlighted in Fig. 22, which shows the position of accelerometer no. 7 on the deformed 3D structure according to mode 2.

Furthermore, by way of example (Fig. 23), the graph of the PSD relating to accelerometer number 7 is shown, with indication of the experimental frequency associated with the second vibrating mode equal to 6.067 HZ.

The various accelerometers have been installed in position and direction of reading such as to be able to provide, on the whole, information on the torsional mode. Therefore, as far as the third way of vibrating is concerned, it is expected that, more or less, all accelerometers will be able to provide adequate information. Despite this, around the theoretical frequency of 9.19 HZ, the PSD charts with the most noticeable peaks are those relative to accelerometers 5, 7 and 8. For the third mode of vibration, an experimental frequency of 8.869 HZ was determined. By proceeding in the same way for all the other significant vibrational modes, the experimental frequencies were determined. The comparison is made by calculating the percentage error (Table 3),

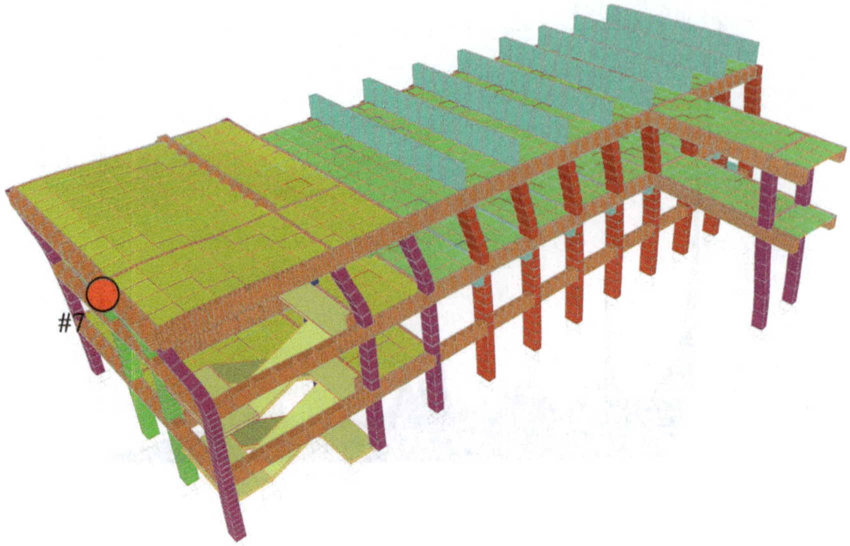

Fig. 22 accelerometer position no. 7

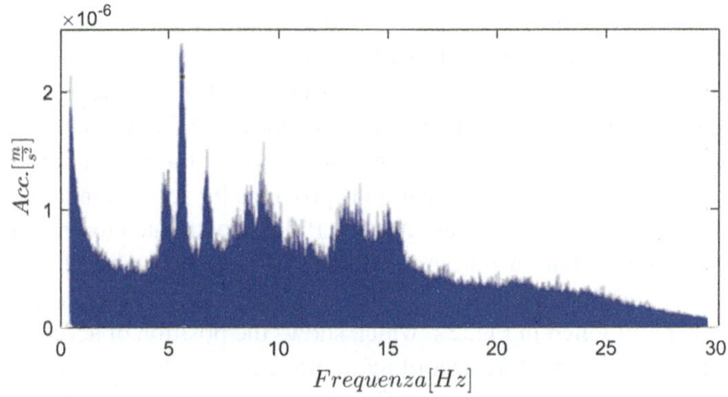

Fig. 23 Indication of the second frequency (accelerometer no. 7)

from which the frequencies provided by the finite element model are close to those calculated experimentally.

The comparison is also carried out by plotting the two estimates on a Cartesian plane (Fig. 24), which shows a good correlation between the two groups of frequencies as close to the line passing through the origin of the axes and with a slope of 45°.

Therefore, the FEM modeling and its analysis can be considered valid and used in future applications, since it is sufficiently representative of the real conditions of the structure from the point of view of the dynamic behavior. On the other hand, theoretical frequencies had been obtained that were clearly different from the experimental

Table 3 Percent error

Modes	f_{teo} (Hz)	f_{sp} (Hz)	Error (%)
1	4,9001	4,852	0,98
2	6,5951	6,067	8,01
3	9,1927	8,969	2,43
4	11,519	10,96	4,85
5	14,759	14,14	4,19
6	15,335	–	–
7	15,747	–	–
8	17,003	17,65	3,81

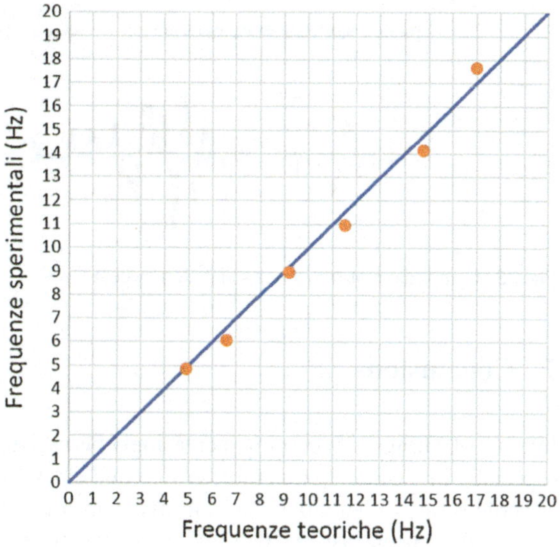

Fig. 24 Frequency comparison plotting

ones (white noise) and the FE modeling was carried out by making changes in the discretization of the structural elements to obtain a model with frequencies similar to the experimental ones.

The previous results obtained from the geometrical and geomatic analyses, from the mechanical characterization, from the theoretical and experimental analyses for the definition of the vibrating modes and the principal frequencies of the structure that can be interpreted as undamaged structure, allow at any time a comparison between the latter data and those instantly detectable by the accelerometric network constantly present on the structure. A difference, over a predefined tolerance, between modes and frequencies in the case of an undamaged and damaged structure, will trigger the transmission of an alert and the consequent actions of maintenance or rescue intervention (Fig. 25).

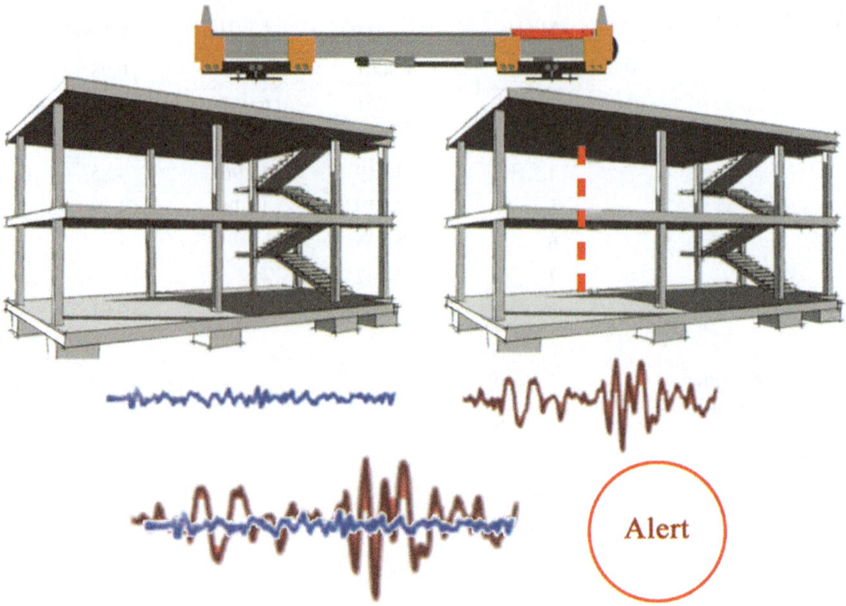

Fig. 25 Comparison of frequencies and alert system

7 Conclusions

Internet of Things is becoming increasingly pervasive in all aspects of everyday life, especially in the aid of vulnerable groups and elderly or disabled people, facilitating a whole series of activities that improve comfort and reduce energy consumption. Among the various aspects that IoT facilitates there are also the safety and security of the home, intended on the front of the usability of the building. With the study presented in this chapter, the building itself was monitored and it was able to self-diagnose its structural state of health and be able to detect dangers for users. It has been demonstrated how the integration with all the other IoT systems is possible and above all useful and efficient. It was even possible to show the operation on a real prototype: ITI Monaco of Cosenza. It can be concluded that the results are more than satisfactory and further possible future developments can be glimpsed in order to pass from simple self-diagnosis and warning of structural danger to a cognitive capacity of the building on the structural aspect, as well as to studies that allow greater cost-effectiveness and ease of installation. solutions, to allow a widespread diffusion of the proposed system.

References

1. F. Cicirelli, A. Guerrieri, G. Spezzano, A. Vinci, An edge-based platform for dynamic Smart City applications. Future Gen. Comput. Syst. **76**, 106–118 (2017)
2. G. Fortino, P. Trunfio (eds.), *Internet of Things Based on Smart Objects: Technology, Middleware and Applications* (Springer International Publishing, 2014)
3. A. Guerrieri, V. Loscri, A. Rovella, G. Fortino (eds.), *Management of Cyber Physical Objects in the Future Internet of Things*. (Springer International Publishing, 2016)
4. G. Deruyter, M. Hennau, V. De Wolf, N. De Wulf, Approach for comparing design and as built models based on data acquisition using a 3rd terrestrial laser scanner, a case study. In: *4th International Workshop on 3rd Geo-Information*, pp. 101–116 (2009)
5. I. Lubowiecka, J. Armesto, P. Arias, H. Lorenzo, Historic bridge modelling using laser scanning, ground penetrating radar and finite element methods in the context of structural dynamics. Eng. Struct. **31**, 2667–2676 (2009)
6. S. Artese, J.L. Lerma, G. Zagari, R. Zinno, The survey, the representation and the structural modeling of a dated bridge, in *Proceedings of the 8th International Congress on Archaeology, Computer Graphics, Cultural Heritage and Innovation 'Arqueológica 2.0' in Valencia (Spain)*, 5–7 Sept 2016
7. I. Detchev, A. Habib, M. El-Badry (2011) Case study of beam deformation monitoring using conventional close range photogrammetry. In: ASPRS 2011 Annual Conference. ASPRS. Milwaukee, Wisconsin, USA
8. S. Artese, V. Achilli, R. Zinno (2018) Monitoring of bridges by a laser pointer: dynamic measurement of support rotations and elastic line displacements: methodology and first test. Sensors **18**(2), 338
9. G. Vosselman, B.G. Gorte, G. Sithole, T. Rabbani, Recognising structure in laser scanner point clouds. Int. Arch. Photogramm. Remote Sens. Spat. Inf. Sci. **46**(8), 33–38 (2004)
10. H.M. Zogg, H. Ingensand, Terrestrial laser scanning for deformation monitoring—load tests on the felsenau viaduct (CH). Int. Arch. Photogramm. Remote Sens. Spat. Inf. Sci. **XXXVII**(B5), 555–562 (2008)
11. EN 12504-4:2005: Testing concrete in structures—Part 4: Determination of ultrasonic pulse velocity
12. EN 12504-2:2012: Testing concrete in structures—Part 2: Non-destructive testing - Determination of rebound number
13. EN 12504-3:2005: Testing concrete in structures—Part 3: Determination of pull-out force
14. ASTM C900—06: Standard Test Method for Pullout Strength of Hardened Concrete
15. EN 12504-1:2009: Testing concrete in structures—Part 1: Cored specimens—Taking, examining and testing in compression
16. R.D. Borcherdt, Effects of local geology on ground motion near San Francisco Bay. Bull. Seismol. Soc. Am. **60**(1), 29–61 (1970)
17. Y. Nakamura, A method for dynamic characteristics estimation of subsurface using microtremor on the ground surface. Quaterly Rep. Railw. Tech. Res. Inst. **30**(1), 25–30 (1989)
18. A. Madeo, R. Casciaro, G. Zagari, R. Zinno, G. Zucco, A mixed isostatic 16 dof quadrilateral membrane element with drilling rotations, based on Airy stresses. Finite Elem. Anal. Des. **89**, 52–66 (2014)
19. D.J. Ewins, *Modal testing: theory and practice*, vol. 15 (Research Studies Press, Letchworth, 1984)
20. R. Potter, M. Richardson, Mass, stiffness and damping matrices from measured modal parameters, in *ISA International Instrumentation-Automation Conference, New York, New York*, Oct 1974
21. M. delle Infrastrutture, Norme tecniche per le costruzioni. Min. Inf **14** (2008)
22. B. Peeters, G. De Roeck, Stochastic system identification for operational modal analysis: a review. J. Dyn. Syst. Meas. Control **123**(4), 659–667 (2001)

A Smart Air-Conditioning Plant for Efficient Energy Buildings

Roberto Bruno, Natale Arcuri and Giorgio Cuconati

Abstract The spread of renewable energy technologies in the building sector has produced the new figure of "*prosumer*", able to consume and produce energy simultaneously. In this context, a correct management of the energy fluxes is required to increase user remuneration. All of this, paired with the use of the emergent IoT technologies, allowed the realization of a Smart Ecosystem devoted to make effective the process of producing, storing and consuming energy. Considering PV generators, the self-produced electricity surplus has to be transferred with advantageous conditions, alternatively it has to be stored. Air-conditioning plants employing heat pumps represent a useful option for the rational management of renewable electricity because the same system can be used as an alternative to "electric storage", cheaper and more reliable than traditional batteries. Heat pumps can be exploited to produce thermal or cooling energy and store it in a suitable tank, though the building does not require them, and to conciliate the time shift between energy demand and offer. In presence of a saturated storage tank, the same building could be used as a further thermal accumulator by exploiting radiant emission systems to activate its thermal mass, by means of either overheating or undercooling strategies. The combination of these solutions allows for noticeable energy and economic savings and a rational use of renewable sources. However, a smart control system is required to make all the various involved devices communicating and coordinating among each other. A smart air conditioning system and the correspondent control strategies adopted for its management, based on the employment of PV driven heat pumps with thermal storage connected to a radiant emission system, is introduced.

R. Bruno (✉) · N. Arcuri · G. Cuconati
Mechanical, Energetic and Management Engineering Department, University of Calabria,
Via P. Bucci 46/ C – ZIP 87036, Arcavacata di Rende (CS), 87036 Rende, Italy
e-mail: roberto.bruno@unical.it

N. Arcuri
e-mail: natale.arcuri@unical.it

F. Cicirelli et al. (eds.), *The Internet of Things for Smart Urban Ecosystems*,
Internet of Things, https://doi.org/10.1007/978-3-319-96550-5_11

1 Introduction

In Italy, the primary energy demand in 2015 was approximately 156 Mtoe (million tons of equivalent oil), an increase of 3.4% compared to the prior year; in particular, in the residential sector the increment was approximately 10% reaching 32 Mtoe [1]. For the latter, a fraction of 70% was employed to satisfy the building air-conditioning demands. Therefore, the building sector represents a key element to improve energy efficiency because 75% of the whole building stock can be classified as "energy inefficient" [1]. A possible solution to reduce the primary energy requirements, deriving from fossil sources, is the correct management of a combination of reliable and existing technologies [2]. These plants involve different primary energy sources, both renewable and non-renewable, therefore their rational administration has to be attained in order to favour the minimization of fossil fuel exploitation [3]. An appropriate auxiliary or integration system, usually supplied by fossil sources, is required to find a solution to the aleatory nature of the renewable energies. The latter are able to limit the building energy demands, especially those connected to the air-conditioning plants, directly and indirectly involving the exploitation of solar radiation [4]. This research field is still largely investigated, because the technologies linked to the so-called "Solar Air-Conditioning" have to be further improved appearing very promising in terms of primary energy saving. The simplest air conditioning system, exploiting solar radiation, is represented by electrical heat pumps connected with a photovoltaic field (PV driven heat pump) [5]. PV electricity represents the direct exploitation of solar energy, whereas the thermal energy embodied in the heat pump sources is indirectly associated to the action of the solar radiation. However, in these systems the solutions of two evident problems have to be attained. First, the asynchrony existing in winter between solar radiation availability and thermal load request, successively heat pumps may not guarantee the coverage of the energy demands, therefore an auxiliary device has to be considered [6]. The first problem can be overcome by means of the sizing of a suitable storage system, in order to accumulate thermal energy when heating or cooling are not required. Moreover, the heat pump allows to transform PV surplus in thermal energy, though a heating load is not requested, acting as an alternative energy storage system, permitting to use the stored thermal energy in differed times. The second problem is solved by an integrative system made by a more reliable technology that assures the coverage of the energy demands. Therefore, such an air-conditioning plant represents a "*hybrid*" system, where different primary sources are involved to provide heating and cooling loads. These typologies of air-conditioning plant, in fact, can be used for the whole year and suitable emitters, such as fan-coils or radiant panels, can be used for cooling applications too. The presence of PV generators and of auxiliary systems makes these devices quite expensive if compared to traditional air-conditioning plants, however the payback period reduces with the self-consumed PV electricity growth [7]. Regarding the storage of thermal energy, insulated tanks represent a consolidated and cheaper technology compared to expensive batteries for the electric energy accumulation. However, a smart management of the building-plant system that considers the strong interaction of different devices (PV generator, heat pumps, tank and emission system), has to be developed. The considered devices were augmented by exploiting IoT-based technologies with the aim of realizing a Smart Ecosystem devoted to make effective the process of producing, storing and consuming energy. In order to attain energy savings and appropriate indoor comfort conditions, the smart control system has to allow simultaneously the achievement of different targets: the maximum exploitation of the renewable source, the correct use of the storage system, and the accurate management of the emission system. For the latter, the employment of radiant panels is suggested to increase the energy efficiency indexes of heat pumps [8]. However, in summer the surface temperature of radiant systems has to be monitored continuously to avoid condensation of the water vapour by means of appropriate inlet temperature. The same emission system can be used to activate the building thermal mass, therefore the control system has to consider the structure as a further thermal energy accumulator to employ in the presence of saturated storage tanks. Thus, an eventual PV electricity surplus can be exploited by means of overheating/undercooling strategies of the indoor environment, avoiding selling the self-produced electricity if its price is not competitive.

2 The Heat Pump Technology

By exploiting the free energy available in renewable sources such as external air, water or ground, air-conditioning plants equipped with heat pumps represent an easy and reliable system to reduce operational costs and the environmental impact for heating, domestic hot water production and cooling in buildings. Thermal energy is transferred from a thermal source to another one by means of the provision of electric energy. The heat transfer occurs by an inverse thermodynamic cycle evolving with low-boiling fluids. In Fig. 1 a typical thermodynamic cycle describing the operation of an electric heat pump, is shown. Initially, the low-boiling fluid in gaseous form is compressed to a higher pressure level by a compressor, absorbing electric energy ($A \rightarrow B$). In these conditions, the fluid tends to pass spontaneously from the gaseous to liquid form, but the phase transition requires the release of thermal energy to a first thermal source ($B \rightarrow C$). When the low-boiling fluid is in liquid phase, it is subjected to a drastic pressure drop ($C \rightarrow D$) and successively it tends spontaneously to become gaseous. During this transformation ($D \rightarrow A$), the low-boiling fluid has to absorb heat from a second thermal source, returning to the initial point. With the transformations BC and DA, the thermal energy is exchanged with the correspondent thermal sources that can be represented by external air or water. For instance, in winter an air-air heat pump extracts thermal energy from the external air to heat an air flow-rate, whereas in presence of an air-water heat pump a water flow-rate suppling appropriate hydronic emitters (fan coils or radiant panels), is employed. During summer, an air-air heat pump extracts heat from the indoor air to "*pump*" the thermal energy to the outdoor air. If hydronic systems are employed, the thermal energy is extracted by the water flow rate circulating inside the emitters. Electric heat pumps connected to the ground

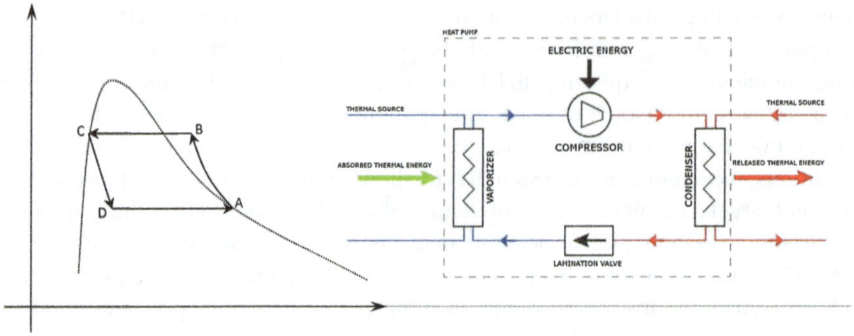

Fig. 1 Thermodynamic cycle and functional sketch of an electric heat pump

as thermal source use apposite heat exchangers to transfer thermal energy between a water flow rate and the soil. Heat pumps provide elevated energy efficiency indexes because the largest amount of heat is absorbed from the renewable sources, allowing a limitation of about 50% concerning the operational costs [8].

The European Directive RES "Renewable Energy Source" fully contemplates the electric heat pumps as devices exploiting renewable energy [9]. The technological evolution has allowed for the production of modern heat pumps with energy efficiency indexes 10 times greater than analogous products made 10 years ago, due to the employment of innovative technologies such as compressors equipped with inverters and refrigerant circuits operating with variable flow rates [6]. The efficiency indexes are named in different ways in accordance with the provided service; however, both are defined as the ratio between the "pumped" heat and the absorbed electric load (see Fig. 2). In winter, the COP (Coefficient of Performance) at seasonal level can be equal to 3, therefore the supply of 1 kWh of electricity (W_e) allows for the production of 3 kWh of thermal energy (H_2). Similar values are detected in summer and the correspondent energy efficiency index is called EER (Energy Efficiency Ratio). Other performance indexes such as SEER and SCOOP, are usually used: the first refers to cooling application, the second to heating application, both calculated by considering the involved energies during a whole year and in three different climatic contexts. The energy performances of electric heat pumps, in fact, cannot be generalized: they depend strongly on the temperature of the external source from which the heat is absorbed (in winter) or transferred (in summer). By setting the temperature of the supplied fluid, COP indexes increase with the external source temperature growth and EER increases with the external source temperature decrement. If external conditions are too severe, heat pumps are not able to operate. Electric heat pumps are not able to provide flow rates with high temperatures, therefore radiant emitters are more suggested because flow rates at 35 °C are sufficient for heating applications. Moreover, flow rates at a temperature of 15 °C are suitable in summer allowing for the achievement of more elevated EER, corresponding to a reduction of absorbed electric load at parity of supplied thermal energy. In Fig. 3, typical trends concerning

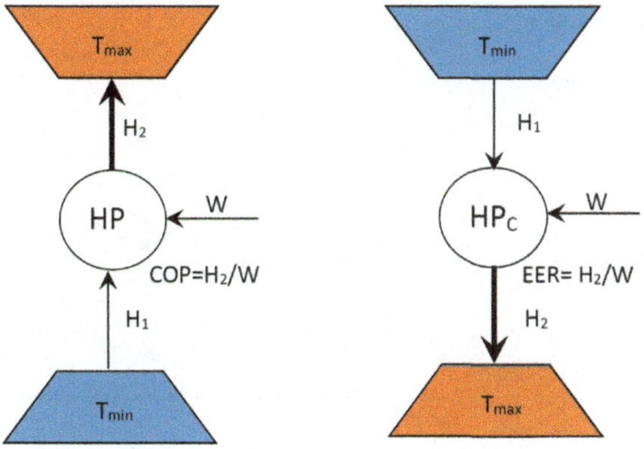

Fig. 2 Heat pump operational scheme for heating (HP_H) and cooling (HP_C)

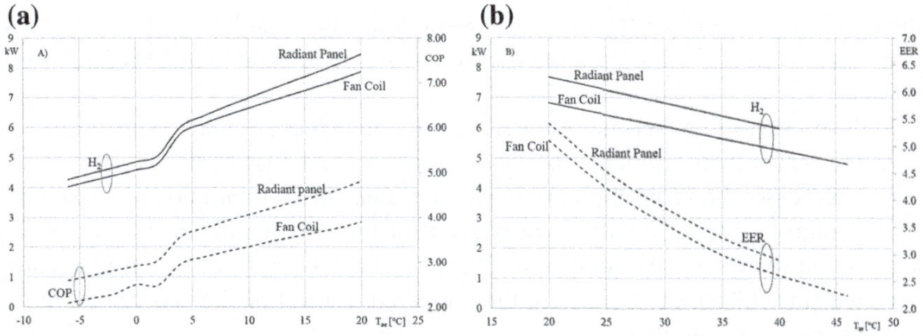

Fig. 3 Thermal power and COP **a** and cooling power and EER **b** trends in function of the outdoor air temperature and for two different types of emitters

the supplied power and the performance indexes, in function of the external air temperature, are shown for a commercial air-water heat pump considering radiant panels and fan coils as emission systems.

The size of heat pumps has to be chosen carefully in relation to the required thermal and cooling loads: if the latter are lower than the supplied one, in absence of thermal storage systems, heat pumps have to operate in partial mode with correspondent worsening of the performance indexes. Usually, the magnitude of heating and cooling loads is strongly different; therefore, the device has to be chosen in function of the greatest power. Thus, the heat pump chosen for the winter period could be oversized than the summer period or vice versa. A possible solution is represented by a generation system constituted with more heat pumps activated in cascade. In Fig. 4, the required and supplied powers of a generator system with 4 heat pumps for heating applications, is shown. At the outdoor temperature t_{oa}, the power P has to be provided, therefore the heat pumps HP1 and HP2 are switched on and operate at the maximum power, with the best COP, whereas only heat pump HP3 has to

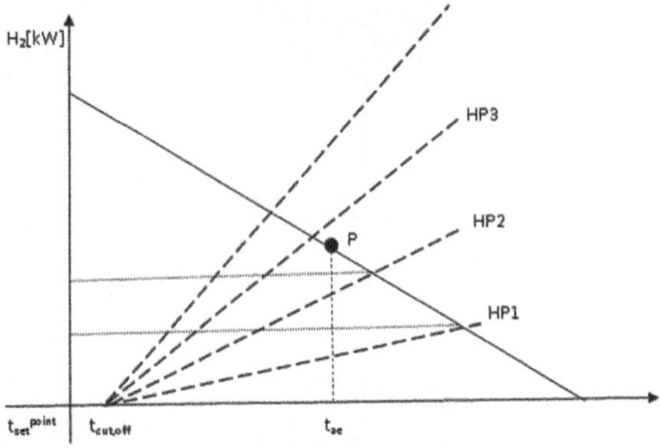

Fig. 4 Example of operation of more heat pumps in cascade to limit partial mode operation

operate in partial mode, avoiding the partialised operation of a sole and greater heat pump. Therefore, an intelligent control system has to consider the actual thermal load required from the building in order to determine the number of heat pumps to activate.

In the presence of a thermal storage, if the supplied thermal load compensates the extracted one, the tank temperature level remains constant. Contrarily, its values could rise or decrease, therefore the tank temperature has to be continuously monitored by the control system. Finally, improvement of the energy efficiency indexes can be achieved increasing the temperature level of the source from which the heat is extracted (in winter), or decreasing the temperature level of the source where the heat is pumped (in summer) [10].

3 Heat Pumps Assisted by Solar Energy: SAHP Systems

Active solar systems are classified in two categories: thermal solar collectors to produce thermal energy (heated air or water flow rates) and photovoltaic collectors to produce electricity. Heat pumps can be connected with active solar devices to form SAHP systems (Solar Assisted Heat Pump). By using thermal collectors, the energy indexes are increased during winter operation by the temperature level growth of the coldest source [11]. Thus, a COP growth can be observed, reducing the electric absorption, and in some circumstances the thermal level is sufficient to supply the emission system directly. In summer, instead, the available thermal energy cannot be used to improve EER, but it could be used to produce domestic hot water. However, an appropriate size of the solar collector field has to be identified in order to avoid a thermal energy availability that exceeds the summer demand of the domestic hot water. PV generators provide a fraction of the electric load required by heat pumps and can be employed for the whole year. These systems do not produce an actual improvement of the performance indexes, but the possibility to satisfy partially the electric loads by using solar radiation as primary source makes the system environmentally more sustainable [12]. From the economic point of view, the cost of PV driven heat pump systems is similar to those required for SAHP equipped with thermal solar collectors, because the recent financial incentives on PV technologies have produced a substantial cost reduction [13]. At parity of solar radiation availability, thermal collectors provide a higher system efficiency reaching elevated temperatures, but can be used only to accumulate hot water useful for DHW production or for heating applications. SAHP systems employing PV panels are less efficient, more limited in term of temperatures levels but allow for satisfying both heating and cooling demands. SAHP system with PV/T panels are more attractive: the latter, for the same occupied space, provides both hot flow rate to use as thermal source in heat pump systems and electricity to cover part of their electric demands [14]. However, these systems are still more expensive and thermal energy has to be opportunely dissipated in summer to avoid an electric efficiency decrement.

4 The Thermal Energy Storage Systems (TES)

The role of thermal energy storage systems (TES) in air-conditioning plants is significant because they are largely recognized as the most reliable systems for the integration of renewable sources in the coverage of building energy demands [15]. In the presence of generation systems supplied by electricity, the TES capacity is that of converting and absorbing any instantaneous electrical surplus (peak production) into thermal energy to use in a deferred manner. Therefore, TES systems can be employed also as an effective tool for electricity demand managing programs (DSM, demand-side management) rationalizing the aleatory electricity production provided by renewable technologies [16]. TES systems, in fact, cover the role of managing the mismatching between the availability of renewable energy and the users energy demand. Moreover, TES systems are simple, safe, efficient, cheaper, of high environmental quality and they are suggested for their versatility, technical and economic feasibility and for the possibility to employ different technologies in relation to the connected plant typology. The intelligent management of electricity fluxes provided by renewable technologies in a TES allows to compensate the decoupling between energy demand and supply. When the production fluctuations typical of renewable sources are shifted in function of the energy demand, the punctual surplus is inevitably lost. In presence of a TES system, instead, this surplus can be transformed in thermal energy and stored in the accumulator that acts as a buffer. An integrated smart control system allows for exploitation of the storage potential and the maximization of the production from renewable sources. TES can be involved not only when the renewable electricity exceeds requirements, but also to store it when external electricity from the grid (whose cost is continuously variable) is cheaper. Thus, the smart management of the electric fluxes increases the production from the renewable source and allows for a rationalized distribution on a large scale favouring the cooperative energy exchange typical of smart districts. In order to qualify a TES system, some aspects such as long life span, large storage capacity, reduced thermal losses through the accumulator envelope, have to be considered. The different solutions of thermal storage have to be chosen in relation to technical and economic factors, however other aspects such as the period requested for the accumulation of thermal energy (daily, monthly or also seasonal), the economic feasibility and the actual operative conditions, have to be considered. The best renewable technology to adopt in connection with TES depends especially on the demand magnitude. Currently, the TES market offers different types of accumulators, which can be classified in three principal categories: sensible TES, latent TES and thermo-chemical TES [15]. The first usually employs water as storage medium in stratified tanks, but other technologies based on the employment of concrete, rocks or mixed system constituted by water and rocks also are available. Latent storages can be made by water ice or with changeable phase materials (PCM) in the desired temperature range. Thermo-chemical storages still represent a limited application and their operation is based on the activation of particular chemical reactions between reagent substances.

5 Radiant Panel for Heating/Cooling Applications

Radiant emitters offer good performances in term of energy savings and comfort conditions, because of the advantageous inlet temperatures and the prevalent infrared radiant exchange. Generally, radiant panels allow for a better exploitation of generation systems, such as condensation boilers or heat pumps, obtaining higher performance indexes. The prevalent radiative thermal exchange involves radiant panels and the indoor in view surfaces first, successively the convective thermal exchange produces the set-point attainment. Thus, the internal air temperature change occurs with delay due to the thermal inertia of the building fabric. A simple ON/OFF control, that acts on the supply flow rate, produces elevated times for the set-point temperatures achievement, therefore it is not suggested and could provide a worsening of the comfort conditions [17]. A key element to increase the plant response to the internal variations is represented by the modality of radiant panel installation: it is possible to distinguish between integrated and non-integrated panels in the structure, mounted in the floor, in the ceiling or in the vertical walls. Integrated panels, such as radiant floors, are considered as an active surface installed inside the structure characterized by an elevated thermal inertia, providing a high response time and a thermal "*flywheel*" effect. A radiant system not integrated in the structure is represented by chilled ceilings, or suspended radiant ceilings, because they are physically disassembled from the building with reduced time response. Both the mentioned systems can be used for heating and cooling applications, however they can act only on sensible loads leaving the control of latent loads to natural or mechanical ventilation, or to other systems [18]. Radiant panels installed in the floor or in the vertical walls can be penalized by the presence of furnishing, therefore the ceiling installation is the most suggested. A typical section of a radiant floor is represented in Fig. 5a: under the floor covering a lightweight concrete slab with additives (to increase the thermal conductivity) is located, in which the pipes are installed to transport hot or cold flow

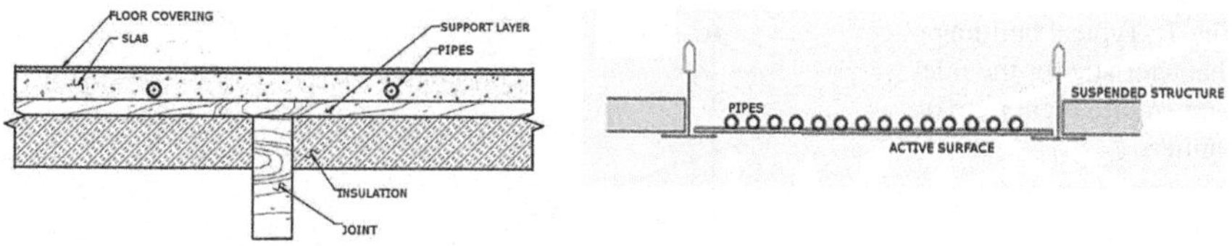

Fig. 5 Sections of a radiant floor **a** and a non-integrated chilled ceiling **b**

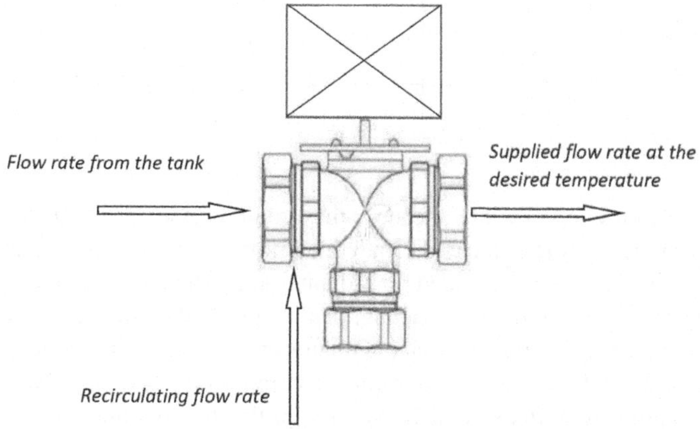

Fig. 6 Scheme of a three-way motorized valve for the temperature regulation

rates, then an insulate layer to avoid the thermal losses downward and finally suitable joints to compensate the thermal dilatations. In Fig. 5b a scheme of a chilled ceiling is represented: a suspended structure supports the active surface supplied by a network of pipes located in the upper part. The active surface can be made with different materials such as mineral fibres or metallic plates, the latter more performant but also more expensive. In order to accelerate the emission system time response, a suitable control strategy varies the inlet temperature in relation of the boundary conditions [19]. Usually, the inlet flow rate is extracted from a storage tank equipped with a motorized three-way valve to obtain the desired temperature, using the returning flow rate from the emitters as mixing fluid (see Fig. 6).

A typical control strategy implemented in air-conditioning plants is the "*climatic*" regulation: in function of the outdoor air temperature, a precise relation calibrated in function of the building characteristics, sets the inlet temperature. For instance, in winter the inlet temperature is increased with the outdoor temperature decrement in order to accelerate the emitters' response, depending on the weather conditions. Every building can be described, in a simplified way, by a linear trend characteristic, to be determined by means of experimental tests, where the slope is related to the mean heat loss coefficient of the building envelope, as shown in Fig. 4 and in a more detailed way in Fig. 7. The inlet temperature regulation has to consider another crucial aspect, because it fixes the panels surface temperatures. The latter cannot

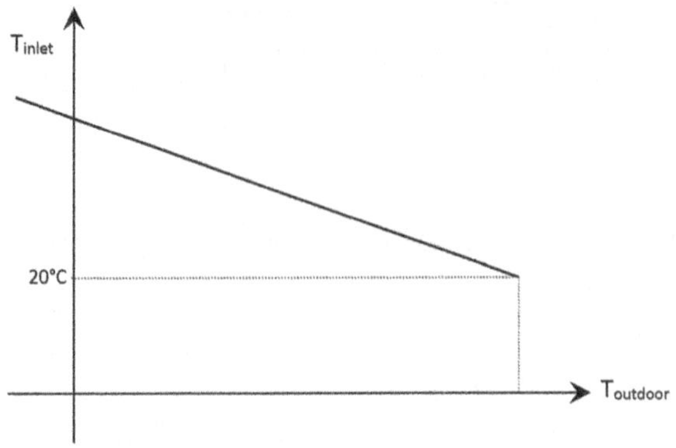

Fig. 7 Typical building characteristic for the inlet temperature regulation of emitters

be excessively elevated in winter, because they could provide discomfort conditions and damage to the emitters due to thermal dilatations. Usually, surface temperatures higher than 29 °C have to be avoided for radiant floors, whereas temperatures slightly more elevated can be admitted for radiant panels installed on the ceiling.

In summer, when the panel surface temperature is next to the indoor dew-point temperature, the risk of water vapour condensation on the active surface has to be considered carefully. It does not depend only on the emitters inlet temperature but also on the indoor humidity level, therefore a prudential control strategy to employ during cooling application has to be developed. Usually, the inlet temperature is set equal to the indoor dew point temperature: thus, the thermal resistances between the supplied flow rate and the panel surface provide a temperature increment that assures a surface temperature that is always greater than the dew point, avoiding the condensation phenomena [20]. However, in the presence of an elevated humidity level in the indoor environment, inlet temperatures could be insufficient to remove thermal loads, thus the indoor set point temperature cannot be guaranteed. Considering the employment of chilled ceilings, theoretical evaluations have demonstrated that it is possible to decrement the dew point temperature by 4 °C in the presence of panels made of mineral fibres or by 2 °C for metallic active surfaces, permitting to respect the set point [21]. Contrarily, a suitable control of the internal latent loads is required: often natural or mechanical ventilation systems are sufficient to contain the indoor humidity levels [18]. In Fig. 8, a thermographic image of a chilled ceiling, where surface temperatures are slightly variable around the value of 17 °C in proximity of the serpentines, is reported.

6 A Smart Air-Conditioning Plant and Control Strategy

Considering the advantages achievable by employing different devices in a combined way, an opportune management allows the attainment of higher performance indexes with a significant use of renewable sources. Thus, energy and economic savings can

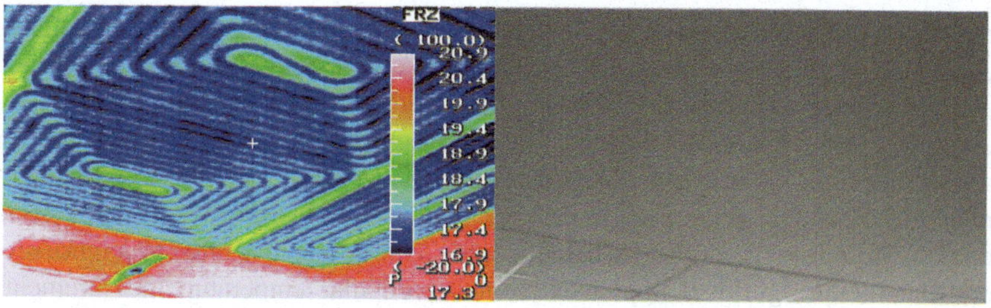

Fig. 8 Thermographic image of a chilled ceiling during a cooling application

be achieved with acceptable payback periods. An efficient and smart air conditioning plant, employed to satisfy heating and cooling requirements, can be constituted by:

- a precise number of air-water heat pumps, in order to allow a rational exploitation of the generation system in relation of the required thermal load;
- a PV plant, to provide a fraction of the electric load absorbed by heat pumps;
- a storage tank, to accumulate sensible energy, in order to manage the mismatching between required thermal energy and renewable energy availability;
- an integration system to guarantee a minimal temperature level in the storage tank in winter;
- a mixing valve, to regulate the emitters inlet temperature;
- internal thermostats, to manage indoor overheating/undercooling strategies;
- radiant ceilings, to supply thermal loads and to exploit the building structure as further thermal storage system.

The control system can be divided in two levels: the first to manage the generation system (primary control), the latter to regulate the emission system (local control). The control strategy can be opportunely improved if different parameters involved in the analysis of the building-plant system are known in advance. Analogously to the electric loads, in the presence of multi-generation systems for air-conditioning applications, the possibility to know the heating and cooling trend of the building-plant system in a predictive manner, allows for the electric energy planning. The predictive knowledge of the required thermal load, in fact, allows to define a suitable management of the generation units and, consequently, a projection of the required electric loads. Moreover, in a future scenario of Smart Cities development, the planning of electricity consumptions allows the coordination of air-conditioning plants with other electric loads in order to make the aggregated users less dependent on the external grid. Regarding the building thermal loads, in literature different predictive models conceived in function of the external weather conditions, geometrical and thermo-physical building characteristics, provide prediction of these loads generally with an advance of 24 h [22, 23].

The control system is an articulated device, where different types of technologies are involved. Preliminarily, an embedded computer is required in order to implement appropriate algorithms able to provide the building heating/cooling loads and the performance indexes concerning heat pumps and integration system. A suitable dashboard is required to set different parameters, concerning the thermo-physical and the geometrical properties of the building and the characteristics of the devices employed in the generation system. The same computer has to be predisposed for an internet connection in order to exploit different services, such as the provision of meteorological data or the situation concerning the electric energy costs on the local market. Moreover, the embedded calculator has to be arranged in order to acquire several data for the algorithms implementation, in particular temperature measurements involving the thermal storage system, the emitters inlet temperature, the outdoor and indoor air temperatures. Finally, the decisions taken by the calculator are executed by means of apposite actuators that not only intervene on the generation system (devices switched on or switched off), but which are able to modify the set-point values concerning thermostats, tank thermal level and emitters inlet temperature.

6.1 Primary Control for Heating Applications

In winter, the smart air-conditioning plant requires both electric energy (to supply heat pumps) and a primary source for the integration system. As the thermal storage system offers a finite capacity, the planning program has to consider its state, in particular the attained thermal level. Therefore, the control system has to be equipped with apposite probes (RTD PT 100, for instance) in order to measure the tank thermal level. In particular, a resistance temperature detector has to be installed in the superior part of the tank for heating applications, due to the temperature stratification effect. The temperature measurements allow for considering the thermal energy availability in the storage tank.

The presence of a PV generator allows for absorption of the self-produced electricity by the heat pump system or, if the latter is not sufficient, to integrate it with that provided by the external grid. If PV electricity surplus is not available (overcast conditions, night hours), the electric energy has to be absorbed completely from the external network. However, the presence of the integration system could make the purchase of its primary source more competitive than electricity acquisition, therefore the thermal cost of the primary source used by the integration system (in €/kWh$_{th}$) has to be compared with the thermal cost provided by the heat pumps. The thermal costs of the generation system are continuously changeable, due to the daily variability of the electricity cost (CE) and the variability of the performance indexes of the employed devices (in function of weather conditions and/or the operation mode), therefore different economic frames have to be analyzed. Supposing the presence of an integration system supplied by methane as primary source, characterized by a lower heating value (LHV) of about 10 kWh/m^3 and an average cost (C_M) of 1 €/m^3, the control system has to carry out the following evaluations. Preliminarily, by means of apposite correlations derived from the manufacturer data sheet, the punctual efficiency (η) of the integration system has to be determined in function of the required heating load; if the efficiency is equal to 80%, the integration system thermal cost (C_I) is calculated as:

$$C_I = C_M/(LHV \cdot \eta) = 1\ €/m^3/(10 \cdot 0.8) = 0.125\ €/kWh_{th} \quad (1)$$

whereas the thermal cost connected with the heat pumps (C_{HP}), hypothesizing a punctual COP of 2.5 and an electricity cost (CE, provided by the national institutions concerning the management of the electric market) of 0.30 €/kWh_{el}, is:

$$C_{HP} = CE/COP = 0.30/2.5 = 0.12\ €/kWh_{th} \quad (2)$$

Therefore, C_{HP} is lower than C_I and heat pumps have to be used if this situation occurs. In a different frame, where the COP is equal to 2 and the integration system efficiency is 95%, the integration thermal cost (0.105 €/kWh_{th}) becomes more competitive than the heat pump thermal cost (0.15 €/kWh_{th}), therefore the employment of the first device is determined. The control system has to be able to evaluate the instantaneous performances indexes of the several devices constituting the generation system, depending on precise parameters such as the outdoor air temperature (T_{oa}), and the heating load requested by the building (Q_h). With reference to Fig. 3a, in fact, by setting the supplied water temperature, the COP dependency from the outdoor air temperature can be determined with good agreement by a second order relation:

$$COP = a \cdot T_{oa}^2 + b \cdot T_{oa} + c \quad (3)$$

where the three coefficients *a*, *b* and *c* can be evaluated by a polynomial regression in function of the data provided by the manufacturer. Generally, a winter set-point value of 50 °C for the thermal level of hot water contained into the tank is suitable in order to exploit the thermal storage capacity. Later, the motorized three-way valve will regulate the emitters inlet temperature at the desired value. The heating load requested from the building allows for determining:

- the number of heat pumps to activate in the generation system;
- the capacity ratio (CR) of the devices that affects the performance indexes.

In a simplified way, the latter can be evaluated as the ratio between the requested heating load and the nominal power of a single heat pump. Thus, if CR = 1 only a heat pump operates in nominal conditions without a worsening of the COP; if CR > 1, other heat pumps have to be activated in cascade. For instance, if CR = 1.7, two devices will be switched on, with the first heat pump operating in nominal conditions, the second one in partialisation mode at 70%. Generally, modern heat pumps equipped with inverters offer a non-penalized COP if the CR is greater than 0.5 [6]. Regarding the required heating load, lumped parameters models depending on few geometrical and thermo-physical parameters are available in literature [24]. However, the boundary initial conditions at the instant $t = 0$ have to be known, whereas in predictive models are not required because they consider the thermal "*history*" of the building structure.

In presence of the PV plant, the electricity surplus ($El,_{PV,s}$) is another crucial parameter to be considered by the control system, supposing that the self-produced electricity is absorbed by other electric loads in a priority way (domestic appliances, artificial lighting system). The air-conditioning plant, in fact, is considered as an alternative electricity storage device. The control system has to manage different situations adequately, depending on energy and economical parameters. In particular, when heating is required ($Q_h > 0$) if a PV surplus is not available ($El,_{PV,s} = 0$), the thermal costs of the devices employed in the generation system have to be compared in order to identify the cheapest solution. If a PV surplus is present, heat pumps will be activated and, if the electricity is not sufficient to maintain the tank thermal level, it can be eventually integrated with that provided by the external grid. If the tank is saturated, the generation system is switched-off and the eventual PV surplus can be transferred to the external environment. Instead, when heating is not required, if a PV surplus is not available the generation system is switched-off, otherwise the thermal level of the storage tank has to be considered. If the latter is lower than the set-point, then the heat pump system is activated anyway (the produced thermal energy is stored in the tank). Contrarily, heat pumps will also be used in the presence of a saturated tank by acting on the zone thermostats with an indoor air temperature set-point raising (T_{ia}, for instance from 20 to 21 °C). Thus, the thermal energy is transferred from the tank to the structure, activating heat pumps exclusively with the PV electricity surplus and exploiting the building as further thermal storage by the radiant ceilings. Indoor air temperatures greater than 22 °C are not suggested in order to avoid discomfort conditions. Consequently, the control system has to be able to modify the indoor set-point temperatures acting directly on the single zone thermostats. Only if the storage tank is saturated and the indoor air temperature set point is already too elevated, the PV surplus can be transferred to the external grid. A flow chart of the described control strategy is shown in Fig. 9.

6.2 *Primary Control for Cooling Applications*

The summer management of the generation system is easier than heating application because the integration system will be never used. Therefore, the sole involved energy is the electricity provided by PV generator ($El,_{PV,s}$) or by the external network to supply the heat pumps system; considering that the self-produced electricity cost is generally lower than the electricity purchase cost from the external grid, PV surplus has to be used in priority way. Though cooling loads (Q_C) are not requested from the building, PV electricity surplus will be equally used considering the possibility to store the cooling energy in the storage system. This time, a temperature set point of the storage tank (T_{tank}) of about 10 °C is appropriate for its rational exploitation; successively, the emitters inlet temperature will be regulated by means of the motorized three-way valve [25].

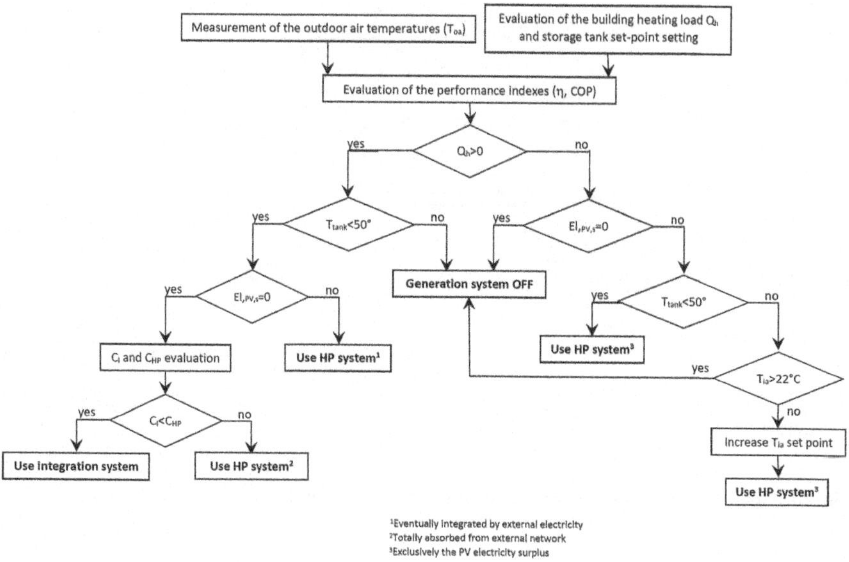

Fig. 9 Control strategy employed by the smart air-conditioning plant during heating applications

In the presence of a saturated tank, the set point temperature of the heat pumps will not be modified, but an undercooling strategy of the indoor environments will be carried out. Analogously to the winter case, the control system acts on the zone thermostats in order to decrease the indoor air temperature set point (T_{ia}, setting 25 °C rather than 26 °C), thus the cooling load will be equally extracted from the storage tank and transferred to the building structure by the radiant emitters. Only in the presence of a saturated tank and a reduced indoor set point temperature, the PV electricity surplus can be delivered to the external grid. The flow chart of the control strategy developed for cooling applications is shown in Fig. 10.

6.3 *Local Control*

The management of the emission system is delegated to a local control that adjusts prevalently on the inlet flow rate and temperature of the radiant panels. It acts prevalently on an electronic hydraulic pump and on the motorized three-way valve; whenever the indoor air temperature drops below the set-point value, the zone thermostat, through the control system, activates the hydraulic pump that operates with variable flow rate to activate the emitters. The electronically controlled hydraulic pump is required to manage the emission system opportunely, because the water flow rate increases with the number of activated radiant ceiling circuits. A PLC, after polling the zone thermostats, sends an ON/OFF control signal to actuate each electric zone

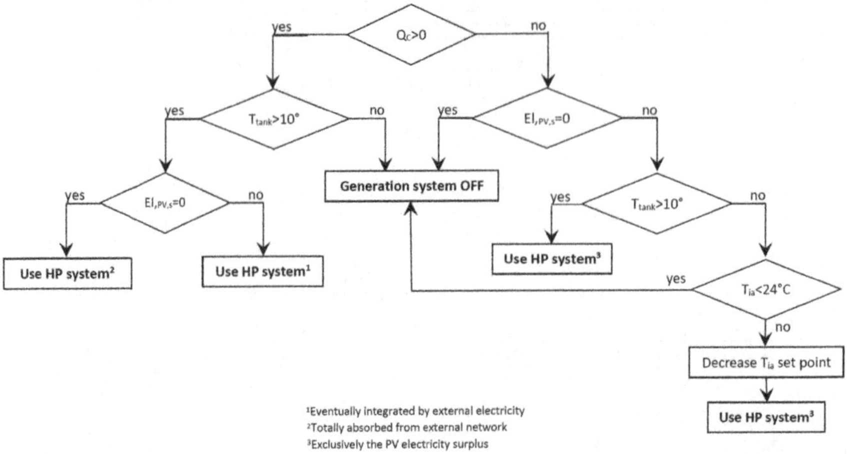

Fig. 10 Control strategy of the smart air-conditioning plant during cooling applications

valve in order to activate the proper hydraulic path. The motorized three-way valve regulates the inlet temperature both in winter and in summer to provide a precise set-point value. The latter can be set as a fixed value or can be set as variable depending on the indoor or outdoor conditions: in the first case, when the indoor temperature is far from the set point, the inlet temperature is increased in winter, whereas the opposite occurs in summer. Generally, the motorized valve uses a PD feedback system to obtain the desired outlet temperature quickly. Finally, in order to avoid the condensation risk of the water vapour on the active surfaces of the ceiling, the control system decides the inlet temperature depending on the indoor dew point temperature.

7 A Demonstrative Smart Air-Conditioning Plant

A demonstrative smart air-conditioning plant was installed at the University of Calabria (Italy) suppling the locals that host the "DOMUS" research district, consisting in a generation and emission systems similar to those prior described. In addition to the several features already cited, the intelligent air conditioning plant was inserted into a wide research project that involves the solution of the so called "electric prosumer problem" for the rational exploitation of the electric loads [26]. In particular, in function of the predicted electricity, the prosumer problem identifies which electric load can operate, also involving the air-conditioning plant. The latter, in fact, is constituted by:

- N° 2 air-water heat pumps equipped with inverter and with a PV generator of
- 4 kW_p tilted of 30° and exposed toward south, installed on the building roof;
- a pellet cogenerative boiler with Stirling engine as integrative system;

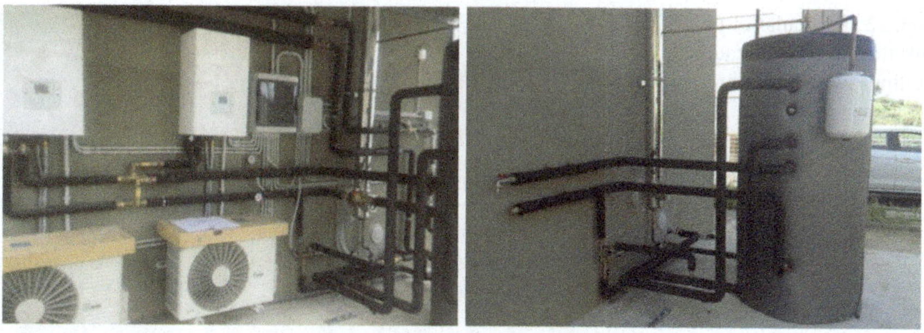

Fig. 11 Particular of the heat pumps, of the correspondent external units and of the vertical storage tank of the demonstrative smart air-conditioning plant

- a 800 L thermal storage tank employed both for heating and cooling;
- radiant ceilings made by mineral fibres panel as emission system.

The generation system was placed outside the building with heat pumps installed on an external wall by means of anti-vibrating bars, whereas the other components were placed in adjacency inside a delimited space (see Fig. 11). Only the integration system was installed inside for obvious operation reasons and for the device connection to an internal electric *microgrid* that manages the electric fluxes [27]. The thermal storage system is represented by a vertical insulated tank equipped with two heat exchangers: the first is connected at the bottom with the heat pumps, whereas the second, at the top, is connected to the integration system. The water flow rate for the emission system is extracted from the storage tank without the employment of a heat exchanger, in order to limit the tank temperature stratification. A motorized three-way valve regulates the emitters inlet temperature and was installed in the upper part of the tank on the extraction side. The radiant panels decoupled from the structure and installed on the ceiling have an active surface of about 125 m^2 that provide heating and cooling loads to 4 different rooms, whereas the remaining surface is equipped with usual suspended ceiling (Fig. 12). In Fig. 13 the radiant system layout, for the different rooms, with the particular of the water distribution network and the correspondent ON/OFF electro-valves employed to activate the emitters hydraulic circuit, is shown. Globally, eight radiant panel circuits have been used: two rooms use a single hydraulic circuit (A), one room two circuits (B) and one room four circuits (C).

In Fig. 14 the sketch plan concerning the generation system with the position of the probes employed for the measurement of the main energy parameters, is shown. A suitable data acquisition system (DAQ), in fact, was provided for the monitoring of the generation system, evaluating different parameters also at energy level. In particular, energy flowmeters mounted on the hydraulic circuits connected to the storage tank, allow to measure the supplied and the extracted thermal energies for the evaluation of the system thermal balance. Flow rates and temperatures can also be monitored in real time by means of different manometers installed in the main

Fig. 12 Particular of the installed decoupled radiant ceilings in mineral fibre

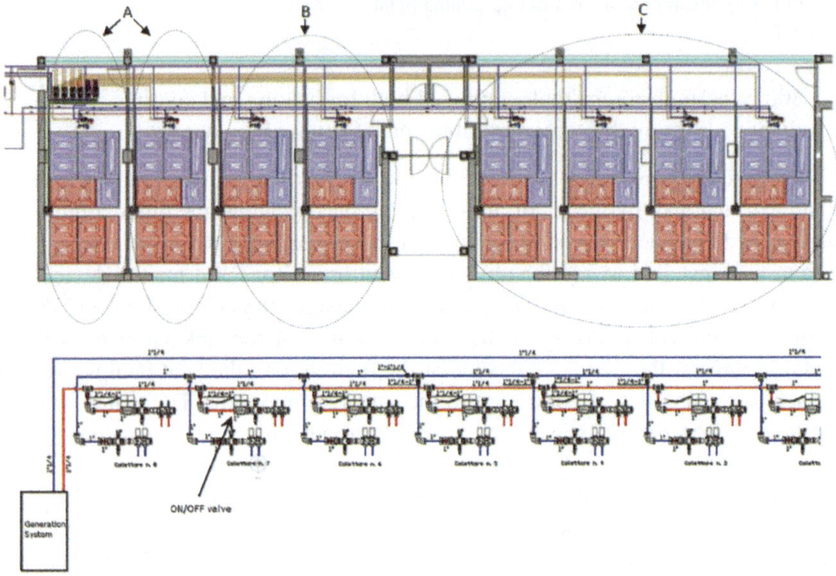

Fig. 13 Radiant panel layout and particular of the main hydraulic distribution system

interest points of the hydraulic circuits. Apposite modules have been provided to measure the electric energy fluxes, in particular those absorbed by heat pumps and provided by the PV generator. Thus, performance indexes concerning heat pumps can be evaluated in the actual operation conditions, as well as the electric fraction provided by the renewable source. The set point for the boiler (or the water supplied by heat pumps) was set in 10 °C in summer and 55 °C in winter, whereas the inlet emitters' temperature was set constant and equal to 35 °C in winter and to the indoor dew point temperature in summer. The nominal power of a single heat pump is 5 kW in heating operation (for a supplied water temperature of 35 °C and an external air temperature of 7 °C), and of about 4 kW in cooling operation (chilled water at 15 °C and external air temperature at 30 °C). The three coefficients of Eq. (3) were evaluated in $a = 0.038$, $b = 0.144$ and $c = 3.42$. In winter, a sole heat pump is activated when

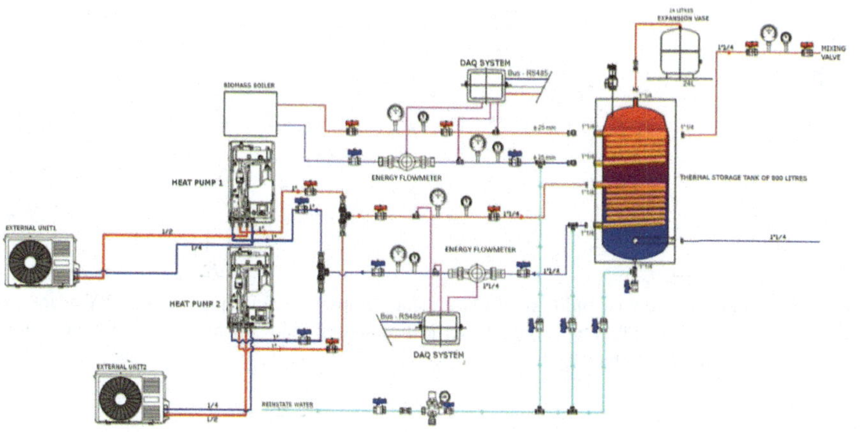

Fig. 14 Sketch plan of the generation system employed for heating and cooling application

the tank temperature ranges between 45 and 55 °C, both operate with temperatures below 45 °C. If the heat pump operation is not sufficient to guarantee the minimal temperature level of 35 °C, the boiler intervenes until the tank temperature reaches the set-point, maintaining the heat pumps switched off.

The integration system is characterized by a thermal power of 14 kW and an electric power of 1 kW (in nominal conditions), managed by the internal *microgrid*. In summer, two heat pumps are employed for a tank temperature greater than the set point, a sole device for temperatures between 7 and 10 °C. The air-conditioning plant has been functioning since the end of the summer 2017.

7.1 Energy Performances of the Demonstrative Air-Conditioning Plant

The air-conditioned building is occupied occasionally, therefore the PV electricity production is prevalently employed for the heat pump system. Thus, PV production is almost coincident with the PV surplus to employ in the air-conditioning plant if the tank temperature and the indoor air temperature allow for the accumulation of produced thermal energy, otherwise the internal *microgrid* carries out its rational management by means of energy and/or economic criteria. In Fig. 15, for the period ranging from the 1st November to the 31st December 2017, the trends concerning the electricity absorbed by the heat pumps, distinguished between that self-produced and that absorbed from the external grid, of the PV surplus exploited with the overheating strategy and the storage tank temperature, are shown. The employment of the external electric energy (blue line) is limited for few days during December, only to guarantee the minimal thermal level in the tank (due to the large thermal energy extraction to

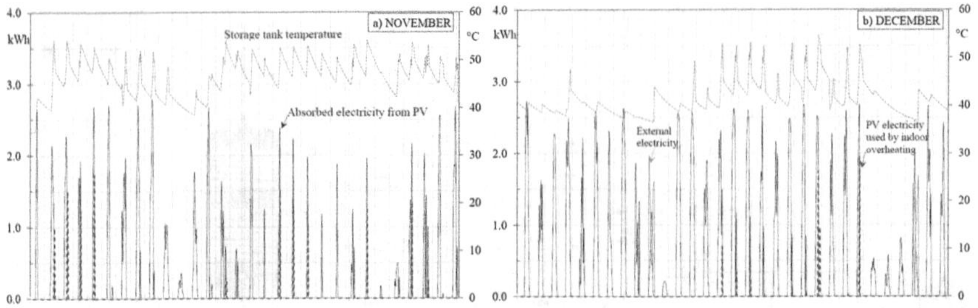

Fig. 15 Demonstrative air-conditioning plant: heat pumps electricity provided by PV plant and external network, PV surplus employed for overheating strategy and tank temperature for the months of November **a** and December **b**

cover the heating loads), and in those circumstances the correspondent cost was cheaper than biomass cost. The red line highlights the PV electricity surplus used for the overheating strategy, because the thermal level of the storage tank is already too elevated. During the monitored period, the indoor set-point air temperature was elevated to 21 °C for 18 days, with a major frequency during November due to the limited heating loads that produce a higher temperature in the storage tank. The latter presents temperature values always contained in the interval of 35–55 °C, suitable for the emission system, observing in days with limited solar radiation (low PV electricity) a consistent decrement of the thermal level of the stored water and no employment of the external electricity or of the integration system.

In Table 1, the main monitored energies in the smart air-conditioning plant, are listed. In Italy, 19% of the electricity absorbed from the external network is produced by renewable sources, analogously 80% of the biomass primary energy is considered renewable, therefore the control system assures a percentage of renewable source employment equal to 92.4% in November and 85% in December. Exclusively from the point of view of electric energy, 100% is provided by renewable sources in November and a percentage of 94.7% was assessed in December. Regarding the thermal energy provided to the storage system, that connected to the heat pumps prevails especially in November due to a major availability of solar radiation and to reduced heating loads. The ratio between the thermal energy produced by renewable sources and the thermal energy required by the building was evaluated as 121% in November and 83.5% in December; for the first, the surplus was opportunely stored in the tank or in the building, contributing to increase their internal thermal energy. The operation of the two heat pumps in cascade allows for the achievement of excellent COP values for the first device; the second one is more penalized because it has to operate in partial mode to cover the power peaks. However, the worsening of the second heat pump occurs when renewable electricity is available, therefore it does not actually represent an inefficient energy employment. The exploitation of the self-produced electricity in the smart air-conditioned plant has allowed for the attainment of an economic saving of about €115 in two months of the heating period. The cost of the consumed pellets was around 50 €, corresponding to approximately

Table 1 Main energy parameter monitored in the smart air-conditioned plant during November and December 2017: electricity provided by PV generator (A) and by external network (B), the PV surplus employed for thermal storage in the building (C), thermal energies provided by heat pumps (D) and integration systems (E), supplied heating energy by emitters (F), average tank temperature (G)

	A (kWh_{el})	B (kWh_{el})	C (kWh_{el})	D (kWh_{el})	E (kWh_{el})	F (kWh_{el})	G (°C)	COP1 (–)	COP2 (–)
November	169.3	0	14.3	558.7	103.0	503.2	46.5	4.12	1.56
December	323.6	22.5	6.3	1120.4	669.9	1703.4	41.4	3.56	2.4

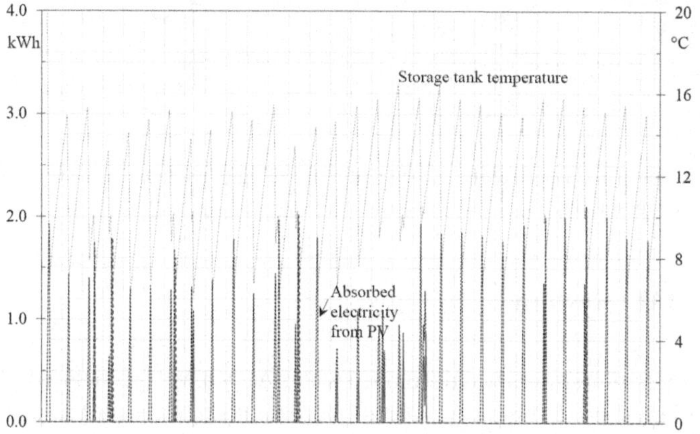

Fig. 16 External network (blue line) and PV electricity (black) absorbed by heat pumps, PV electricity stored under the form of thermal energy in the building structure (red) and tank temperature (violet) in the months of September for cooling applications

150 kg of biomass. Similar evaluations have been carried out for September (cooling application), with trends reported in Fig. 16: tank temperature levels range from 7 to 15 °C, with a reduced heat pump operation time that provides a rapid temperature decrement in the storage system due to the limited requests of cooling loads.

In some circumstances, with PV surplus availability, an undercooling strategy was undertaken reducing the indoor air set-point temperature from 26 to 25 °C, avoiding the transfer of the produced PV electricity to the external grid. However, in 7 days the absorption of external electricity was requested to avoid excessive temperature level in the storage tank, that could not guarantee the indoor air set-point. From the energy point of view, the absorbed PV electricity amounts to 89 kWh, from which 11 kWh were employed to produce thermal energy to store inside the building structure. Moreover, a further 7 kWh have been absorbed by heat pumps from the external network, producing an overall cooling energy of 288 kWh; therefore, the mean monthly EER was evaluated as equal to 3.

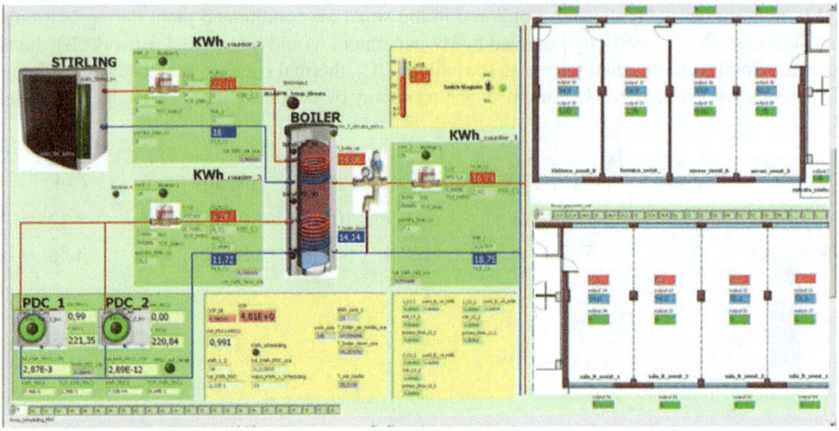

Fig. 17 Labview front panel for the visualization of the main parameters involved in the smart air-conditioned monitoring

8 The DAQ System

The data acquisition system was developed with LABVIEW®, namely a dataflow programming language (Laboratory Virtual Instrument Engineering Workbench) suitable for data acquisition, data analysis and data presentation. A chassis that can hold eight possible I/O modules interfaced with temperature, analog, digital and counter sensors, makes the I/O hardware. In Fig. 17, the LABVIEW® front panel of the smart air-conditioning system, is shown. The energies involving the storage tank are also measured at hourly level, with the exception of the thermal energy losses through the tank shell. Two energy flowmeters are located on the heat pumps and on the integration system, a third one is installed on the extraction side to determine the thermal energy transferred to the emission system.

Each energy flowmeter is made by a water flow rate measurement sensor and by two RTD PT100 for the evaluation of the temperatures in the inlet and outlet pipes. The probes signals are transferred to a signal conditioning apposite module, elaborated for communication through an Ethernet cable, and sent to a personal computer, where they are analysed, visualized and stored. The same DAQ acquires the temperature at the bottom and top of the storage tank, as well as the temperature inside the conditioned rooms. Moreover, it acquires the ON/OFF state of the electrovalves to identify the operation of a precise radiant ceiling system. Other modules and sensors are used for the measurement of the electric powers absorbed by heat pumps; thus, the data analysis involving the provided thermal energy and the absorbed electric energy allows for the COP/EER calculation.

The DAQ is integrated with a control system in order to allow the operation of the heat pumps in accordance to a suitable strategy. Because the prosumer problem is able to determine 24 h in advance the electric energy available for the heat pumps, the

DAQ system acquires this information and the successive day actuates the heat pumps until the absorbed electric energy (measured by the same DAQ system) reaches the values indicated by the prosumer problem solution. The latter information is usually stored in a power cloud where the DAQ system connects to retrieve data.

References

1. ENEA, Italian agency on the energy efficiency, Analysis and results on the energy efficiency policy in Italy, Executive summary, 2017
2. A. Aswani, N. Master, J. Taneja, A. Krioukov, D. Culler, C. Tomlin, Energy-efficient building HVAC control using hybrid system LBMPC, in *The Proceedings of 4th IFAC Conference, International Federation of Automatic Control*, 2012
3. G. Nicoletti, N. Arcuri, R. Bruno, G. Nicoletti, A technical and environmental comparison between hydrogen and some fossil fuels. Energy Convers. Manag. **89**, 205–213 (2015)
4. G. Oliveti, N. Arcuri, R. Bruno, A. Mazzuca, Energy performances of an absorption chiller supplied by solar collectors in mediterranean area, in *The Proceedings of the ISES Solar Word Congress 2005*, Florida, USA, 2005
5. R.G. Morgan, Solar Assisted Heat Pumps. Sol. Energy **28**(2), 129–135 (1982)
6. K.J. Chua, S.K. Chou, W.M. Yang, Advances in heat pump systems: a review. Appl. Energy **87**(12), 3611–3624 (2010)
7. Castillo-Cagigal, E. Caamano-Martın, E. Matallanas, D. Masa-Bote, A. Gutierrez, F. Monasterio-Huelin, J. Jiménez-Leube, PV self-consumption optimization with storage and active DSM for the residential sector. Sol. Energy **85**, 2338–2348 (2011)
8. H.J. Sauer, R.H. Howell, *Heat Pump Systems* (Wiley, New York, NY, 1983)
9. Directive 2009/28/EC of the European Parliament and of the Council of 23 April 2009 on the promotion of the use of energy from renewable sources and amending and subsequently repealing Directives 2001/77/EC and 2003/30/EC
10. A. Dikici, A. Akbulut, Performance characteristics and energy–exergy analysis of solar-assisted heat pump system. Build. Environ. **43**(11), 1961–1972 (2008)
11. T.T. Chow, A review on PV/T solar technology. Appl. Energy **87**(2), 365–379 (2010)
12. J. Jie, L. Keliang, C. Tin-tai, P. Gang, H. Wie, H. Hanfeng, Performance analysis of a photovoltaic heat pump. Appl. Energy **85**(8), 680–693 (2008)
13. D. Feldman, G. Barbose, R. Margolis, R. Wiser, N. Darghouth, A. Goodrich, Photovoltaic (PV) pricing trends: historical, recent, and near-term projections, Lawrence Berkeley National Laboratory, 2014, https://escholarship.org/uc/item/06b4h95q
14. G. Pei, J. Ji, K. Liu, H. He, A. Jiang, Numerical study of PV/T-SAHP system. J. Zhejiang Univ. **9**(7), 970–980 (2008)
15. I. Dincer, M. Rosen, *TES: Systems and Applications* (Wiley edition, New York, NY, 2002)
16. A. Arteconi, N.J. Hewitt, F. Polonara, Domestic demand-side management (DSM): role of heat pumps and TES systems. Appl. Therm. Eng. **51**(1–2), 155–165 (2013)
17. N. Arcuri, R. Bruno, S. Ruffolo, Radiant floor system supplied by solar collectors. Thermal and economic analysis, in *The Proceedings of the 5th ISES Europe Solar Conference*, Freiburg, Germany, vol. I, pp. 86–95, 2004
18. S.P. Corgnati, M. Perino, G.V. Fracastoro, P.V. Nielsen, Experimental and numerical analysis of air and radiant cooling systems in offices. Energy Build. **44**, 801–806 (2009)
19. N. Arcuri, R. Bruno, Energy performances of radiant ceiling heating system supplied by solar collectors, in *The Proceedings of the Second Mediterranean Congress of Climatization—Climamed*, Madrid, 2005
20. G. Oliveti, N. Arcuri, R. Bruno, Resa termica di soffitti radianti che impiegano tubi capillari per il riscaldamento e il raffrescamento degli ambienti, in *The Proceedings of the 62° A.T.I. Conference*, Cuzzolin Edition, Naples, pp. 119–123, 2007

21. N. Arcuri, R. Bruno, Prestazioni termiche di sistemi di raffrescamento a soffitto radiante e relative strategie di controllo, in *The Proceedings of the 60° ATI Conference*, Rome, 2005
22. R. Bruno, N. Arcuri, G. Pizzuti, The prediction of thermal loads in building by the EN ISO 13790 dynamic model: a comparison with TRNSYS. Energy Proc. **101**, 192–199 (2016)
23. R. Bruno, N. Arcuri, G. Pizzuti, A simplified hourly calculation code to evaluate the buildings heating load: a case study for Italian conditions. Simul. Series **48**, 174–180 (2016)
24. ISO 13790, Thermal performance of buildings—calculation of energy use for space heating and cooling, International Organization for Standardization, Geneva, 2005
25. N. Arcuri, R. Bruno, P. Bevilacqua, Influence of the optical and geometrical properties of indoor environments for thermal performances of chilled ceilings. Energy Build. **88**, 229–237 (2015)
26. G. Belli, A. Giordano, C. Mastroianni, D. Menniti, A. Pinnarelli, L. Scarcello, N. Sorrentino, M. Stillo, A unified model for the optimal management of electrical and thermal equipment of a prosumer in a DR environment. IEEE Trans. Smart Grids, https://doi.org/10.1109/tsg.2017.2778021
27. D. Menniti, N. Sorrentino, A. Pinnarelli, A. Burgio, G. Brusco, G. Belli, In the future smart cities: coordination of micro smart grids in a virtual energy District, in *International Symposium Power on Electronics, Electrical Drives, Automation and Motion (SPEEDAM)*, pp. 676–682, 2014, https://doi.org/10.1109/speedam.2014.6872123

A Comprehensive Approach to Stormwater Management Problems in the Next Generation Drainage Networks

Patrizia Piro, Michele Turco, Stefania Anna Palermo, Francesca Principato and Giuseppe Brunetti

Abstract In an urban environment, sewer flooding and combined sewer overflows (CSOs) are a potential risk to human life, economic assets and the environment. In this way, traditional urban drainage techniques seem to be inadequate for the purpose so to mitigate such phenomena, new techniques such as Real Time Control (RTC) of urban drainage systems and Low Impact Development (LID) techniques represent a valid and cost-effective solution. This chapter lists some of the recent experiences in the field of Urban Hydrology consisting in a series of facilities, fully equipped with sensors and other electronical component, to prevent flooding in urban areas. A series of innovative numerical analysis (in Urban Hydrology research) have been proposed to define properties of the hydrological/hydraulic models used to reproduce the natural processes involved.

1 Introduction

During the last few decades, the area of impervious surfaces in urban areas has exponentially increased as a consequence of demographic growth. This long-term process has altered the natural hydrological cycle by reducing the infiltration and evaporation capacity of urban catchments, while increasing surface runoff and reducing groundwater recharge. Moreover, the frequency of extreme rainfall events, characterized by high intensity and short duration, is expected to increase in the near future as a consequence of global warming [34, 38]. In addition, these processes have led to an increase in the frequency and magnitude of two undesired phenomena which negatively affect human life, economic assets and the environment: (i) local flooding and (ii) combined sewer overflows (CSOs) [44, 47]. Urban flooding occurs when the urban drainage system (UDS) overload during extreme rainfall events, causing untreated combined sewage and storm water to back up into basements and to over-

---

P. Piro · M. Turco (✉) · S. A. Palermo · F. Principato · G. Brunetti
Department of Civil Engineering, University of Calabria, Rende, CS 87036, Italy
e-mail: michele.turco@unical.it

F. Cicirelli et al. (eds.), *The Internet of Things for Smart Urban Ecosystems*, Internet of Things, https://doi.org/10.1007/978-3-319-96550-5_12

flow from manholes onto surface streets. This phenomenon is generally worsened by obstructions in conduits and manholes due to an infrequent maintenance.

CSO [42, 43] takes place when the wastewater treatment plant (WWTP) is not able to treat the wastewater delivered by the UDS. Specifically, the sewage and wet weather flows that exceed the WWTP treatment capacity. Specifically, the sewage and wet weather flows are conveyed through the UDS to the WWTP until the maximum treatment capacity is reached. The exceedance of the water flows is discharged directly into the receiving water bodies, such as rivers or lakes, without receiving any treatment. As a consequence, CSO is one of the major contributors to water pollution experienced in rivers, lakes, etc.

This work proposes two innovative alternatives to manage stormwater in urban areas:

1. Direct management by using offline storage facilities with decentralized Real Time Control (RTC) system;
2. Pervasive management by using Low Impact Development techniques (LID).

The offline storage facilities such as storage tanks, which have the goal to temporarily accumulate stormwater volumes, are widely used, even though they are often overly expensive due to the high construction and maintenance costs. In contrast, approaches aiming at temporarily accumulating stormwater volumes directly in the existing UDSs have also been developed thus avoiding large investments [5, 7, 52]. These approaches are supported by the fact that the UDSs are typically designed by taking into account a set of safety factors. In particular, conduits are intentionally designed to be larger than required in the case of typical network working conditions. Basically, the UDS is managed by a real-time control (RTC) system which requires the network to be embedded with sensors and actuators permitting the network to be real-time monitored and regulated so as to adapt to the different rainfall events [1, 19].

Previous studies in literature was focused on RTC based on a centralized approach. In the study of Pleau et al. [45] a sewer networks global optimal control (GOC) scheme with a two-level architecture has been designed. The upper level was composed of a central station, which computed flow set points, whereas the lower level was composed of local stations, which are used for monitoring, flow computation, data validation and feedback control. The real-time computer was dedicated to all RTC operations and supports a supervisory software, a GOC software, a non-linear hydrologic–hydraulic model and a non-linear programming algorithm. The site was controlled automatically under a flow set point computed by the GOC scheme. The optimization problem was defined by a multi-objective (cost) function and a set of equality and inequality constraints, based on the following control objectives: minimizing overflows, minimizing set point variations and maximizing the use of WWTP capacity. In Fu et al. [23] a multi-objective optimization genetic algorithm was proposed which is used to derive the Pareto optimal solutions, which can illustrate the whole trade-off relationships between objectives. In Schütze et al. [52] a global optimal predictive real time control system has been implemented, which involves solution of a multi-objective optimization problem. The control objectives were the minimization of overflows, the maximization of the use of the treatment plant capacity, the minimization of accumulated volumes and, finally, the minimization of variations of the setpoints. The real time control system was implemented at a central station and used flow monitoring and water level data, rainfall intensity data, radar rainfall images and 2 h rain predictions. Set-points were translated into moveable gate positions at local stations by Programmable Logic Controllers (PLC). In this work we illustrate the Distributed Real-Time Control (DRTC) system already proposed in previous studies [16, 24, 25]. A multi-agent paradigm and specifically a gossip-based algorithm has been exploited. The UDS was equipped with electronically moveable gates and a set of water level sensors spread across the network. All the gates are locally controlled by Proportional Integrative Derivative (PID) controllers which are globally orchestrated by the mentioned gossip-based algorithm thus achieving an optimal hydrodynamic behaviour in terms of CSO and flooding reduction. The case study is the UDS of the city of Cosenza (Italy), which is modelled by using the StormWater Management Model (SWMM) simulation software. SWMM is an open-source computer model widely used by the hydraulic engineering community for simulation of hydrodynamic water and pollutant transport in sewer systems. It is provided by US EPA [50] and permits an accurate simulation of the hydrological and hydraulic behaviour of the UDS during both dry and wet weather conditions. SWMM simulation software has been customized in order to allow it to be integrated with an external real-time control module. Experiments, conducted using a set of selected extreme rainfall, show a substantial reduction of both CSO and flooding when the proposed approach is exploited.

The other innovative approach presented here consist of implementation of pervasive technique. This approach to land development known as low-impact development (LID) has gained increasing popularity. LID systems consist of a series of facilities whose purpose is to reproduce the site's pre-developed hydrological processes using design techniques that infiltrate, filter, store, evaporate, and detain runoff close to its source. Low-impact development practices consist of bioretention cells, infiltration wells or trenches, stormwater wetlands, wet ponds, level spreaders, permeable pavements, swales, green roofs, vegetated filter and buffer strips, sand filters, smaller culverts, and water harvesting systems. In recent years, researchers have focused their attention on applying and developing empirical, conceptual, and physically based models for LID analysis. In their review, Li and Babcock [36] reported that there were >600 studies published worldwide involving green roofs, with a significant portion of them related to modeling.

Benefits of LIDs in terms of runoff reduction and pollutants removal have been widely discussed in the literature [14, 26, 29, 32]

For example, Kamali et al. [32] investigated the performance of a permeable pavement under sediment loadings during its life span by evaluating the temporal and spatial clogging trends of this facility and by finding its vulnerability to sediment loadings during rainfalls. Zhang and Guo [68] developed an analytical model to evaluate the long-term average hydrologic performance of green roofs. Local precipitation characteristics were described using probabilistic methods, and the hydrological behavior of the system was described using mass balance equations. Carbone et al. [15] proposed a conceptual model to predict the hydraulic behaviour of a full-scale physical model of a vegetated roof, located at University of Calabria, Italy. The model idealized the vegetated roof as a system consisting of three individual components in series. A mass balance equation was applied to each block, taking into account the specific physical phenomena occurring in each module. The model was validated using dataset observed from the monitoring campaign carried out on the prototype of a full-scale vegetated roof. She and Pang [53] developed a physical model that combined an infiltration module (based on the Green–Ampt equation) and a saturation module (SWMM). The model calculates the water content in a green roof in a stepwise manner from the initiation of precipitation until saturation. In simulating the hydraulic response of green roofs to precipitation, an infiltration module is used before field capacity is reached and when no drainage is produced, while a saturation module is used after field capacity is reached and when drainage is produced. However, because runoff and infiltration can occur simultaneously during heavy precipitation, this stepwise approach may not be appropriate for a wide range of precipitation events.

Carbone et al. [12] developed a physically-based model using the explicit Finite Volume Method (FVM) for the infiltration process during rainfall events in green roof substrates. The model solves a modified version of the Richards equation which considers neglected the soil water diffusivity.

In another work, Huang et al. [29] proposed a numerical model for permeable pavements and also proved its applicability by applying it to simulate both hydraulics and water quality. The results of this study demonstrated a good agreement between field measurements and modeled results for three types of pavement in terms of hydraulics and water quality variables including peak flow, time to peak, outflow volume and TSS removal rates. The sustainable management of water resources requires the identification of procedures to optimize the use and the management of resources [17, 37]. As pointed out by several authors (e.g., [22, 65]), there is a strong demand for predictive models that can be applied across a range of locations and conditions to predict the general performance of a range of stormwater treatment measures. In addition, the heterogeneity of the materials that compose LIDs (concrete, gravel, soils, etc.) and their strongly unsaturated hydraulic behaviour, pose significant modelling challenges. In this way, several studies demonstrated that physically-based models can provide a rigorous description of various relevant processes such as variably-saturated water flow, evaporation and root water uptake, solute transport, heat transport, and carbon sequestration [10].

Although analytical and conceptual models represent a viable alternative to the numerical analysis of green roofs, their use suffers from several limitations. Conceptualization of the physical processes involved often leads to simplification of the system and a reduction in numerical parameters. While in a physical model each parameter has its own meaning, in conceptual models, lumped parameters often incorporate different components of the described process. These lumped parameters are case sensitive and need to be calibrated against experimental data, implying a lack of generality of the model itself. These drawbacks could represent a barrier to the use of modeling tools among practitioners who need reliable and generally applicable models.

For these reasons, in this work some techniques/procedures on how to interpret the hydraulic behaviour of several LIDs (green roof, permeable pavement) part of the "Urban Hydraulic Park" of the University of Calabria, south Italy, have been presented. It will suggest experimental and mathematical procedures for model calibration, which consists of: (a) experimental design (system construction, and number and character of measured transient flow data); (b) methods for independently evaluating of material hydraulic properties; (c) additional analysis of material hydraulic parameters using the transient flow data; and (d) model validation.

2 Real Time Control Approach

In urban areas with Combined Sewer Systems, stormwater and wastewater are collected in the same conveyance pipes. During heavy rainfall events, due to eventual obstructions in the pipes or to a poor maintenance, the surface runoff may overburden existing storm-water management facilities and cause flooding or combined sewage overflow into receptive water bodies.

Since the current drainage systems will fail to control and manage a constantly increasing runoff volume, and given the potential risk to human life, economic assets and the environment, an efficient adaptation strategy able to improve the flood resilience for the future urban environment is needed.

In this context, the main objective of this section is to illustrate a Distributed Real Time Control (DRTC) approach, proposed in previous studies [16, 24, 25], as a solution to mitigate CSO and reduce flooding at catchment scale. To illustrate the advantages of the flood alleviation strategy focused on DRTC, the drainage network of the city of Cosenza (south Italy), was chosen as a testbed. The proposal consists in instrumenting the existing urban drainage network with sensors and a series of movable gates that, by monitoring water level and the filling degree in each conduit, self-adjust in real time to optimize the storage capacity of the pipelines and reduce the CSO. In the following subsections will be described the components of the DRTC system and the approach used, then will be provided details regarding the implementation in the network and some results.

2.1 *Components and Implementation of the DRTC System*

In order to achieve the proposed goals and be controlled in Real Time, an urban drainage network needs to be instrumented with a series of components, whose conceptual organization is basically structured in control loops which can be implemented by hardware components including sensors, actuators, controllers and telemetry systems (Fig. 1). Sensors, by monitoring the process evolution, collect

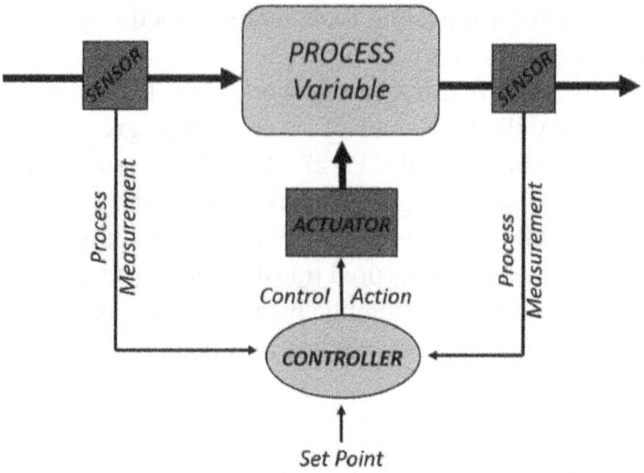

Fig. 1 Control Loop Scheme. Simple arrows indicate data flows, bold arrows the actions. Bold letters indicate hardware components and italic letters indicate transferred information

information about the current state of the system, actuators modify the monitored process and influence it, controllers adjust actuators with a certain objective and the telemetry system supports the data transmission among the different devices.

With regard to urban drainage networks, most of the time RTC implementations are based on water level measurements. The selection of the correct equipment, together with the choice of an adequate communication system and the proper software to be used, is thus crucial for a durable and reliable installation of RTC systems in UDSs. More in detail, to improve the performance of existing drainage networks and balancing water level throughout the conduits of the network, so as to reduce water level in the more overloaded conduits, the drainage network needs to be implemented with the following equipment:

(i) Water level sensors, which measure the water level in each conduit and the flow on the outfall;
(ii) Moveable gates, functioning as actuators, which can be real-time regulated electronically;
(iii) Computational nodes, which can host and execute the distributed control algorithm.

Sensors, which monitor water level and, hence, the degree of filling in each conduit, are positioned into pipes to evaluate flow depths and to monitor surcharge conditions during rain events. Since the mechanism cannot work properly if the increase of water level is not correctly perceived, it is important to correctly position the water level sensor, because if it is placed in the "underloaded" part the gate-agent would perceive a decreased water level instead of an increased one. In the proposed approach, this issues is addressed by deploying more than one sensor per conduit and taking the maximum sensed value as the water level value for the conduit.

Using the information acquired by the sensors, electronically Moveable Gates—functioning as actuators—can self-regulate to intelligently manage the storage capacity of the pipelines. The moveable gates are made up of mobile plates rotating around a horizontal hinge placed on the bottom of the conduit, as shown in Fig. 2. The gates

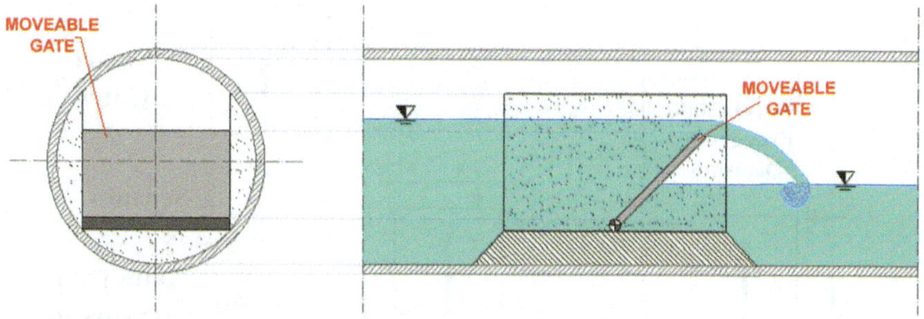

Fig. 2 The down-hinged movable gate

are dynamically regulated in order to utilize the full storage capacity of the pipeline by accumulating the excess stormwater volume in the less overloaded conduits thus preventing CSO.

The gate is completely closed when the plate rotates in a perpendicular position with respect to the flow direction. Conversely, the gate is fully open when the plate is parallel to the flow. When the gate is closed, the opening area is null and no flow rate is delivered from the node. An intermediate position of the gate corresponds to a partial opening degree. The gates, as actuators, are the regulators elements of the RTC system that are used to adjust flows and water levels in the controlled system.

The manipulation of actuators in RTC systems is performed by control units (controller). A certain number of computing nodes are spread throughout the drainage network in order to cover all the points of interest. In particular, these devices read data received from local sensors through wireless connection, and collectively elaborate the acquired information in order to provide—according to the objectives—output adjustments to actuators and thus supply the gates with an "intelligent" behaviour (Fig. 3).

The gates are located at the points of the network where subnetworks are connected to the main channel. Figure 4a shows the logical places for inserting the gates, while Fig. 4b shows the gates insertion in a case of a realistic network.

Each computational node has a partial view of the network as it can read only from sensors which are located in its spatial neighbourhood and can actuate only gates it can physically reach. Thus, these devices dynamically regulate the gates according to the information acquired by the sensors in the neighbour areas. On the basis of the previous considerations the idea proposed lies in using a distributed agent-based architecture [66]. The agent paradigm has several important characteristics:

- Autonomy: Each agent is self-aware and has a self-behaviour. It perceives the environment, interacts with others and plans its execution autonomously;
- Local views: No agent has a full global view of the whole environment but it behaves solely on the basis of local information;
- Decentralization: There is no "master" agent controlling the others, but the system is made up of interacting "peer" agents.

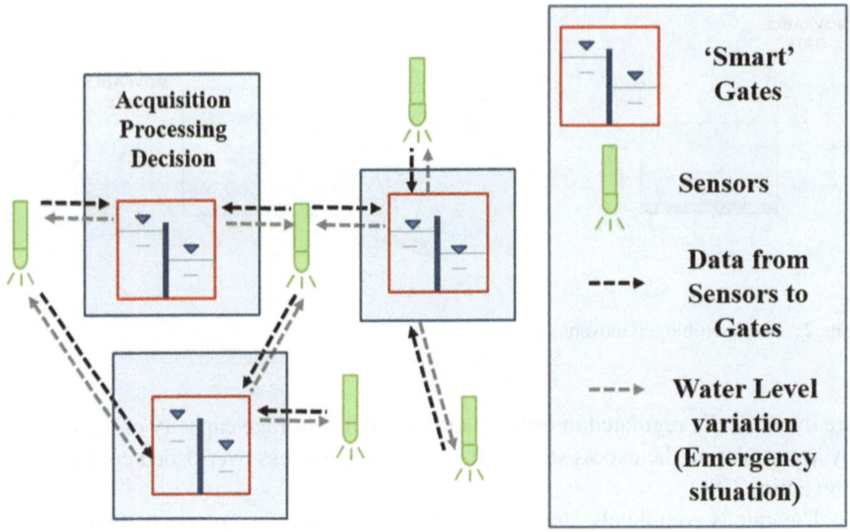

Fig. 3 Schematisation of the DRTC system

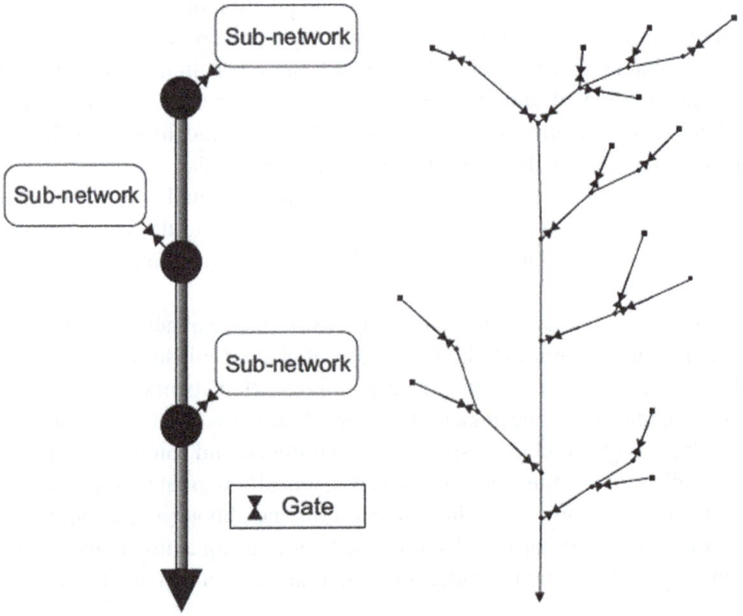

Fig. 4 Gate position

From the point of view of software architecture, the approach requires one agent per gate. Each gate-agent runs on one of the computational nodes covering the specific gate, it can perceive the local water level and communicate with the neighbouring gate-agents in order to elaborate a proper actuation strategy for its gate. In addition to the gate-agents, another agent, called outfall agent, is logically associated with the outlet node, it behaves the same as other agents except for the actuation part, indeed, it is not associated with any gate. For each generated network, an optimization algorithm, executed on computing nodes in a distributed fashion, leads to balance—in real-time—the water level perceived by the agents and aims to distribute equally the degree of filling of the conduits, thus preventing overcharge phenomena as far as possible. The proposed goal has been achieved by means of agents continuously executing two tasks:

1. Figuring out collectively the average of the water level in the generated network;
2. Each agent triggers its specific gate in order to bring the water level closer to that average.

Task 1 is accomplished by exploiting a Gossip-based aggregation [30] for dealing with the global aspect of the drainage network, while Task 2 is accomplished exploiting locally a PID controller [4]. Once an agent knows the global water level through the "gossip-based aggregation", there remains the problem of appropriately tuning its gate so as to reach that "desired" level. This issue is addressed using the PID logic which, sending Proportional Integral and Derivative control action signals to actuators, can be used when you do not know an exact mathematical model of the system you want to control. A PID controller is a control loop feedback mechanism where an error value is computed as the difference between a measured output of a process and the desired value (setpoint). In the case studies proposed by Giordano et al. [24] and Garofalo et al. [25], each gate of the drainage network is controlled by a PID implemented by the gate-agent.

Once established the optimization computational operations to be carried out, the drainage network of Cosenza was simulated using a customized version of SWMM software, built for permitting it to Real-Time communicate with a separate Java controller which implements the algorithm described before. Further hardware and software details of this approach, which allows a dynamic management of the drainage network, are given elsewhere [24, 25].

2.2 *Case Study: The Drainage Network of Cosenza*

The test site chosen for the proposed approach and the hydrological modelling of the conversion scenarios was the Liguori catchment, a highly urbanized catchment in Cosenza. The catchment has a population of 50.000 inhabitants and a total surface area of 414 ha, of which almost one-half (48%) is densely urbanized and highly impervious, while the other 52% (202 ha) is pervious, occupied by natural areas (Fig. 5a).

The catchment is drained by a combined sewer system that collects sanitary sewage and stormwater runoff in a single pipe system, conveying the entire water flow directly to the wastewater treatment plant (WWTP). During the most intense

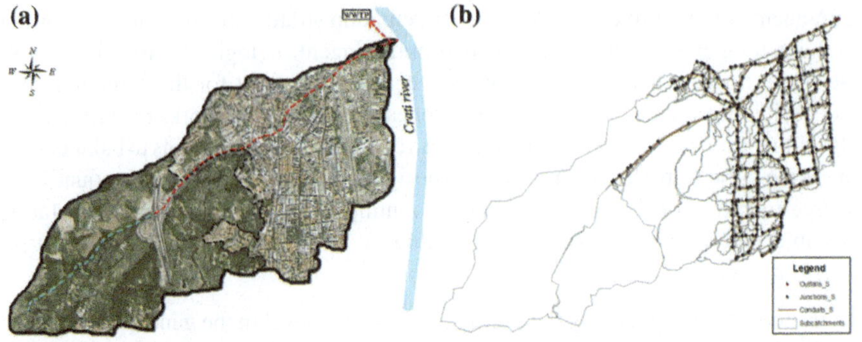

Fig. 5 **a** The Liguori catchment of Cosenza; **b** SWMM model of the drainage network of Cosenza

rainfall events, wet-weather flows occasionally exceed the capacity of the sewer system and the excess flows escape from the sewer system, via an overflow structure, as a combined sewer overflow (CSOs) [43]. Such overflows are directly discharged, without receiving any treatment, into the receiving water body, the Crati River.

In order to simulate the response of the LC drainage network to storm events, the sewer dataset and the physiographic characteristics of the sub-catchments, were imported into EPA-SWMM software [50] for the next hydraulic modelling of the drainage system. The software SWMM, provided by EPA, is an open-source computer model that allows a dynamic rainfall-runoff simulation for predicting hydrological and hydraulic behaviour of urban drainage systems and watersheds.

More in detail, in SWMM, the Liguori Catchment area was simplified in 296 sub-catchments, of which 258 are mostly urbanized (%Imp > 0.7) and 23 average urbanized (0.3 < %Imp < 0.7). The urban drainage network modelled in SWMM (Fig. 5b), instead, consists of 324 conduits with different shapes and sizes and a slope varying from 0.5 to 6%. Some pipes are circular and egg-shaped with diameters varying from 0.3 to 1.5 m and others are polycentric pipes. Finally, there are in total 326 nodes (with Outfall and Junctions functions) which represent the catch basins. The model used in this study was previously calibrated on the basis of several measurement campaigns [42]. Calibration parameters taken into consideration in flow modelling were surface roughness of the conduit (n), the impervious (N-Imperv) and pervious (N-Perv) surfaces in the catchment, and the depths of surface depressions on impervious (Dstore-Imperv) and pervious (Dstore-Perv) areas.

As previously said, for the purposes of the proposed studies, the drainage network was simulated using a customized version of SWMM built for the purpose for permitting it to communicate in real time with a separate Java controller which implements the algorithm described before. More in detail, in this version, the moveable gates are modelled as a transverse weir with the opening area equal to the conduit section area.

2.3 Experiences and Results

In the following section, the finding obtained from the DRTC application to the case study, will be described and discussed.

At first Giordano et al. [24] used and applied a totally decentralized RTC approach to a drainage network. In this study the experiments were carried out using a simplified network, which consists in a main channel of 1 m diameter and a total of 35 pipes inside the sub-networks, instrumented with a series of moveable gates and sensors which monitor water level and, hence, the degree of filling in each conduit, during severe rainy events. The water level in the pipes is balance by a combination of a Gossip-based algorithm, which ensures a global correct behaviour, and a PID controller used for each gate, so as to maintain locally its related water level as close as possible to the "suggested" value computed by the algorithm. The results have demonstrated that when the RTC is applied, the filling degree of conduits are much closer to each other as the load on the entire network is more balanced. This means that the network does properly exploit the residual water capacity of the undercharged conduits and the latter implies an improvement in the behavior of the critical conduit that reaches the overcharge condition later. The proposal provides positive effects on the overall hydraulic performance of the network as it is able to prevent (or delay) flooding events that would occur in the original (not instrumented) network. Following works, instead, focused on extending the algorithm and validating the DRTC approach in real drainage networks.

In the study carried out by Carbone et al. [16], the DRTC investigations have been extended to the real drainage network of Cosenza. To evaluate the effect of the moveable gates on reducing the storage capacity of the conduits, a part of the drainage network of Cosenza was investigated by comparing two scenarios of the CSS in SWMM: the existing configuration (Scenario 0) and a new one (Scenario 1) where six sluice gates were placed in the secondary conduits of the system. To investigate the response of the new configuration of the drainage system, three extreme events (dated March 1st, June 16th, October 8th) occurred in 2011 and which put in crisis the system, were analysed.

Specially, in Scenario 0 were identified 2 most overloaded pipes in the main conduit and used in this study to demonstrate the beneficial influence of gates in alleviating the most critical sections of the system. Two example results obtained for the Conduit 2 are reported in Fig. 6.

Figures show the distribution of storage capacity for two of the analysed events and for both considered scenarios; as it can be noticed the storage capacity in Scenario 0 varies up to 100%; this mean that in some time points the selected conduits get completely full. Instead the filling degree is lower when sluice gates are used to control stormwater volumes. In Table 1 is reported the average storage capacity reduction for each event; the reduction varies from 22 to 77% showing lower value for the event with higher rainfall volume. However, as the authors have pointed out, the beneficial effect may be dependent upon the storm characteristics (such as hydrographs, intensity and duration).

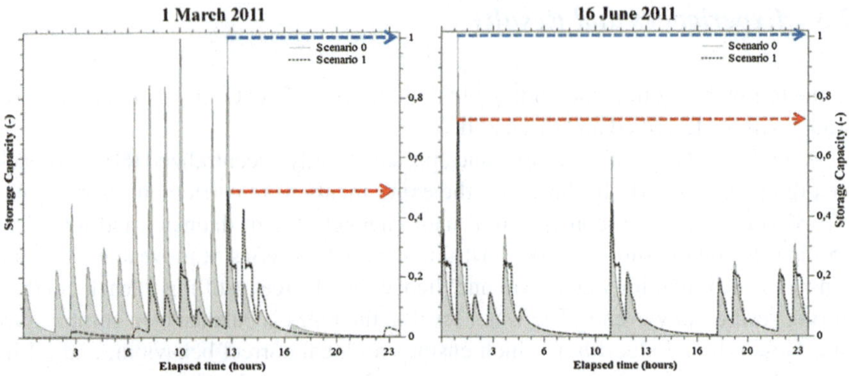

Fig. 6 Temporal distribution of storage capacity for Conduits 2

Table 1 Selected rainfall events

Rainfall event	Rainfall volume (mm)	Average storage capacity reduction (%)
1 March 2011	97.40	22
16 June 2011	20.10	60
8 October 2011	48.60	77

The findings show that a series of devices inside the urban drainage systems are actually able to control the flow rate to drop the storage capacity to a reasonable value.

In Garofalo et al. [25], later, different scenarios have been analysed to evaluate the performance of the DRTC as a function of the number of the moveable gates placed in the system. As for the previous studies, the scenario without DRTC, which corresponds to the actual UDS, is called Scenario 0. The other scenarios, controlled by the DRTC, and differ according to the number of secondary pipes equipped with moveable gates: 91 for Scenario 1 (S1), 107 for Scenario 2 (S2), 214 for Scenario 3 (S3) and 322 for Scenario 4 (S4). The response of the UDS for all these scenarios is modelled for 15 independent rainfall events recorded in the weather station in Cosenza during the years 2010–2015.

In this work, for each selected rainfall event, were evaluated both CSO reduction (computed as the relative percentage difference between the CSO volume in the scenarios with DRTC and the Scenario 0), and the local flooding reduction (as the relative percent difference between the total flooding volumes from the conversion scenarios and the reference one). Referring to the Scenarios 1 and 4, was observed a CSO reduction which varies respectively between 2.7 and 83% (S1), and from 13 to 99% (S4), according to the rainfall events. The consistently higher CSO drop in the S4, demonstrates the beneficial effect provided by using a larger number of moveable gates. At the local level, thanks to the DRTC, the temporary stormwater detention provided in the less overloaded conduits, utterly prevents the UDS from

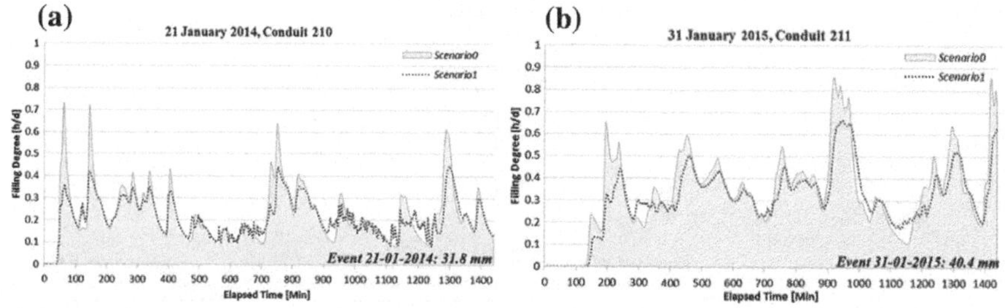

Fig. 7 Temporal distribution of Filling Degree (h/d) of two conduits during two rainfall events

local flooding in S1, where a drop of 100% is obtained for all the events. Otherwise, the S4 the risk of flooding is solely mitigated, with reductions varying from 2.4 to 13.4% for all the events, except for one event (dated 23 Nov. 2013), where a drop of 100% is obtained. As highlighted by the authors, the reasons why S4 offers a limited flooding reduction are strongly related to the high number of gates adopted. The high number of gates involved, able to exploit all the possible storage capacity of the network, makes the Scenario 4 the best choice in order to prevent the CSO, but it performs quite badly in terms of flooding reduction with respect to the other controlling scenarios. The reason for this behaviour lies in the fact that when the whole storage capacity is exploited, no additional water can be stored temporarily, and so a growth in incoming water flows produces unavoidable flooding phenomena. Therefore, these findings suggest that S2 and S3 are the most convenient solutions, since they offer the highest overall performance in terms of reduction of local flooding and perform well also with respect to the CSO reduction. Summing up, this study clearly demonstrated that the DRTC produced beneficial effects on the management of the UDS by substantially mitigating the risk of flooding and CSO.

More recently, Principato et al. [46], evaluated the potential of an integrated and sustainable approach for a better management of the drainage network. The main objective of this study was to assess the mitigation of CSO's impact when dynamic (RTC) and static (LID) measures are simultaneously adopted to cope with greater stormwater volumes. With this purpose different conversion scenarios have been applied on a portion of the Liguori Catchment (LC) of Cosenza: Scenario1, emulates the behavior of the drainage network regulated by moveable gates controlled in Real Time, while Scenario2 investigate the hydrologic response of the network considering Green Roofs (GRs) implementation, in replace of impervious rooftops, in a portion of the LC. A last scenario (Scenario3) has also been developed to analyze the combined effect of RTC and GRs implementation, in the same portion of LC considered for the Scenario2. As a confirm of the studies already analyzed, also these model results revealed that the RTC of urban drainage system, equipped with a series of gates, provides beneficial effects to the overall hydraulic performance of the network. Results reported in Fig. 7, show that the filling degree of two selected conduits is lower when smart gates are used to control stormwater volumes: the peak reduction is around 29% for conduit 210 and 19% for conduit 211.

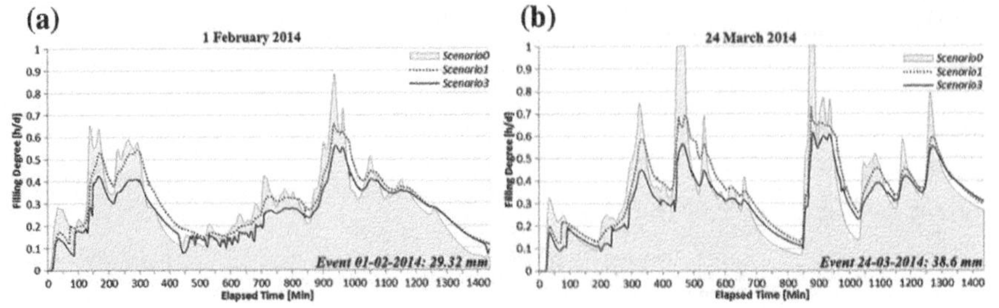

Fig. 8 Temporal distribution of Filling Degree (h/d) of the last conduit of the network during two rainfall events

A novelty, compared to the studies already analyzed, are the results obtained from Scenario3, which prove the importance of an integrated approach on the overall hydraulic performance of the network, as a valid solution for controlling flooding in urban areas. In particular Fig. 8, which refers to the last conduit of the network, reveals that the coupling of distributed (RTC) and source control (LID) solutions, leads a further reduction of the filling degree also compared to Scenario1, for the most part of the event chosen.

In conclusion, the use of smart moveable gates provides beneficial effects to the overall hydraulic performance of the network during critical rainfall-runoff events, offering an alternative and valid solution for controlling flooding in urban areas. The advantage of this system is to utilize the full storage capacity of the pipeline by accumulating the excess volumes of rainwater that otherwise would be spilled out in the pipes with a low water level. The advantage is also from the economic point of view because it takes advantage of the existing sewer system.

3 LID Approach

Despite the hydrological benefits of LIDs are already studied in literature, these techniques are not yet widespread probably because modelling tools often used simplified methodologies, based on empirical and conceptual equations, which do not take into account hydrological processes in a physical way. In addition, the hydraulic properties of LIDs materials have not been investigated in a comprehensive manner, limiting the investigation only to specific properties [59]. In this way, Brunetti et al. [8, 9] proposed an innovative approach to investigate the hydraulic behaviour of several LIDs using a mechanistic model coupled with specific numerical analysis to explore the hydraulic properties of LIDs techniques. The LID systems considered in this chapter are all implemented in the "Urban Hydraulic Park," which includes a permeable pavement, a stormwater filter, and a sedimentation tank connected to a treatment unit.

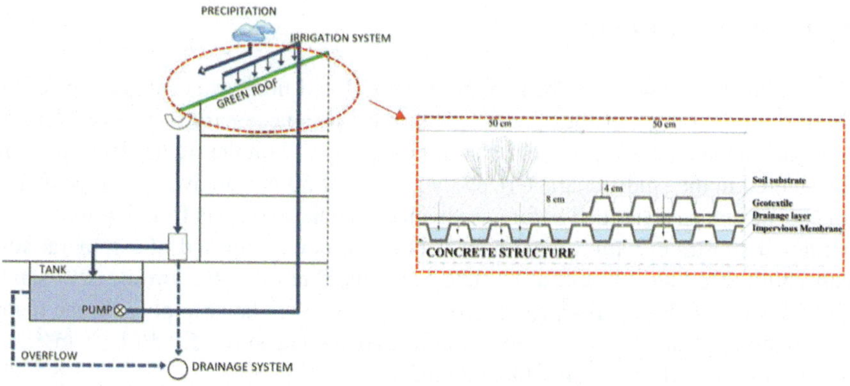

Fig. 9 Schematic representation of the RWH system to collect the rainwater from green roof (left) and the inset of a cross-section of the GR (right)

3.1 Green Roof Experience

To analyse the green roof and LIDs hydraulic behaviour, as discussed in the Introduction sectin, different models from the conceptual and analytical to the mechanistic ones have been developed and widely used, but very few studies have focused on a comprehensive analysis of the hydrological behavior of a green roof. Starting from this assumption Brunetti et al. [8] carried out an accurate and comprehensive analysis of Variably Saturated Hydraulic behavior of the experimental green roof installed at University of Calabria by using HYDRUS 3D. The experimental green roof (GR), considered in the study, was built on a fifth-floor terrace of the Department of Mechanical, Energy and Management Engineering (DIMEG) at the University of Calabria (Italy), in Mediterranean climate region. The area of an existing roof was parcelled into four sectors: two sectors are vegetated with the same native Mediterranean species (Carpobrotus edulis, Dianthus gratianopolitanus, and Cerastium tomentosum), but present different drainage layers; a third sector is mostly characterized by spontaneous vegetation; while the last sector is the original roof, considered as the reference compartment for experimental data analysis.

The water supply of GR is guaranteed by reusing the green roof's outflow, collected in a specific storage tank and distributes through a drip irrigation system during drought periods. The Rainwater Harvesting (RWH) system (Fig. 9, left), designed ad hoc for the site specific, consists of: (1) a system for collecting rainwater from the experimental site; (2) a storage tank of 1.5 m^3 with a pump to relaunch the irrigation system; (3) a connection system with the water supply to ensure the full satisfaction of irrigated demand in any condition. When the storage capacity of the tank is reached, the overflow is directly discharged into the sewer system. The drip irrigation system is currently actioned by an electric valve at predetermined time, and the irrigation rate is recorded by a water counter with a frequency of 1 min.

3.1.1 Modeling Theory

To investigate the hydraulic behaviour of the GR and finally to evaluate a possible optimization strategy of the specific green roof, it is necessary to proceed first of all with a detailed description of its stratigraphy (Fig. 9 on the right). Thus, the GR considered in the study, characterized by an area of 50 m^2, an average slope of 1% and vegetated, as described before, with native species, consists from top to bottom of: a soil substrate, with a maximum depth of 8 cm, composed of a mineral soil consisting of a hetero-disperse Particle Size Distribution (74% gravel, 22% sand, and 4% silt and clay); a permeable geotextile; a drainage layer in polystyrene foam with a storage capacity of 11 L/m^2 and a drainage capacity of 0.46 Ls^{-1} m^{-2}; an anti-root layer and a waterproof membrane.

A weather station located directly at the site collect precipitation, velocity and direction of wind, air humidity, air temperature, atmospheric pressure, and global solar radiation [11]. Rainfall data are measured every minute by using a tipping bucket rain gauge with a resolution of 0.254 mm. While the outflow from each sector is recorded at the base of the building by a flowmeter device composed of a PVC pipe with a sharp-crested weir and a pressure transducer (Ge Druck PTX1830) to estimate the water level inside the pipe.

In the work of Brunetti et al. [8] two-months rainfall data from 2015-09-01 and 2014-10-30 were used and the Penman-Monteith equation [2] was implemented to estimate the hourly reference evapotranspiration. To model the water flow in unsaturated soils by the Richards equation, and, thus, estimate the water retention function $\theta(h)$ and hydraulic conductivity function K(h), the evaluation of unsaturated hydraulic properties of GR substrate was carried out by implementing a simplified evaporation method with the extended measurement range (down to –9,000 cm) [60]. While to simultaneously fit $\theta(h)$ and K(h) to the experimental data obtained by the evaporation method, HYPROP-FIT [41] numerical optimization procedure was used. For the description of soil hydraulic properties, first of all the unimodal van Genuchten–Mualem (VGM) model [62] was implemented; next, since the unimodal VGM model couldn't always describe the full complexity of measured data, the bimodal model of Durner [20] was taken into account. Results of the experiments are reported in Table 2.

To describe the complex physical features of the experimental green roof, HYDRUS-3D software [54], which solves the Richards equation for multi-dimensional unsaturated flow, was used implementing the parameters obtained with the evaporation method. Finally, to evaluate the agreement between measured and modeled hydrographs the Nash-Sutcliffe Efficiency (NSE) index [40] was evaluate.

The results concerning the estimated soil hydraulic parameters with their confidence intervals, reported in this study revealed that the bimodal function presents a more accurate description of the retention curve. While the findings obtained during the validation process showed that: the unimodal and bimodal models are both able to accurately describe the GR hydraulic behavior; a higher precision is achieved by the bimodal model; both model slightly overestimate the outflow.

Table 2 Estimated soil hydraulic parameters and their confidence intervals (CIs) for the unimodal and bimodal hydraulic functions

Parameter	Unimodal	CIs	Bimodal	CIs
Residual water content, θ_r (–)	0	0.05	0.070	0.007
Saturated water content, θ_s (–)	0.551	0.01	0.562	0.003
Air-entry pressure head index for the first pore system, α_1 (1/cm)	0.13	0.03	0.843	0.07
Pore-size distribution index for the first pore system, n_1 (–)	1.25	0.06	1.24	0.04
Saturated hydraulic conductivity, K_S (cm/day)	4700	3500	12,600	3700
Air-entry pressure index for the secondary pore system α_2 (1/cm)	–	–	0.01	0.001
Pore-size distribution index for the secondary pore system n_2 (–)	–	–	1.97	0.08
Weight coefficient w_2 (–)	–	–	0.422	0.01
Tortuosity and pore connectivity parameter, L (–)	0.53	0.02	0.5	–

In addition, starting from the assumption that critical rainfall events occur in a very short time [8, 13] have investigated the hydrological response of the GR to single precipitation events. The results obtained for four rainfall events with different total precipitation volume (V_{prec}) in terms of peak flow reduction P_{red} (%) and volume reduction V_{red} (%), for both modeled and measured outflow, are shown in Table 3 and Fig. 10. By the analysis of this results the authors concluded that the green roof hydraulic performance was affected primarily by the antecedent substrate moisture and secondly by the precipitation pattern.

Table 3 Analysis of the hydrological performance of the green roof during single precipitation events

Rainfall events	V_{prec} (mm)	Modeled outflow		Measured outflow	
		P_{red} (%)	V_{red} (%)	P_{red} (%)	V_{red} (%)
9 September 2015	100	5	12	7	16
7 October 2015	42	40	27	60	31
10 October 2015	69	5	5	45	9
21 October 2015	120	5	17	7	17

3.1.2 Future Perspective: GR and RHW from a Smart and Innovative Point of View

The integration of GR and RWH system, like in the case of the experimental site of University of Calabria, allows considerable benefits in terms of rational management of water resource. Furthermore, different studies have considered the RWH systems as a good strategy to limit environmental impacts that the on-going urbanization produce on the drainage network and receiving water bodies [48].

However, so far, these techniques (GR and RWH) have been studied from a purely hydrological-hydraulic point of view, there are no studies that consider these ones as smart objects for an integrated management of the water resource and urban drainage system.

In light of this, an innovation in the field of Urban drainage is look at the single techniques GR and RWH as smart objects, optimizing them with ICT technologies, based on the IoT (Internet of Things) paradigm.

A new aspect, in fact, could lie in the integration of GRs and RWH techniques through a complex network equipped with: sensors, which allow rapid quantitative assessments from a hydraulic, energy and environmental point of view; regulators or actuators, able to modify the processes in progress; transducers that allow the conversion of the data detected in command; control units that report the variables to the pre-established threshold values.

According on what was previously discussed about Green Roof experimental site of University of Calabria, the rainwater collected in the storage tank is re-introduced through the irrigation system at fixed time and in quantity set by the operator of the experimental site. The optimization of this system could be achieved by considering the smart automation through the estimation of water content and the evaluation on the wheatear situation. More in detail, when the water content, monitored by specific sensors dislocated within the layers of the green roof, falls below a threshold value that causes the plant water stress, the smart system sends a command to activate the irrigation withdrawing the water from the storage tank in the needed rates to re-reach

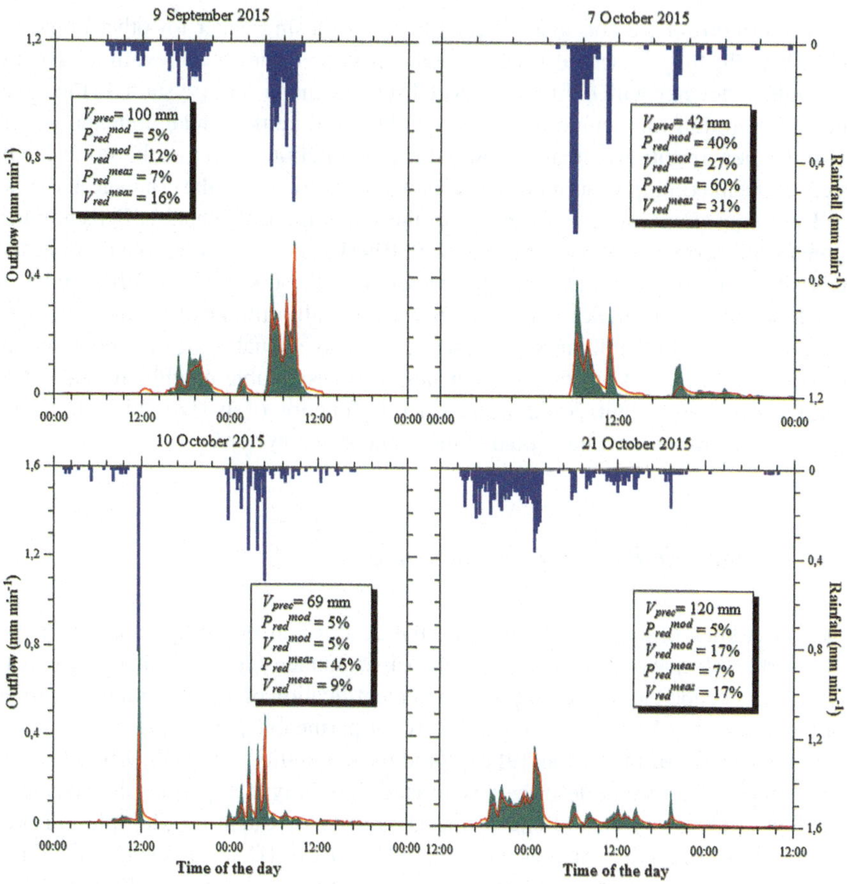

Fig. 10 Rainfall (blue area) and modeled (cyan area) and measured (red line) outflow for four selected rainfall events in the analysis of the hydrological performance of the green roof during single precipitation events; Pred meas and Pred mod are the measured and modeled peak flow reductions, respectively, and Vred meas and Vred mod are the measured and modeled volume reductions, respectively

the appropriate water content. The Weather Station located at the experimental site offers also the opportunity to make preventive estimates, based on the rainfall regime and solar radiation recorded on the site, so as to better calibrate the operation of the irrigation system.

Through these innovative strategies, in fact, not only a hydraulic benefit would be obtained, optimizing the reuse of the rainwater and reducing the flow to the drainage system, but also a thermo-energetic one. The activation of the irrigation system could also be carried out following the temperature measurements in the rooms below in order to improve the summer thermal comfort of the building.

Furthermore, if we consider the possibility of reusing water for other domestic uses (WC flushing, machine washing, ect.), it is possible to achieve total system hydraulic efficiency with minimum runoff discharge in the sewer system. In this case, the collected rainwater could be reused totally, avoiding that, in autumn and winter season, when irrigation demand is lower and precipitations increase, the storage tank exceeds the maximum value and the overflow is directly sent in the drainage network.

This innovation could be extended to the integration of others LID techniques (green wall, permeable pavement, etc.) with RWH systems in order to maximize the hydraulic, environmental and energy efficiency of these solutions. Furthermore, it would be appropriate to develop of an optimization algorithm that not only includes local actions for the building-scale system, but also evaluates the efficiency on the district and basin scale to favour smart and eco-sustainable neighbourhoods. The smart system, so thought, could be classified in function of its Hydraulic Efficiency Class, i.e. the rainwater rate spilled in the urban sewer system.

3.2 Permeable Pavement Experience

Lack of studies in literature focused on the description of the hydraulic behavior of a permeable pavement in a comprehensive manner suggested that research is particularly needed in the development and identification of accurate modeling tools for the analysis of LID practices, especially for permeable pavements.

In their work, Brunetti et al. [9] explored the suitability of the HYDRUS mechanistic model to correctly describe unsaturated flow in typical permeable pavement, installed at the experimental site of the University of Calabria. Multiple uniform and nonequilibrium flow models included in HYDRUS-1D, such as single and dual-porosity models, are used to define the hydraulic behavior of the permeable pavement. The problem was addressed by combing a Global Sensitivity Analysis (GSA), used to prioritize the hydraulic parameters and identify those that are non-influential, with a Monte Carlo filtering approach, used to investigate the parameter space and identify behavioral regions. Results from these analysis are then used in the calibration process conducted with the Particle Swarm Optimization (PSO) algorithm. Finally, the calibrated model was validated on an independent set of measurements.

The studied permeable pavement has an area of 154 m^2, an average slope of 2%, and a total depth of the profile of 0.98 m. Figure 11 shows a schematic of the permeable pavement, consisting of 5 layers.

The surface wear layer consists of porous concrete blocks characterized by high permeability (8 cm depth). Base (35 cm depth), sub-base (45 cm depth) and bedding layers (5 cm depth) were constructed by following the suggestions of the Interlocking Concrete Pavement Institute (ICPI), which recommends certain ASTM stone gradations.

A weather station located directly at the site measures precipitation, wind velocity and direction, air humidity, air temperature, atmospheric pressure, and global solar radiation. Rain data are measured by a tipping bucket rain gauge with a resolution

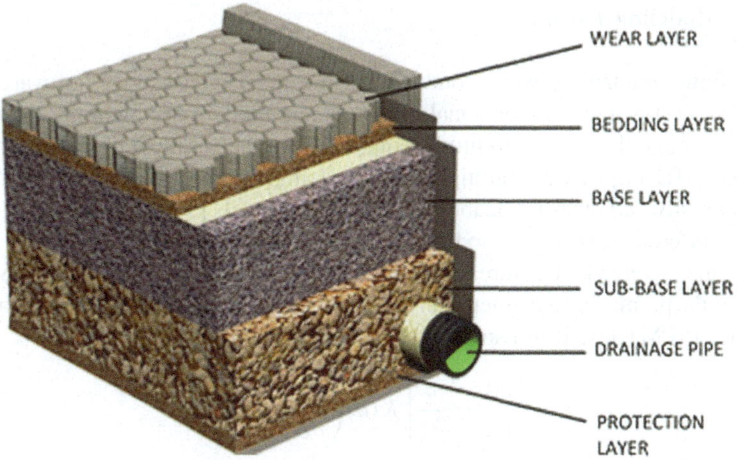

Fig. 11 A Schematic of permeable pavement

of 0.254 mm and an acquisition frequency of one minute. Climatic data are acquired with a frequency of 5 min. Data are processed and stored in the SQL database.

Outflow from the pavement is measured by two flux meters, composed of a PVC pipe with a sharp-crested weir and a pressure transducer. The pressure transducers were calibrated in the laboratory by using a hydrostatic water column, linking the electric current intensity with the water level inside the column. The exponential head-discharge equations for the two PVC flux meters were obtained by fitting the experimental data with a coefficient of determination R2 = 0.999 for both devices. No measurements of pressure heads or volumetric water contents inside the pavement were taken.

Two month-long data sets were selected for further analysis. The first data set, which started on 2014-01-15 and ended on 2014-02-15, was used for parameter optimization and sensitivity analysis. Total precipitation and total potential evapotranspiration for the first data set were 274 and 43 mm, respectively. The second data set, which started on 2014-03-01 and ended on 2014-03-31, was used for model validation. Total precipitation and total potential evapotranspiration for the second data set were 175 mm and 81 mm, respectively. The second data set was selected so that it had significantly different meteorological data than during the first period. The optimization set is characterized by multiple rain events with few dry periods. The validation set has fewer rain events, which are concentrated at the beginning and end of the time period and separated by a relatively long dry period between. Surface runoff was not observed during these time periods.

Potential evaporation was calculated using the Penman-Monteith equation [2]. The permeable pavement was installed in 2013 and has been constantly exposed to atmospheric conditions and traffic since then that has altered the surface roughness and color. For these reasons, an albedo of 0.25 was used as suggested by Levinson and Akbari [35] for weathered gray cement.

3.2.1 Modeling Theory

Water flow simulations were conducted using the HYDRUS-1D software [57]. HYDRUS-1D is a one-dimensional finite element model for simulating the movement of water, heat, and multiple solutes in variably-saturated porous media. HYDRUS-1D implements multiple uniform (single-porosity) and nonequilibrium (dual-porosity and dual-permeability) water flow models [56].

Two different conceptual models have been used to study the unsaturated water flow in the pavement structure. Scenario I assumed that water flow in all five soil layers of the permeable pavement can be described using the classical single-porosity approach (SPM) described using the one-dimensional Richards equation:

$$\frac{\partial \theta}{\partial z} = \frac{\partial}{\partial z}\left[K(h)\left(\frac{\partial h}{\partial z} + 1\right)\right] \tag{1}$$

where θ is the volumetric water content [–], h is the soil water pressure head [L], K(h) is the unsaturated hydraulic conductivity [LT^{-1}], t is time [T], and z is the soil depth [L]. The soil hydraulic properties are described by the van Genuchten–Mualem relation [62].

Scenario II assumes a single-porosity model for the wear layer, the bedding layer, and the protection layer, and a dual-porosity model for the base and sub-base layers. This configuration was selected in order to consider the occurrence of preferential flow in the coarse layers of the pavement that are composed of crushed stones, with particle size diameters ranging from 2.5 to 37 mm in the base layer and from 20 to 75 mm in the sub-base layer. Crushed stones were washed before installation in order to remove fine particles. From a physical point of view, the structure of the base and sub-base materials closely resembles fractured aquifers [6].

In this way, the classical approach to model water flow in fractured porous media is the so-called "dual-porosity" or "mobile-immobile water" (MIM) approach [6, 63, 64]. This approach assumes that flow occurs only in the mobile fracture domain, while water in the matrix domain is immobile with a coefficient Γ_w that represents the mass transfer between two domains, which is assumed to be proportional to the difference in effective saturations of the two regions [55, 56].

$$\Gamma_w = \omega \cdot (S_\theta^m - S_\theta^{im}) \tag{2}$$

In this, Scenario II thus includes 20 parameters (additionally also ω and θs of the immobile domain for the base and subbase layers).

3.2.2 Global Sensitivity Analysis

A sensitivity analysis (SA) can identify the most influential parameters and their interactions and how these parameters affect the output [51].

Most SAs performed in the literature of environmental sciences are the so-called 'one-at-a-time' (OAT) sensitivity analyses, performed by changing the value of parameters one-at-a-time while keeping the others constant [18, 28, 49].

One of the most widespread algorithms for the GSA is the variance-based Sobol' method [58]. Variance-based methods aim to quantify the amount of variance that each parameter contributes to the unconditional variance of the model output. For the Sobol' method, these amounts are represented by Sobol's sensitivity indices (SI's). These indices give quantitative information about the variance associated with a single parameter or related to interactions of multiple parameters. For a more complete explanation about the Sobol' method, please refer to Sobol' [58].

In order to assess the accuracy of estimations of the sensitivity indices, the bootstrap confidence intervals (BCIs) [21] were estimated. The rationale of the bootstrap method is to replace the unknown distribution with its empirical distribution and to compute the sensitivity indices using a Monte Carlo simulation approach where samples are generated by resampling the original sample used for the sensitivity analysis. In our case, the q samples used for the model evaluation were sampled 1000 times with replacement, whereby Sobol's indices were calculated for each resampling. In this way, 95% confidence intervals are constructed by using the percentile method and the moment method [3].

The sensitivity analysis was conducted using the programming language Python and in particular, the Sensitivity Analysis Library (SALib) [61]. An elaborated script overwrites the input file containing the parameters for different materials at each iteration. The script then executes HYDRUS-1D, which usually runs less than one second.

As stated before, the GSA was also coupled with a basic Monte Carlo filtering in order to identify behavioral regions in the parameter space and to reduce the uncertainty in the following parameter estimation step by using the same sample and runs of the GSA. Potential solutions are divided into two groups depending on the value of the objective function calculated: behavioral, solutions with NSE$>$0.0, and non-behavioral, solutions with NSE$\leq$0.0. Particle Swarm Optimization Inverse modeling is a procedure to estimate unknown parameters of the model from experimental data. In this work a global search method based on Particle Swarm Optimization (PSO) [33] have been used. PSO has been used in multiple studies involving inverse modeling with complex environmental models [27, 31, 67]. In PSO, collections of "particles" explore the search space, looking for a global or near-global optimum.

For the optimization process, a modified version of the PySwarm Python Library has been used.

3.2.3 Experimental Results

Results from SA indicated that only two parameters exhibit a significant direct influence on the output's variance, the pore-size distribution index n1 and the air-entry pressure parameter a1. The third most influential parameter, the saturated hydraulic conductivity Ks1, has the effect, which is only half of the second most influential parameter, a1. Ten parameters have a first-order index lower than 1%, which indicates that their main effect on the output variance is negligible.

In addition, SA showed that almost 75% of variance in simulated outflow is caused by n1, either by the variation of the parameter itself (30%) or by interactions with other parameters. Together with a1 (51%) and Ks1 (42%), it is the most influential parameter for simulated flow. It can be noted that the saturated hydraulic conductivity, Ks1, has a relatively low main effect but a relatively high total effect. That indicates that this parameter has a limited direct effect on the variance of the objective function, but it has an effect in interactions with other parameters.

The effect of the sub-base layer on the output is less significant, while the wear layer strongly conditions the output.

A Monte Carlo Filtering procedure was applied to the runs of the GSA. The threshold value of NSE = 0.0 produced a filtered sample composed of 1,452 behavioral solutions.

Also, for Scenario II, parameters a1 and n1 exhibit the highest main effects on the output's variance (about 35%). For both scenarios, modeling results are most sensitive to the wear layer, which strongly influences the output's variance. However, in Scenario II, the influence of the wear layer is partially reduced and redistributed to other layers. It is evident that the adoption of the dual-porosity model for the unsaturated hydraulic properties significantly affects the influence of the base and sub-base layers on the model's output. The dynamics of sensitivity indices between the two scenarios suggest that the physical description of unsaturated flow in the sub-base layer is an important element in numerical simulations.

A Monte Carlo Filtering procedure was again applied to the runs of the GSA. The filtered sample now consisted of 28,107 behavioral solutions. The filtered sample of behavioral solutions for Scenario II was considerably larger than for Scenario I. This indicates that the implementation of the dual-porosity model leads to higher values of the objective function.

Figure 12 compares measured and modeled hydrographs for the two scenarios. The PSO for Scenarios I and II resulted in NSE values of 0.43 and 0.81, respectively. Both NSE values of the objective function are higher than zero and thus admissible [39]. However, the implementation of the dual-porosity model for the base and sub-base layers in Scenario II provides a more accurate description of the hydraulic behavior of the permeable pavement.

In order to evaluate the reliability of the estimated parameters, the model has been validated on another independent set of experimental data. Figure 13 shows a comparison between measured and modeled hydrographs for the two scenarios during the validation period.

The value of the objective functions is NSE = 0.43 for Scenario I and NSE = 0.86 for Scenario II. For Scenario I, the value of the objective function remains the same, which confirms the reliability of the calibrated model. Although the simulated hydrograph provides an overall sufficiently accurate description of the hydraulic behavior of the pavement, it is less accurate during rainfall events, which may be a time period of main interest. For Scenario II, the value of the objective function

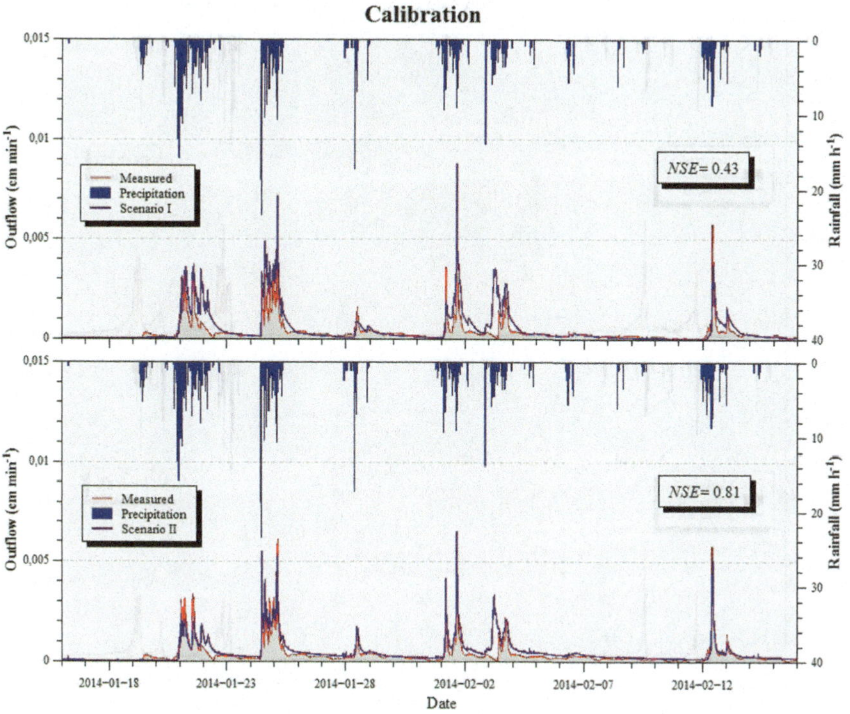

Fig. 12 Comparison between the modeled and measured hydrographs for Scenarios I (top) and II (bottom) for the optimization process

actually increased and reached the value NSE = 0.86, which is very high and reflects the accuracy of the modeled hydrograph.

4 Conclusion

The aim of this chapter was to presents the recent experiences in the Urban Hydrology field able to manage stormwater in a correct way. The cited works have shown how the smart management of the drainage networks and the application of urban regeneration facilities such as green roof or permeable pavement can help in knowledge of flooding phenomena. In particular, findings from Garofalo et al. [25] showed how DRTC algorithm proposed was able to balance the hydraulic capacity of the conduits within the system by utilizing the storage capacity of the less overwhelmed conduits during intense rainfall events. In other words, the DRTC algorithm was able to control the water level within the UDS successfully, ensuring a full utilization of the actual storage capacity of the system. The findings clearly demonstrated that the DRTC produced beneficial effects on the management of the UDS by substantially

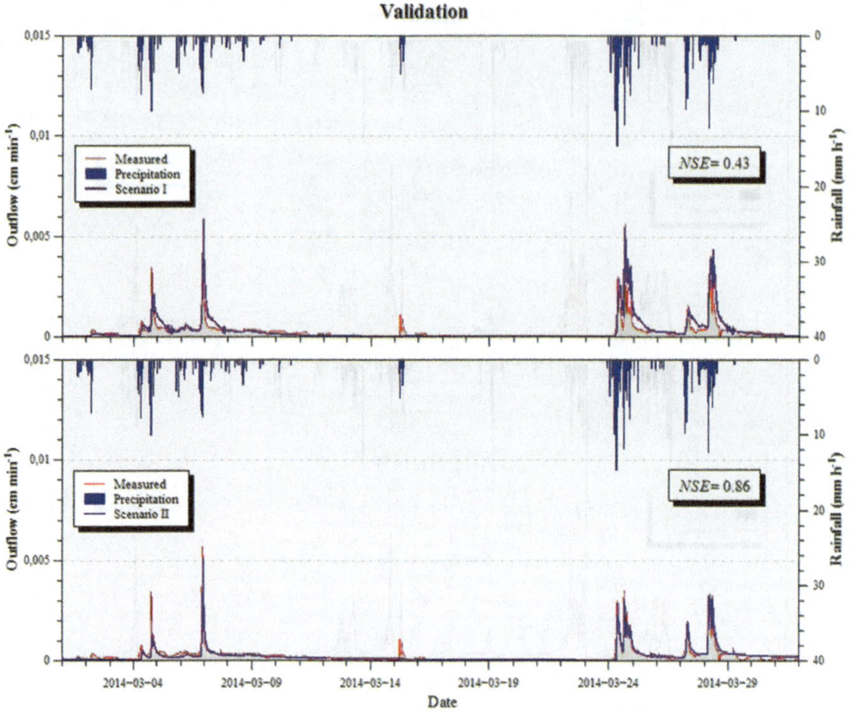

Fig. 13 Comparison between the modeled and measured hydrograph for the two scenarios for the validation period

mitigating the risk of flooding and CSO. Results from Brunetti et al. [8, 9] showed that the implementation of a model aimed at soil systems, together with accurate experimental and numerical procedures, has been able to accurately describe the hydraulic behaviour of systems of multiple layered materials that are not really soils. Future perspectives are oriented in the smart optimization of LID and RWH systems by the IoT advanced innovations in order to maximize the hydraulic efficiency of these techniques and mitigate the urban flooding risk.

References

1. S. Achleitner, M. Möderl, W. Rauch, CITY DRAIN ©—an open source approach for simulation of integrated urban drainage systems. Environ. Model. Softw. **22**, 1184–1195 (2007). https://doi.org/10.1016/j.envsoft.2006.06.013
2. R.G. Allen, L.S. Pereira, D. Raes, M. Smith, FAO Irrigation and Drainage Paper No. 56: Crop Evapotranspiration, FAO. Rome (1998)
3. G.E.B. Archer, A. Saltelli, I.M. Sobol, Sensitivity measures, ANOVA-like techniques and the use of bootstrap. J. Stat. Comput. Simul. **58**, 99–120 (1997). https://doi.org/10.1080/00949659708811825

4. K. Astrom, PID controllers: theory, design and tuning. Instrum. Soc. Am. (1995). ISBN 1556175167
5. P.M. Bach, W. Rauch, P.S. Mikkelsen, D.T. McCarthy, A. Deletic, A critical review of integrated urban water modelling—urban drainage and beyond. Environ. Model. Softw. (2014). https://doi.org/10.1016/j.envsoft.2013.12.018
6. G. Barenblatt, I. Zheltov, I. Kochina, Basic concepts in the theory of seepage of homogeneous liquids in fissured rocks [strata]. J. Appl. Math. Mech. **24**, 1286–1303 (1960). https://doi.org/10.1016/0021-8928(60)90107-6
7. T. Beeneken, V. Erbe, A. Messmer, C. Reder, R. Rohlfing, M. Scheer, M. Schuetze, B. Schumacher, M. Weilandt, M. Weyand, Real time control (RTC) of urban drainage systems—a discussion of the additional efforts compared to conventionally operated systems. Urban Water J. **10**, 293–299 (2013). https://doi.org/10.1080/1573062X.2013.790980
8. G. Brunetti, J. Simunek, P. Piro, A comprehensive analysis of the variably-saturated hydraulic behavior of a green roof in a mediterranean climate. Vadose Zo. J. **15** (in press) (2016a). https://doi.org/10.2136/vzj2016.04.0032
9. G. Brunetti, J. Šimůnek, P. Piro, A comprehensive numerical analysis of the hydraulic behavior of a permeable pavement. J. Hydrol. **540**, 1146–1161 (2016). https://doi.org/10.1016/j.jhydrol.2016.07.030
10. G. Brunetti, J. Šimůnek, M. Turco, P. Piro, On the use of surrogate-based modeling for the numerical analysis of low impact development techniques. J. Hydrol. **548**, 263–277 (2017). https://doi.org/10.1016/j.jhydrol.2017.03.013
11. M. Carbone, F. Principato, G. Garofalo, P. Piro, Comparison of evapotranspiration computation by FAO-56 and Hargreaves methods. J. Irrig. Drain. Eng. **142**(8), 06016007 (2016). https://doi.org/10.1061/(ASCE)IR.1943-4774.0001032
12. M. Carbone, G. Brunetti, P. Piro, Modelling the hydraulic behaviour of growing media with the explicit finite volume solution. Water (Switzerland) **7**, 568–591 (2015). https://doi.org/10.3390/w7020568
13. M. Carbone, M. Turco, G. Brunetti, P. Piro, A cumulative rainfall function for subhourly design storm in mediterranean urban areas. Adv. Meteorol. **2015**, 1–10 (2015). https://doi.org/10.1155/2015/528564
14. M. Carbone, M. Turco, G. Nigro, P. Piro, Modeling of hydraulic behaviour of green roof in catchment scale, in *14th SGEM GeoConference on Water Resources. Forest, Marine and Ocean Ecosystems* (2014a), pp. 471–478. https://doi.org/10.5593/sgem2014/b31/s12.061
15. M. Carbone, F. Principato, G. Nigro, P. Piro, Proposal of a conceptual model as tool for the hydraulic design of vegetated roof, in *Applied Mechanics and Materials*, vol. 641 (Trans Tech Publications, 2014b), pp. 326–331. https://doi.org/10.4028/www.scientific.net/AMM.641-642.326
16. M. Carbone, G. Garofalo, P. Piro, Decentralized real time control in combined sewer system by using smart objects. Procedia Eng. 473–478 (2014c). https://doi.org/10.1016/j.proeng.2014.11.237
17. M. Carini, M. Maiolo, D. Pantusa, F. Chiaravalloti, G. Capano, Modelling and optimization of least-cost water distribution networks with multiple supply sources and user. Ricerche Mat. **2017** (2017). https://doi.org/10.1007/s11587-017-0343-y
18. B. Cheviron, Y. Coquet, Sensitivity analysis of transient-MIM HYDRUS-1D: case study related to pesticide fate in soils. Vadose Zo. J. **8**, 1064 (2009). https://doi.org/10.2136/vzj2009.0023
19. G. Dirckx, M. Schütze, S. Kroll, C. Thoeye, G. De Gueldre, B. Van De Steene, Cost-efficiency of RTC for CSO impact mitigation. Urban Water J. **8**, 367–377 (2011). https://doi.org/10.1080/1573062X.2011.630092
20. W. Durner, Hydraulic conductivity estimation for soils with heterogeneous pore structure. Water Resour. Res. **30**, 211–223 (1994). https://doi.org/10.1029/93WR02676
21. B. Efron, R. Tibshirani, Bootstrap methods for standard errors, confidence intervals, and other measures of statistical accuracy. Stat. Sci. **1**, 54–75 (1986)
22. A.H. Elliott, S.A. Trowsdale, A review of models for low impact urban stormwater drainage. Environ. Model. & Softw. **22**, 394–405 (2007). https://doi.org/10.1016/j.envsoft.2005.12.005

23. G. Fu, D. Butler, S.-T. Khu, Multiple objective optimal control of integrated urban wastewater systems. Environ. Model Softw. **23**, 225–234 (2008). https://doi.org/10.1016/j.envsoft.2007.06.003
24. A. Giordano, G. Spezzano, A. Vinci, G. Garofalo, P. Piro, A cyber-physical system for distributed real-time control of urban drainage networks in smart cities, in *International Conference on Internet and Distributed Computing Systems* (Springer, Cham, 2014), pp. 87–98. https://doi.org/10.1007/978-3-319-11692-1_8
25. G. Garofalo, A. Giordano, P. Piro, G. Spezzano, A. Vinci, A distributed real-time approach for mitigating CSO and flooding in urban drainage systems. J. Netw. Comput. Appl. **78**, 30–42 (2017). https://doi.org/10.1016/j.jnca.2016.11.004
26. G. Garofalo, S. Palermo, F. Principato, T. Theodosiou, P. Piro, The influence of hydrologic parameters on the hydraulic efficiency of an extensive green roof in mediterranean area. Water **8**(2), 44 (2016). https://doi.org/10.3390/w8020044
27. M.K. Gill, Y.H. Kaheil, A. Khalil, M. McKee, L. Bastidas, Multiobjective particle swarm optimization for parameter estimation in hydrology. Water Resour. Res. **42**, n/a–n/a (2006). https://doi.org/10.1029/2005wr004528
28. T. Houska, S. Multsch, P. Kraft, H.-G. Frede, L. Breuer, Monte Carlo based calibration and uncertainty analysis of a coupled plant growth and hydrological model. Biogeosci. Discuss. **10**, 19509–19540 (2013). https://doi.org/10.5194/bgd-10-19509-2013
29. J. Huang, J. He, C. Valeo, A. Chu, Temporal evolution modeling of hydraulic and water quality performance of permeable pavements. J. Hydrol. **533**, 15–27 (2016). https://doi.org/10.1016/j.jhydrol.2015.11.042
30. M. Jelasity, A. Montresor, O. Babaoglu, Gossip-based aggregation in large dynamic networks. ACM Trans. Comput. Syst. **23**, 219–252 (2005). https://doi.org/10.1145/1082469.1082470
31. Y. Jiang, C. Liu, C. Huang, X. Wu, Improved particle swarm algorithm for hydrological parameter optimization. Appl. Math. Comput. **217**, 3207–3215 (2010). https://doi.org/10.1016/j.amc.2010.08.053
32. M. Kamali, M. Delkash, M. Tajrishy, Evaluation of permeable pavement responses to urban surface runoff. J. Environ. Manag. **187**, 43–53 (2017). https://doi.org/10.1016/j.jenvman.2016.11.027
33. J. Kennedy, R. Eberhart, Particle swarm optimization. Eng. Technol. 1942–1948 (1995)
34. Z.W. Kundzewicz, M. Radziejewski, I. Pińskwar, Precipitation extremes in the changing climate of Europe. Clim. Res. **31**, 51–58 (2006). https://doi.org/10.3354/cr031051
35. R. Levinson, H. Akbari, Effects of composition and exposure on the solar reflectance of portland cement concrete. Cem. Concr. Res. **32**, 1679–1698 (2002). https://doi.org/10.1016/S0008-8846(02)00835-9
36. Y. Li, R.W. Babcock, *Green roof hydrologic performance and modeling: A review* (Technol, Water Sci, 2014). https://doi.org/10.2166/wst.2013.770
37. M. Maiolo, D. Pantusa, An optimization procedure for the sustainable management of water resources. Water Sci. Technol.: Water Supply **16**(1), 61–69 (2016). https://doi.org/10.2166/ws.2015.114
38. S.K. Min, X. Zhang, F.W. Zwiers, G.C. Hegerl, Human contribution to more-intense precipitation extremes. Nature **470**, 378–381 (2011). https://doi.org/10.1038/nature09763
39. D.N. Moriasi, J.G. Arnold, M.W. Van Liew, R.L. Binger, R.D. Harmel, T.L. Veith, Model evaluation guidelines for systematic quantification of accuracy in watershed simulations. Trans. ASABE **50**, 885–900 (2007). https://doi.org/10.13031/2013.23153
40. J.E. Nash, J.V. Sutcliffe, River flow forecasting through conceptual models Part I—A discussion of principles. J. Hydrol. **10**, 282–290 (1970). https://doi.org/10.1016/0022-1694(70)90255-6
41. T. Pertassek, A. Peters, W. Durner, HYPROP-FIT Software User's Manual, V.3.0 (2015)
42. P. Piro, M. Carbone, A modelling approach to assessing variations of total suspended solids (TSS) mass fluxes during storm events. Hydrol. Process. **28**, 2419–2426 (2014). https://doi.org/10.1002/hyp.9809
43. P. Piro, M. Carbone, G. Garofalo, Distributed vs. concentrated storage options for controlling CSO volumes and pollutant loads. Water Pract. Technol. **5**, wpt2010071–wpt2010071 (2010a). https://doi.org/10.2166/wpt.2010.071

44. P. Piro, M. Carbone, G. Garofalo, J. Sansalone, Size distribution of wet weather and dry weather particulate matter entrained in combined flows from an urbanizing sewershed. Water Air Soil Pollut. **206**, 83–94 (2010). https://doi.org/10.1007/s11270-009-0088-7
45. M. Pleau, H. Colas, P. Lavallée, G. Pelletier, R. Bonin, Global optimal real-time control of the Quebec urban drainage system. Environ. Model. Softw. (2005). https://doi.org/10.1016/j.envsoft.2004.02.009
46. F. Principato, S.A. Palermo, G. Nigro, G. Garofalo, Sustainable strategies and RTC to mitigate CSO's impact: different scenarios in the highly urbanized catchment of Cosenza, Italy, in *Proceedings of the 14th IWA/IAHR International Conference on Urban Drainage, ICUD2017*, Prague, CZ, 10–15 Sept 2017, Oral Presentation, pp. 587–589
47. A. Raimondi, G. Becciu, On pre-filling probability of flood control detention facilities. Urban Water J. **12**, 344–351 (2015). https://doi.org/10.1080/1573062X.2014.901398
48. A. Raimondi, G. Becciu, Probabilistic modeling of rainwater tanks. Procedia Eng. **89**, 1493–1499 (2014). https://doi.org/10.1016/j.proeng.2014.11.437
49. M. Rezaei, P. Seuntjens, I. Joris, W. Boënne, S. Van Hoey, P. Campling, W.M. Cornelis, Sensitivity of water stress in a two-layered sandy grassland soil to variations in groundwater depth and soil hydraulic parameters. Hydrol. Earth Syst. Sci. Discuss. **12**, 6881–6920 (2015). https://doi.org/10.5194/hessd-12-6881-2015
50. L.A. Rossman, Storm water management model quality assurance report: dynamic wave flow routing. Storm Water Manag. Model Qual. Assur. Rep. 1–115 (2006)
51. A. Saltelli,, S. Tarantola, M. Saisana, M. Nardo, What is sensitivity analysis?, in *II Convegno Della Rete Dei Nuclei Di Valutazione E Verifica*, Napoli 26, 27 Gennaio 2005, Centro Congressi Universitá Federico II, Via Partenope 36 (2005)
52. M. Schütze, A. Campisano, H. Colas, W. Schilling, P.A. Vanrolleghem, Real time control of urban wastewater systems—where do we stand today? J. Hydrol. **299**, 335–348 (2004). https://doi.org/10.1016/j.jhydrol.2004.08.010
53. N. She, J. Pang, Physically based green roof model. J. Hydrol. Eng. **15**, 458–464 (2010). https://doi.org/10.1061/(ASCE)HE.1943-5584.0000138
54. J. Šimůnek, M.T. van Genuchten, M. Šejna, Recent developments and applications of the HYDRUS Computer Software Pac. Vadose Zo. J. **15**, 25 (2016). https://doi.org/10.2136/vzj2016.04.0033
55. J. Simunek, N.J. Jarvis, M.T. van Genuchten, A. Gardenas, Review and comparison of models for describing non-equilibrium and preferential flow and transport in the vadose zone. J. Hydrol. **272**, 14–35 (2003). https://doi.org/10.1016/S0022-1694(02)00252-4
56. J. Šimůnek, M.T. van Genuchten, Modeling nonequilibrium flow and transport processes using HYDRUS. Vadose Zo. J. **7**, 782 (2008). https://doi.org/10.2136/vzj2007.0074
57. J. Šimůnek, M.T. van Genuchten, M. Šejna, Development and applications of the HYDRUS and STANMOD software packages and related codes. Vadose Zo. J. **7**, 587 (2008). https://doi.org/10.2136/vzj2007.0077
58. I. Sobol', Global sensitivity indices for nonlinear mathematical models and their Monte Carlo estimates. Math. Comput. Simul. **55**, 271–280 (2001). https://doi.org/10.1016/S0378-4754(00)00270-6
59. M. Turco, R. Kodešová, G. Brunetti, A. Nikodem, M. Fér, P. Piro, Unsaturated hydraulic behaviour of a permeable pavement: laboratory investigation and numerical analysis by using the HYDRUS-2D model. J. Hydrol. **554**, 780–791 (2017). https://doi.org/10.1016/j.jhydrol.2017.10.005
60. UMS GmbH, UMS (2015): Manual HYPROP, Version 2015-01 (2015)
61. W. Usher, Xantares, D. Hadka, bernardoct, Fernando, J. Herman, C. Mutel, SALib: New documentation, doc strings and installation requirements (2015). https://doi.org/10.5281/zenodo.31316
62. M.T. van Genuchten, A closed-form equation for predicting the hydraulic conductivity of unsaturated soils. Soil Sci. Soc. Am. J. **44**, 892 (1980). https://doi.org/10.2136/sssaj1980.03615995004400050002x

63. M.T. Van Genuchten, P.J. Wierenga, Mass transfer studies in sorbing porous media I. Analytical solutions. Soil Sci. Soc. Am. J. **40**, 473–480 (1976). https://doi.org/10.2136/sssaj1976.03615995004000040011x
64. J.E. Warren, P.J. Root, The behavior of naturally fractured reservoirs. Soc. Pet. Eng. J. **3**, 245–255 (1963). https://doi.org/10.2118/426-PA
65. T.H.F. Wong, T.D. Fletcher, H.P. Duncan, G.A. Jenkins, Modelling urban stormwater treatment—a unified approach. Ecol. Eng. **27**, 58–70 (2006). https://doi.org/10.1016/j.ecoleng.2005.10.014
66. M. Wooldridge, *An Introduction to MultiAgent Systems*, 2nd edn. (Wiley, 2009), ISBN-10 0470519460, ISBN-13 978-0470519462
67. M. Zambrano-Bigiarini, R. Rojas, A model-independent Particle Swarm Optimisation software for model calibration. Environ. Model Softw. **43**, 5–25 (2013). https://doi.org/10.1016/j.envsoft.2013.01.004
68. S. Zhang, Y. Guo, Analytical probabilistic model for evaluating the hydrologic performance of green roofs. J. Hydrol. Eng. **18**, 19–28 (2013). https://doi.org/10.1061/(ASCE)HE.1943-5584.0000593

Cooperative Video-Surveillance Framework in *Internet of Things* (IoT) Domain

A. F. Santamaria, P. Raimondo, N. Palmieri, M. Tropea and F. De Rango

Abstract In this chapter a cooperative heterogeneous system for an enhanced video-surveillance service will be presented. Edge and fog computing architectures make possible the realization of even more complex and distributed services. Moreover, the distribution of sensors and devices gives us the possibility to increase the knowledge of the monitored environments by exploiting *Machine to Machine* (M2M) communications protocols and their architectures. The rapid growth of IoT increased the number of the smart devices able to acquire, actuate and exchange information in a smart way. In this chapter, the main issues related to the design of an architecture for a smart cooperative video-surveillance system will be presented. The end-system shall exploit edge and fog computing for video-analytics services and communication protocols for cameras data exchange. Finally, all systems together realize a cooperative tracking among cameras that involves detection and tracking techniques to work jointly. At the end a detected anomaly can be followed among cameras generating alerting and notifying messages that will be sent to the designed human interaction system without explicit human interactions in the detection, tracking and system managing processes.

A. F. Santamaria (✉) · P. Raimondo · N. Palmieri · M. Tropea · F. De Rango
DIMES - University of Calabria, Pietro Bucci street, Rende, Italy
e-mail: afsantamaria@dimes.unical.it

P. Raimondo
e-mail: p.raimondo@dimes.unical.it

N. Palmieri
e-mail: nunzia.palmieri@dimes.unical.it

M. Tropea
e-mail: mtropea@dimes.unical.it

F. De Rango
e-mail: derango@dimes.unical.it

F. Cicirelli et al. (eds.), *The Internet of Things for Smart Urban Ecosystems*,
Internet of Things, https://doi.org/10.1007/978-3-319-96550-5_13

1 Introduction

In the last few years, computer vision algorithms have acquired even more importance in video surveillance platforms. This happened due to several reasons that have permitted to increase related services and functionalities to make more reliable and affordable systems. Usually, these systems or platforms are composed of several cameras distributed in the areas of interest sending video streaming towards a centralized node that has the main task to elaborate videos and collect meta-data and information on recognized activities. The detection of anomalies and alarms is left to human checks that may commit evaluation errors. Hereby, people designed monitoring tasks are not able to recognize on the fly some important events and only a post-analysis of saved videos can bring up important information about what has happened. Moreover, video surveillance system has become a critical part in the security and protection system of modern cities and smart urban ecosystems. Smart monitoring cameras and smart devices can cooperate for the realization of an intelligent video analytic platform. Efficient techniques can monitor and pre-alarm abnormal behaviors or events spreading these information in the smart-device environment ecosystem. However, with the expansion of the surveillance network, massive data are collected by video-cameras. These data open new and complex challenges to the analytics, storage and retrieval in the Big Data era [1]. In this chapter, the basis for building an enhanced framework for the automatic detection of abnormal behavior of people in a given area of interest will be provided. Moreover, once the anomaly is detected, its source will be tracked in *Line Of Sight* (LOS) of the camera that is performing monitoring of the *Region of Interest* (ROI). Tracked objects/people will be followed in the neighbors camera in case of leaving the LOS of the current camera by adopting cooperation mechanisms that will enroll the whole system. For detecting an object a self-adapting mechanism can be used that exploits the availability of video-streaming. It analyzes frames and through a learning algorithm can carry out meta-data about normal behaviors that people or object may assume in the observed ROI. This complex schema can be realized because the availability of novel technologies that allow the implementation of a distributed architecture. In this way it is possible to decentralize the computational load and distribute in the environment several custom devices keeping the price of infrastructures low [2]. The feasibility and the characteristics of these devices may open the road for the design and development of new services. In fact, smart devices have grown in interest because of the opportunities that they may create in several application domains. One of the major fields of interest is represented by the video-surveillance systems able to cooperate with external services for achieving a smarter system. The whole system will be composed of several heterogeneous sub-systems able to work as independent platforms with their services and their users interactions. In the same way, each platform will cooperate with the others to better know the application domain increasing this way the quality of supplied services. In this chapter, the main goal is to design a partially autonomous system for the video-surveillance tasks. It shall be able to recognize malicious events by cooperating with all sub-systems available in the environment. The whole system could be autonomous or composed of several services and devices exploiting M2M and IoT architectures and communications protocols. This makes possible the realization of complex activities by distributing computation and information exchange. Modular system may be applied in several fields but this chapter will focus on the building of a smart video-surveillance system. The main objective is to create a feasible, simple and efficient cooperation between systems and sub-systems. This cooperation brings up several problems to be addressed such as interfaces, roles and communication rules between heterogeneous systems. Alse some basic concepts about distributed systems will be given. We will introduce communication protocols for M2M and IoT domain and computer vision principles that will be used to detect anomalies in the ROI. Distributed architectures are very useful because allow to distribute computational load and increase system scalability. This makes services more reliable, robust and detailed because of the possibility to sample a huge amount of data from sensors and devices placed in the environments. Nowadays social security is even more important and tracking is a key aspect for smart surveillance systems. These applications must detect and follow position and movement of people, objects and vehicles. The reasons for these tracking tasks can concern, for example, the presence of people inside a restricted area. The knowledge of movements and positions of these individuals are really important security issues in order to avoid dangerous situations. A well known issue of security/surveillance systems is the difficulty to analyze and manage complex scenarios. Inter alia the more the complexity increase, in terms of number of objects or individuals, the more is difficult to successfully detect them and the more is the computational complexity required. This issue becomes harder to treat when we try to analyze crowd scenarios with an high density and it's necessary to profile the behavior of a group of people. The major problem in analyzing a crowd scenario is represented by the difficulty of tracking a single person inside the crowd. Individuals inside a crowd interact among them, the partial occlusion and the complex behavior-driven mechanism that depends on individuals role increase exponentially the difficulty of tracking a single person. To avoid this kind of issues a set of tracking algorithms that ensure a good accuracy learning collective movement patterns has been developed [3–5]. These collective patterns are extracted, analyzed and learned in relation to a specific scene or fixed point of view and are used to predict probable crowd movements detecting abnormalities. Recently data-driven methods for visual recognition [6, 7] have been developed, it is possible to analyze and test videos evaluating a large set of crowd patterns and use these information already learned to enhance efficiency of, for example, an online tracking video technique. Regarding image analysis the extraction of already learned features eases and enhances the tracking operations. Finally a testbed will be designed to bring up the goodness and feasibility of the communication architecture and modular system herein described.

2 Distributed Architecture

Even more solutions use distributed architecture to build off novel applications and services. In particular, when a high number of users, devices or appliances need a communication infrastructure a decentralized solution may help. The scalability of systems and applications is a common issue for complex architectures, in fact, several researches have been made covering several fields of interest such as *Call Admission Control* (CAC) algorithm, routing protocols, multimedia systems and heterogeneous architectures [8–10]. For example, the scalability is one of the most important issues to face in case of large systems because of the number of users that may request access to the network and its services. Moreover, when the mobility of the users has to be considered new challenges raise such as the *Quality of Service* (QoS) or the possibility to keep the connection up. Here the routing protocols have the goal to reduce the side effects of mobility by assuring the delivery of packets and avoiding network resource wastages [11–13]. In these last years, the edge and fog computing makes possible to simplify data filtering and aggregation for further uses on the cloud based services. Usually, data are produced at user or device layers and sent towards the cloud for storage, aggregation, filter or data mining elaborations. This process is complex to be performed and with the increase of accesses the whole system presents performances degradation in terms of time and space allocation. For these reasons the edge/fog computation may help to reduce these performance degradations.

The edge-fog cloud paradigm is characterized by a huge number of networked objects that generate and consume data placed in a large geographic area or even the whole world. This distributed architecture is hierarchical and can be seen in Fig. 1. As showed this architecture is able to support future IoT applications with the help of edge-fog cloud computing facilities deployed. Vehicles, buildings and even people equipped with proper devices can be considered IoT nodes inside a city. Besides these low-level nodes we can find at few hops or at the network edge a set of loosely coupled often human-operated nodes such as tablets, smart-phones, wireless access points and small data centers. These edge devices are equipped with wireless device-to-device connectivity and with a reliable and stable connection to fog nodes. Their purpose is to collect data from the sensors and pre-process them as shown in Fig. 1. In this figure is possible to point out how these edge nodes aggregate and shrink this huge amount of environmental data of orange IoT nodes in order to detect events of interest like fire or collapsing structures. Green nodes instead can be used to detect sudden accidents on the road and alert in soft real-time other reachable cars that are coming in the same dangerous area. Fog nodes are distant from the area of concern compared to IoT objects and are machines with a remarkable computational power with the main task of analyzing a huge amount of data and temporary store them. Even if these nodes have great computational capabilities they are not able to face the increasing complexity of possible future applications. For this reason most of analytics have to still be hosted in a central cloud. So in order to take some of this complex services from the cloud to the edge nodes of the networks some virtualized entities that run both on the nodes and the cloud are necessary. Usually

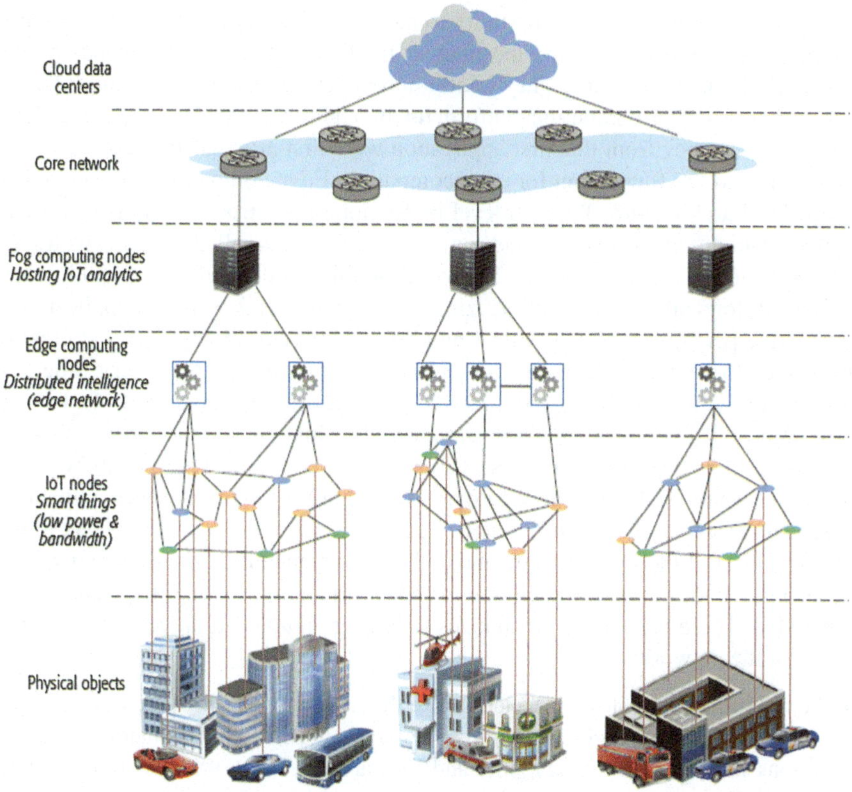

Fig. 1 A whole distributed ecosystem in the modern cloud domain

these devices are virtualized appliances or networking devices called smart gateways with adequate computing capabilities and a fast and stable access to the network. Fog nodes are capable of running on their architecture cloud applications developed by cloud providers. These nodes are mainly used to analyze time-sensitive data and to compute heavy-load tasks to obtain a reduction of latency and traffic. Cloud Computing has been one of the hot topic in the *Information Technologies* (IT) in the past decade with two main innovations. The first is the centralization lowering this way marginal costs of system administrations and operations. The second one is that in this way small organizations can avoid the starting costs for an IT infrastructure using services and resources that are already exploitable through a fast and stable internet connection. The trend in the last years is to externalize all IT needs into the cloud data-centers spread across the world with resulting economic benefits and less risks. Opposite to this centralization trend there is a dispersion trend driven by the appearance of a huge amount of small devices with network and computing capabilities known as IoT. We can consider fog-computing paradigm as a middle way in which micro data centers, cloudlets (also called fog-nodes) are placed on the network's edges in close proximity to mobile devices and sensors. Thanks also to giants like Nokia and IBM the interest for this midway paradigm has grown dramatically in recent years. They jointly introduced the Radio Applications Cloud server that is an edge computing platform for 4G/LTE networks. Data collected about end-to-end latency from this first application were first articulated in 2009 and can be considered the foundation for edge computing. Edge computing is structured as a two-level architecture. The first level is the cloud infrastructure and is still today almost unmodified [14]. The second one consists of devices called cloudlets with state cached from the first level. In 2012 Flavio Bonomi et al defined fog computing as a dispersed cloud infrastructure prioritizing scalability rather than mobile applications' interactive performance [15] with a multilevel hierarchy of fog nodes stretching from the cloud to IoT devices. The chance of accessing internet every time and everywhere with 3G/4G connections led us to ignore the advantages of physical proximities. These advantages are summarized in low latency, low jitter and high bandwidth. Physical proximity ensures low end-to-end latency, high bandwidth at low cost, secure establishment of trust and survivability. Obviously, even with direct hardware access to the network, latency and performances have a limit due to the physical limit of the speed of light and to the heterogeneous nature of networks that needs relay nodes to work. Each hop introduces queuing, routing delay and buffer bloat [16]. There are several advantages in having cloudlets relatively close on the edges of the network:

- the proximity of deployed services make them highly responsive achieving a low end-to-end latency, high bandwidth and low jitter. This is very valuable for applications such as Augmented reality and Virtual reality that offload computation to the cloudlets.
- Easy scalability. Traffic generated from the sensor is summarized and reduced before it is sent to the cloud.
- Privacy-policy enforcement. A cloudlet can enforce the privacy policies on the first contact with IoT nodes before the data release towards the cloud
- Redundancy of services. When a service is not available the cloud can mask the outages by fallbacking to a service on a nearby cloudlet temporarily.

2.1 Scalability in the Edge Computing

Edge computing ensures the reduction of ingress bandwidth into the cloud. Taking into account a video analytic application where several video analysis are need to be performed in a quasi real-time way users continuously send video to the cloud for content analysis. The data in the cloud is received in a cumulative way and the ingress rate is very high. Therefore it is not easy for the cloud to satisfy each single request in a reasonable time. Just considering a certain number of users in a metropolitan area it is easy to achieve a saturation of network in a short time. In Fig. 2 a possible solution is shown. Here users send videos to nearby fog devices that are responsible

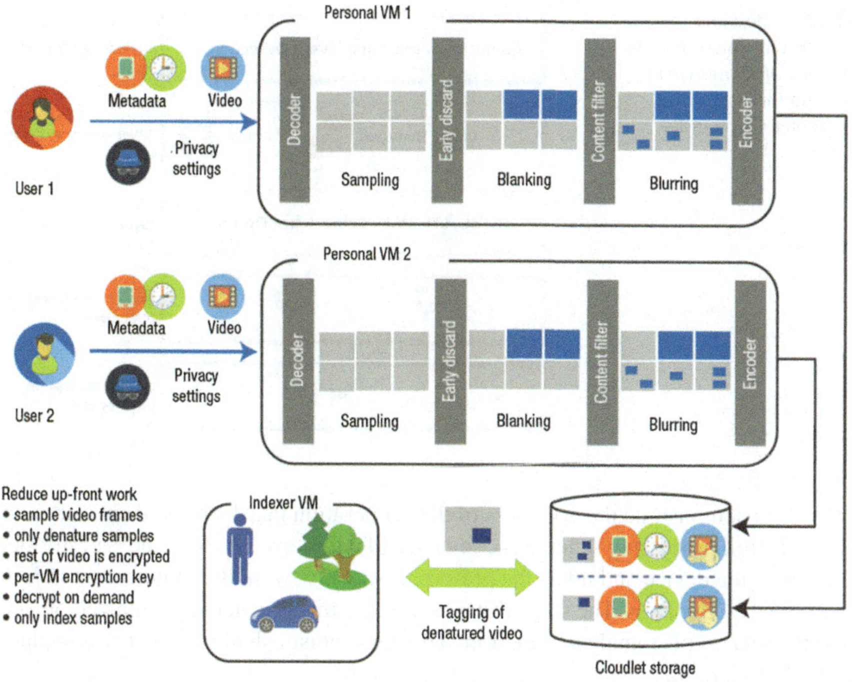

Fig. 2 A possible solution for fog computing in the computer vision an analytics domain. Here video streaming analysis is performed locally in the fog layer and the results of the analysis in terms of metadata are sent to the cloud. This permits to gain in terms of ingress bandwidth into the cloud

of processing data streaming [17]. The results of this video analysis can be obtained in a quasi real-time way because of the available resources and low delays. Finally, analysis results are send back to user. These results are also sent to the cloud in terms of meta-data for further analysis. In the fog layer, it is possible to instantiate processing, filtering or aggregate routines to reduce the amount of data to transfer from user devices to cloud services.

2.2 *Distributed Tasks*

People monitoring and tracking activities in surveillance system usually generate massive amount of data from IoT devices such as cameras. Several issues need to be addressed, including data migration over limited bandwidth and high latency in communication networks. The decentralization of tasks between cloudlets and cloud platforms may help to reduce the overload in the end devices to achieve better performances [18]. The authors in [18] propose to use the architecture layers shown in Fig. 3. This architecture is commonly used to decentralize the tasks and also in this

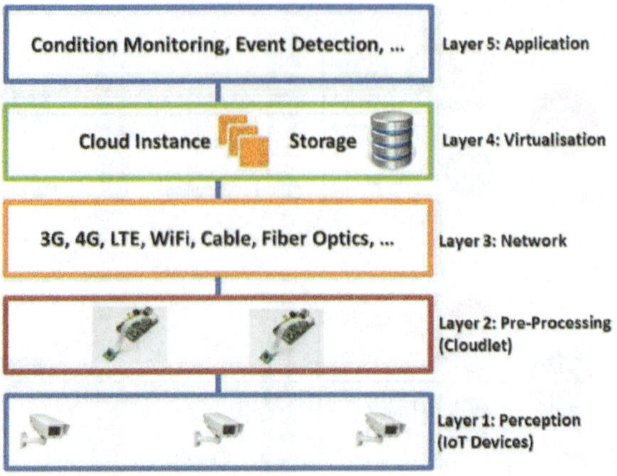

Fig. 3 A possible architecture solution in the *Internet of Things* (IoT) domain for distributing computer vision tasks

chapter to demonstrate the goodness of the cooperation in a distributed environment. Here, the full processing schema is composed of two steps. The first one is performed locally on the device and it has the pre-processing task of finding interesting objects in the analyzed frames. The second step is the classification of the recognized objects and the related post-analysis function. This mechanism shall be better investigated in further sections.

3 Computer Vision

3.1 Introduction

In the last few years there have been a lot of improvements in video surveillance techniques thanks to the application of computer vision algorithms. The word *tracking* is used to describe the process of pursuing one or more object that are moving along sequential frames. This kind of algorithms analyzes each frame of a video returning as output the position of the objects of interest also called targets. There are several application environments in which these computer vision techniques are used like industrial automation, multimedia entertainment, security and so on. The exponential diffusion of small devices able to process video stream fostered the spread of smart applications in which user identity or user activity have to be recognized in order to reach their goals. One of the most challenging scenario is linked to very complex frames, full of not interesting objects or noise, and to high crowded application cases. Most of the techniques that are stated in literature are based on the target model construction and on the tracking method. The target model can be defined as the set of features that trigger a positive individuation. These features are hard to find due to the visual nature of the task in which dimensions, color and shape are very sensible to light, point of view and so on. The most known tracking algorithms come from engineering-mathematic application of image elaboration theory like: Particle Filter, Kalman Filter, Template matching, Background subtraction, object detection and histograms [19]. The particle filter technique has a high computational cost that makes it not really adapt to real time analysis application. On the contrary the Kalman filter technique is really fast but is not suitable to track objects with unpredictable trajectories. Template matching is not suitable when the target object frequently changes its pose or its scale compared to the environment. Background subtraction it's fast and accurate enough in detecting the shape of the target but it's not able to discern between multiple shapes and it cannot be used with mobile cameras. Also Object Detection techniques have problems in discerning between similar objects and has a really high computational cost, moreover they must be trained on a set of very specific features. Histogram based algorithms are very fast when the target dimensions are small. They do not perform well when the environment has multiple features in common with the target object. The most well known histogram based techniques are Meanshift and *Continuously Adaptive Mean Shift* (CAMShift). They both can estimate the position of the target inside the frame finding the local max of the probability distribution but the second one it's also capable of estimating important parameters like target dimension and orientation. CAMShift is a color tracking algorithm developed for the first time by Bradski for facial recognition continuously adapting the searching region inside the frame. Background Subtraction algorithms can extract shapes of human/objects in the foreground detecting differences between the video frame and the background model. This technique has limits: the streaming camera must be fixed, only moving objects are detected and it's really hard to discern human shapes in a crowd.

3.2 *Classifiers*

One of the most important instruments for the object recognition task is represented by the available classifiers for the computer vision framework. One of the most used framework for computer vision is OpenCV thanks to a lot of different classifiers that are chosen on the basis of the case of use. The classifier has the task to assign a label at recognized objects. In order to build and test a classifier it is important to train it by supplying a controlled input. This set of input is called training set. Two kinds of learning are possible:

- supervised learning;
- unsupervised learning.

Supervised learning is where you have input variables (x) and an output variable (Y) and you use an algorithm to learn the mapping function from the input to the output.On the contrary in unsupervised learning you have only input variables (x) and the algorithm has to model the underlying structure or distribution in the data in order to learn more about the data. To verify a classifier goodness a test set is tried and the results are evaluated. OpenCV presents several methods to build new classifiers, some of them are herein reported:

Neural Networks — *Machine Learning* (ML) algorithms based on *Neural Network* (NN) in OpenCV works with the model *Multi Layer Perceptions* (MLP) which is the most commonly used type of neural networks. Its main working schema is depicted in Fig. 4. All neurons in the layers are quite similar. The inputs of a neurons are summed with a certain weight plus a bias term. This sum is then transformed by an activation function that may be different for each neuron. Three common activation functions exist: Identity, Symmetrical sigmoid and Gaussian;

K-Nearest Neighbours — This model is based on the *k-Nearest Neighbor* (KNN), it is one of the most simple classification algorithm and it belongs to the supervised learning set; The basic idea is to find the closest class into a given featured space considering k-neighbors in the nearest space. An example about this classification algorithm is shown in Fig. 5;

Decision tree — A decision tree is a binary tree that can be used either for classification or for regression. When used as a classification model each leaf is marked with a class label. Multiple leaves may have the same label. When used as a regression model to each leaf a constant is assigned. It is important to point out that when this technique is applied to computer vision sometimes certain features of the input vector are missed. An example is the missing color if the tree is operating on a dark image. In this case surrogate splits are used in order to allow the classifier to do not get stuck on a node.

Cascade Classifier — The word cascade refers to several sub-sequential applications of a simple classifier inside an image. The classifier

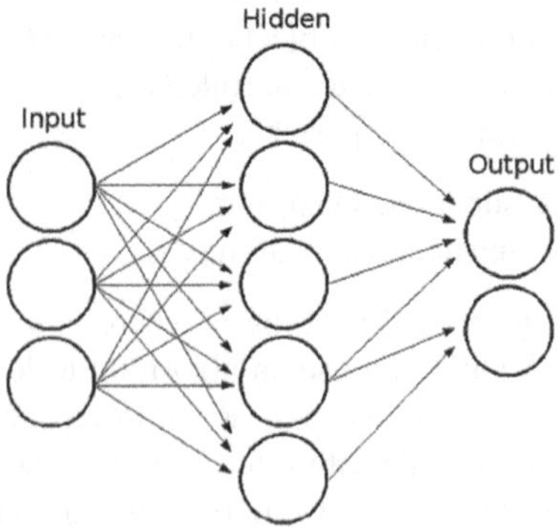

Fig. 4 *Multi Layer Perceptions* (MLP) based schema where an Input layer is connected with the hidden layer. Finally the output layer will supply the results of classification; Each layer is composed of one or more neurons

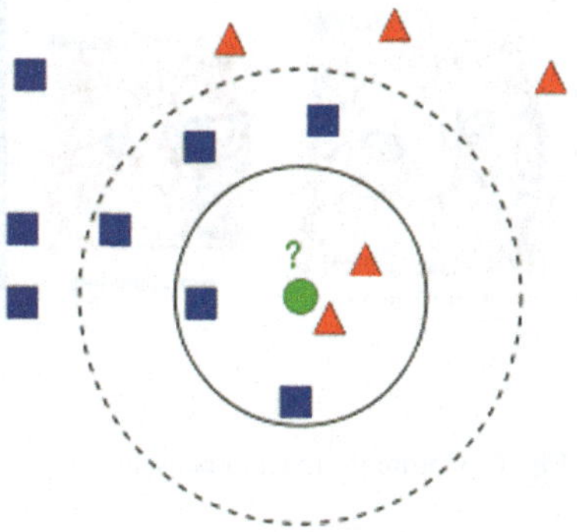

Fig. 5 *k-Nearest Neighbor* (KNN) algorithm example in a 2-D featured space. Here the red triangle, and blue square class has been used to define the featured space in the trained set. The classifier will assign a class to the green circle taking into account the k-nearest neighbors applying sum operator

works with haar-like features and is trained over a few hundred sample views of a particular object. This kind of classifiers use boost technique over very simple classifiers to combine their results and achieve a better accuracy.

Bayesian Classifier — This model is used in the ML problem. The classifier functions are based on statistical laws. This simple classification technique assumes that feature vectors from each class are normally distributed. In this way the whole data distribution function can be considered a Gaussian mixture with one component per class. During training data are used to estimate mean vectors and covariance matrices for every class that are used during the prediction phase.

3.3 *Image Recognition*

Image recognition represents one of the most interesting feature of computer vision. Several algorithms are introduced in this application field year by year outperforming older. Also human detection in video frames represents a great challenge because the difficulty in building efficient classifiers. A feature descriptor called *Histograms of oriented gradients* (HOG) was very useful and changed the way to approach the problem. It was proposed in 2005 by by N. Dalal and B.Triggs with a paper titled *Histograms of oriented gradients for human detection*[20].

3.4 *Object Detection*

Most of computer image classification algorithms to detect objects of interest follow a pipeline of three main steps:

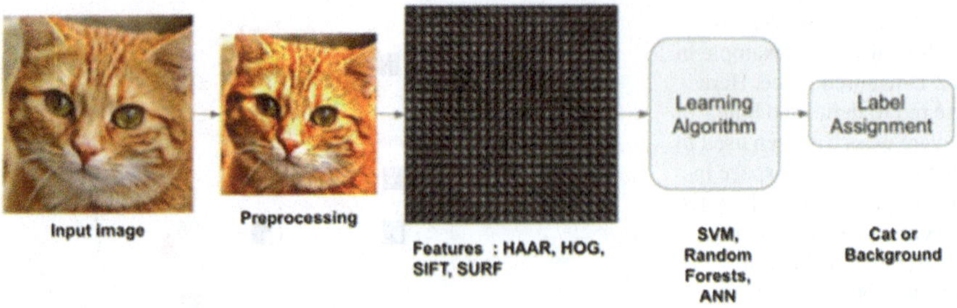

Fig. 6 Common image processing steps

- Preprocessing
- Feature Extraction
- Learning Algorithms.

Techniques that involve deep learning bypass the second step. The preprocessing step consists of normalizing contrast and brightness effects. One of the most common preprocessing techniques is to subtract the mean of image intensities and divide by the standard deviation. It is impossible to know in advance which of the various preprocessing steps will produce good results so this can be considered a trial&phase [21]. Operations like cropping or resize can also be considered as preprocessing techniques. Every image contains not necessary information for detecting objects of interest so the feature extraction step is necessary. For example if we want to detect a button in an image feature like color is unnecessary while edges and shape are relevant. A set of features that are very well known in computer vision environment are the Haar-like features. These features can count HOG, *Scale Invariant Feature Transform* (SIFT), *Speeded Up Robust Feature* (SURF) etc. In particular HOG features will be object of further analysis. The third step takes in input the features extracted during step two and learn how to classify images returning in output a class label. Before a classification algorithm can return good results a training stage is necessary in which thousand of examples are required. An example of a full image processing in Fig. 6 is shown [22].

3.5 Tracking

Tracking can be defined as the process of locating a single or multiple objects over time inside a video stream. A video stream can be considered a set of consecutive frames in which the tracking algorithms have to associate target objects. This task can be really difficult when the objects are moving fast relative to the frame rate. The recognition of objects is a very complex task that a human brain learn to do starting from early years but is really hard to be replicated on machines. The difficulty increases when objects inside frames change orientation or perspective and when brightness is low. Brightness and image resolution affect the ability of the machines to recognize interesting features like colors or specific shapes.

Tracking techniques can be divided into two main groups.

- low level techniques fast and robust;
- high level techniques able to detect complex shapes but slow and less robust;

In the low level set we can count approaches like blob detection, background subtraction, corner detection. Template matching instead can be considered as a borderline technique due to its complexity. In the high level set we can count a lot of very complex approaches and some of the most known are face recognition and activity recognition. When the stream is recorded from a fixed camera one of the most relevant techniques to use is background subtraction. In these scenarios the background is any pixel that does not change between consecutive frames. Then the subtraction is applied and an image with only the changed pixels is calculated. Combining images obtained with background subtraction techniques we can build the *Motion History Image* (MHI). When using this kind of techniques it's required the complete view of the object of interest to work. If the target is not full visible then the recognition is not sure. For good results it is important to recognize which kind of approach is better to use. With low level techniques the identification of interesting features is entwined with the computational capacity of the device ind its low level functions already implemented. These functions are typically really generic and hardly customizable so it's nearly impossible to tune them for a very specific target. On the other hand high level techniques can manage more complex functions. Obviously the computational cost is much higher also because most of them use mathematical approaches with probability distributions and classification. Generally hybrid approaches are used in order to achieve good enough results in a reasonable time.

3.5.1 Blob Detection

The goal of Blob Detection is the detection of points or areas of interest inside a picture. These points or regions are characterized by a certain brightness value that is processed from the algorithm. These set of techniques can be divided into two main groups:

- differential methods based on derivative expressions
- local environmental intensity based techniques

Both approaches have the goal to identify interesting features. An example of this technique results are shown in Fig. 7. As can be seen in Fig. 7 the main task of the algorithm is to detect image areas that are less bright than the others. Areas of interest are highlighted with a green circle. One of the strengths of these approaches is that is able to find points of interest even if the shapes or outlines are not well defined. Descriptors based on Blob Detection have been used to outline the presence of features with high information content for the detection of objects based on local

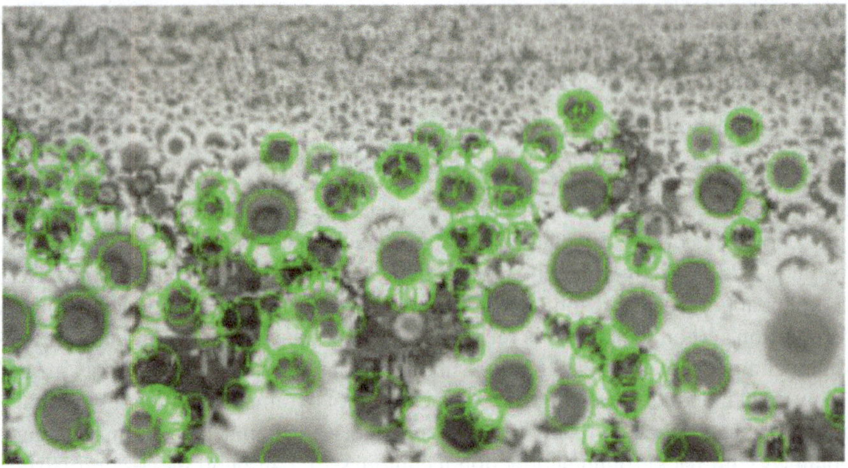

Fig. 7 Example of Blob detection application

image statistics. Moreover this technique is able to recognize dimension and pitch changes.

3.5.2 Corner Detection

Blob Detection techniques are suitable to detect image areas where there are brightness variations of a set of pixels. They are not able to scan points belonging to corners or angle points. An edge is an area of the image with a strong brightness gradient discontinuity only in one direction. A corner instead possesses a strong brightness gradient discontinuity in more than one direction. One of the most important characteristics of corners is that they have a good resistance to lightness change and for this reason are recognizable in dynamic scenarios. Mathematically the gradient is defined as a vector containing partial derivatives of the analyzed function along principal directions:

$$\nabla I[x, y] = \left[\frac{\delta I[x, y]}{\delta x}, \frac{\delta I[x, y]}{\delta y} \right] \tag{1}$$

To explain how this works let's take into account a gray scale image. In regions with a strong depth discontinuity can be detected by the analysis of partial derivatives of the image pointing out this way edges and corners. When there are strong variations the derivative assumes high values instead where the texture is continuous the derivative module is near zero. To enhance noise resistance when the partial derivative are calculated some digital filters are applied like Sobel filter that can be considered as an operator.A weak point of this technique occurs when it's necessary to evaluate the gradient along directions that are different from the one of the target. This problem is also known as the aperture problem and it's bound to the eye perception of objects seen through an aperture. In this situation it can be difficult to understand the direction of the moving object if it has some kind of patterns. For these reasons edges are not taken into account and only angular points are considered in order to not produce ambiguities. Every system must count on an efficient and reliable algorithm for corner extraction. Not all the areas of an image must be taken into account. For example a smooth surface without any texture do not possess information about the movement. An object border gives movement information only when the object moves perpendicularly compared to another object border. There are two groups of algorithms to manage these situations: the first one extracts borders inside the image and search for points with maximum curvature or where borders collide; the second one search for angles analyzing directly the gray scale.

3.6 Human Detection

The task of detecting a human body relying only on the appearance is a more difficult task than detecting rigid objects as cars or still faces. Unlike rigid objects the human body is dynamic and articulated and this implies that can assume a very high range of poses or postures. Furthermore when performing human detection task is not possible to use as key features textures or colors because of the variety of worn clothes. In order to detect a human body a robust feature set is necessary to discriminate cleanly targets even in cluttered backgrounds or in low light conditions. For this purpose a combination of HOG features and *Support Vector Machines* (SVM) are used [23]. HOG features have been chosen as they provide excellent performance over other existing feature sets like wavelets. These features are computed on set of uniformly spaced cells to overlap local contrast normalizations improving this way performances. Using HOG descriptor is advantageous because it shows invariance to geometric and photometric transformations. Simplicity and speed are the desired attributes found in SVM classifiers used as baseline [23]. For human detection one of the feature descriptors that works better, as previously introduced, is the HOG. This technique counts occurrences of gradient orientation in localized area of an image. The computation is performed over uniformly spaced cells using overlapping local contrast normalization to improve accuracy. The theory behind this kind of descriptors is that an object appearance and shape can be described by a distribution of intensity gradients. The key feature of this descriptor is that, since it works on local portion of the image, is unaffected by geometric and photometric transformations. Change of object orientation is still a problem. These qualities make HOG descriptors highly suitable for human detection [24]. In [25] authors propose a technique to identify and track humans in the acquired video. This method initially removes noise from the captured images then segments the images using frame difference and binary conversion techniques and finally tracks the person using a bounding box based on occurrence of high intensity values horizontally and vertically. To enhance these mechanisms we worked on the possibility to store meta-data in the cloud to permits framework to better analyze current frames. An example of achieved results

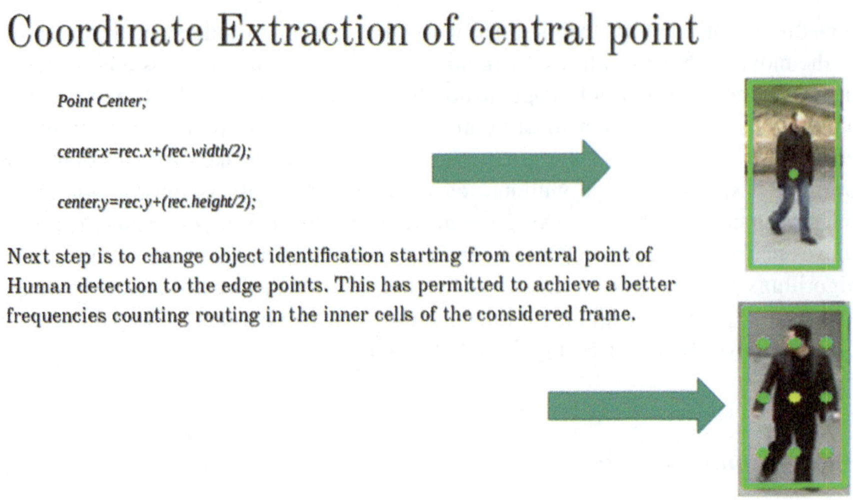

Fig. 8 Human detection based on edge contours to simplify frequency routine counter for building historical meta-data of the *Region of Interest* (ROI)

is shown in the Fig. 8. A detected human figure is then followed by tracking analysis of the incoming video streaming acquired by the smart cameras. Human and object tracking is still an open challenge because of occlusions, illuminations, shadows, motion blur and so on. At present, the existing tracking methods are divided into two categories. The First are generative methods that use the appearance model to estimate the target state in a new frame, such as mean shift [26], L1 [27]. Second instead learn a classifier to discriminate between the target and background, such as KCF [28], SVM [29]. Moreover, nowadays it is possible to work with a combination of both approaches as shown in [30].

3.6.1 Classifier Optimization Using HOG

Using this technique it's possible to considerably reduce false positives during frame analysis. This algorithm splits the frame in a cells matrix of variable size in order to adapt them to the environment composition. This cell division is used to build an occurrence vector in which are stored information about the presence of a human in a specific cell of the frame. Using a based Daimler classifier (Pedestrain classification) it is possible to detect multiple humans inside the frame. If another type of objects must be detected we can simply change classifier and obtain the same results. In order to achieve better results we decided to use three different types of area:

- frequented areas with an high density of targets
- partially frequented areas with an average density of targets
- poorly frequented areas with a low density of targets

Using this approach we achieved a better comprehension of targets behavior building a better occurrences map. In this way we identify the type of scenario finding paths that are more used. If a target is detected inside a poorly frequented area an alarm is raised. To make this approach more adaptive we used a dynamic approach allowing cells to change their type during time taking into account present and past information with different weights.

4 Distributed Computer Vision

One recent evolution of computer vision is the development of distributed multi-camera systems in which object tracking inside images is performed with cooperative techniques. Pioneers work that use this approach are based on video surveillance systems and DARPA monitoring. For example the system developed in [31] use a cooperative tracking technique where tracked objects are processed by multiple cameras; Particularly each sensor (SPU) classifies targets in two main groups: human or vehicle. In [32] Mallet and Bove developed a distributed network of cameras that is able to manage targets in real time. This distributed network is composed by small cameras that are placed on the room ceiling. Cameras are able to move and in this way it is possible to improve their focus on the target depending on the information provided by the system. Lin et al in [33] developed a distributed system for recognition management that integrates data received after a certain number of image processing, this is achieved using a peer-to-peer protocol for communication. A similar approach is used in [34] where the moving target inside the scene is "captured" and followed by every camera of the system. Every camera has a tracking task that depends on the movement of the target. Let's explain more clearly what is a distributed camera system. A distributed camera system is a system that merges data and features that are detected. These data are combined and the final result is subjected to classification. The simplest form of data fusion is pixel combination. For this purpose algorithms like Image Stitching can be used like in [35] to perform merging of different images. Usually a preprocessing of data is performed before merging operation due to the great amount of data that need to be computed. A crucial operation is the features extraction. These features can be extracted using various methods. The SIFT, for example, is a very common technique. First of all a feature vector set is extracted from the image according to the value returned from the subtraction operation applied between a Gaussian function and a smoothing filter applied to sampled image. Then these features are clustered and compared among them. Using distributed algorithms these clustered features are combined and processed. A general schema of a distributed camera system can be summarized in next figures:

Inside this Fig. 9 on the left cameras are placed evenly inside a circular fictitious region in which they belong. A camera group is determined by radial distance between circular areas (usually adjacent to each other). Figure on the right instead shows another way to pace cameras inside the system; in this scenario the camera groups

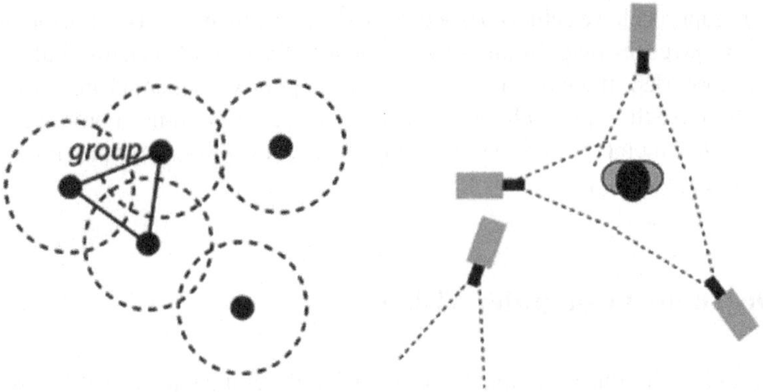

Fig. 9 Distributed camera system where multiple cameras intersect their *Line Of Sight* (LOS)

are determined by the field of view on the target. On the right also is shown how cameras overlap their field of view in order to focus on a common target. Unlike the figure on the left , in this scenario cameras must not be close to each other. A common problem inside distributed camera system is represented by tuning. Camera networks to operate efficiently need to be tuned temporally and spatially. Without an appropriate tuning it is impossible to compare or analyze data processed by each camera. Spatial calibration can be divided in intrinsic and extrinsic. In the first case the instrument is tuned compared to the coordinates of a reference system. In the second one instead the tuning is made compared to parameters like focal distance. Another important aspect is temporal calibration. Cameras must be synchronized among them self and if this synchronization is lost there must be a mechanism to get it again as fast as possible. Often used systems cannot guarantee a good synchronization because of their low quality. When developing a distributed system is crucial the choice of the right trade off between costs and performances. A system that can ensure good performance must posses a good temporal stability (synchronization) even if it means reducing the quality of image processing units.

4.1 Cooperative Video-Surveillance Framework

In this section, the proposal framework will be tested in some case of uses to demonstrate the feasibility and the interoperability of the systems. The main goal is to bring up the cooperative capabilities of the involved sub-systems to accomplish the main tasks. First of all the main scenarios have to be designed to have a suitable case of use avoiding false positive results. The whole testbed system will be composed of a cloud-assisted platform that has the main task to instantiate and manage software agents that implements communication interfaces between managing platforms, field devices and user interaction layers such as notifying subsystems. The

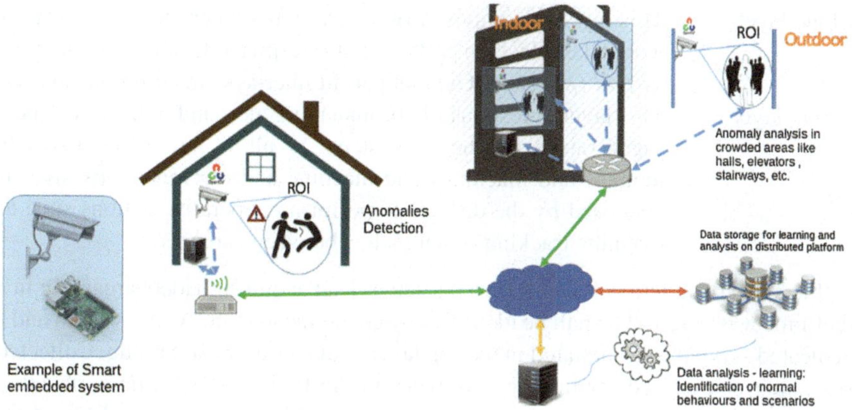

Fig. 10 Reference scenarios used as testbed

resultant system is composed of a distributed and decentralized architecture assisted by edge and cloud computing layers that permits to achieve a scalable and reliable system. In order to evaluate the whole framework two different environments have been realized that are herein summarized:

- Outdoor Environment where the video-analytic and computer vision algorithms work exploiting raw video streaming acquired from a camera devices. Here algorithm works observing the real-time video and taking into account historical meta-data to identify anomalies;
- Indoor Environment, in this environments the system performs a continuous monitoring of the interested areas to detect anomalies;

the reference scenarios are depicted in detail inside Fig. 10. In the depicted scenario different system components work in a cooperative way to exchange heterogeneous information. In particular, embedded devices are used to use a smart camera as field device. Its main task is to acquire video streaming sending it to video-surveillance framework. Fog computing is performed on several devices placed along the edge of monitored area to perform video analysis. Historical data and meta-data are achieved by elaborating key frames on edge layers, alarms and system notifications, instead, are managed by cloud layers exploiting a dedicated agent based framework. Thus the framework is composed of the following layers:

Field data layer — This layer is composed of field devices such as cameras, sensors and smart devices that are responsible of their physical control and management services.

Fog layer — This layer is composed of computational devices that are responsible of the elaboration of data acquired by the field devices. Moreover, it is also responsible of data communication between systems and devices as well as virtualization tasks.

Cloud layer — This layer is composed of several units where the cloud framework takes place. In particular, it is responsible of instantiate some *Virtual Objects* (VOs) that permit inter-system communications.

Notify layer — This layer is responsible of managing alert and notification messages raised by the subsystems to allow interaction between humans and machines and machine and machine. This layer is also used by the devices to perform cooperative actions such as dynamic tracking of anomalies between cameras.

Each smart camera is focused on a particular ROI acquiring video streaming in a real-time way. Each ROI shall be identified by an unique id in the whole system and a dedicated service is instantiated in the fog layer to elaborate its streaming. Collected streaming is analyzed for finding anomalies in the ROI. This specific application anomaly is verified when human actions are detected in some areas of ROI where usually those areas are not involved in human activities. The identification of these anomalies is based on machine learning routines that keep updated identification patterns.

4.2 Smart Camera

In order to allow the aforementioned services and applications, the entire system needs to be equipped with smart devices that are able to work in a distributed architecture and communicate with heterogeneous systems such as other devices, systems or external services. This is possible designing a custom device able to acquire data from field sensors elaborate these data and communicate results to others. In Fig. 11 a possible architecture is presented. It is based on raspberry pi 3 device equipped with OpenCv framework and software API to instantiate cloud interfaces and their agent. Thus the device is virtualized and it is able to communicate with other VOs.

4.3 Anomaly Detection

In this section, we investigate the problem of identifying an anomaly in a given ROI. The first step is to define what the anomaly is in order to train the analytic engine that has to elaborate incoming streaming for rising on alarm notification. In this example the anomaly has been identified as the possibility to recognize human actions in ROI areas where commonly people are not moving on. Two different way could be used, the first one is to train a static engine where the areas to monitor are pre-setted and the engine has the task to identify people and monitoring their action in the ROI. The second one is represented by the possibility to let the engine to dynamically find those area in the ROI. In the proposed framework the second method has been chosen and a learning time-lapse is shown in Fig. 12. Here in the image is shown

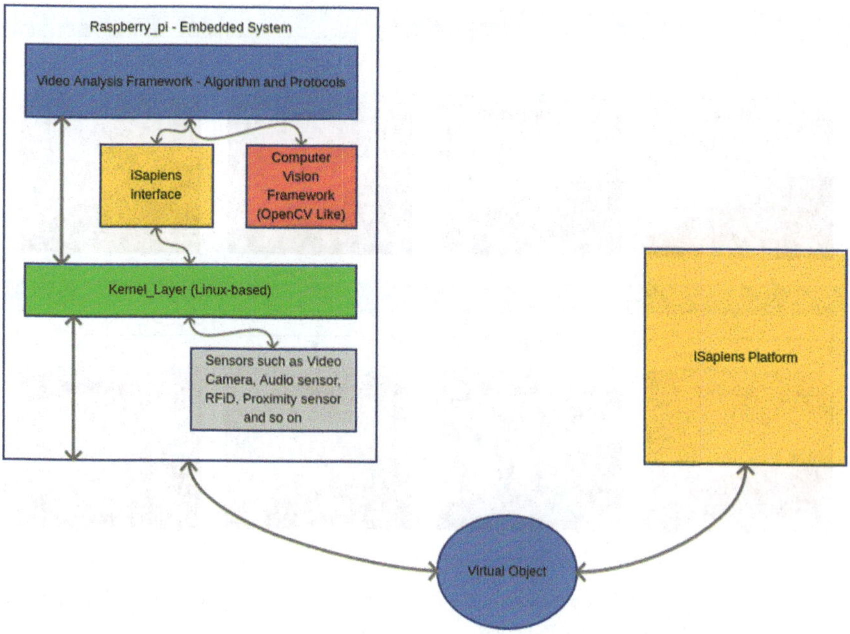

Fig. 11 Logical architecture of the designed smart camera able to cooperate in an IoT domain

how the training is made in the ROI with a step by step sequence. A time based mechanism is used to define an observation window that is exploited to mark the sub-area of the ROI where humans are recognized. The whole ROI is assumed to be composed of several sub-areas where the size of each sub-area is evaluated at run-time and depends on several factors such as average and maximum number of detected activities, light condition, relative speeds of the objects in the camera LOS and so on. Each sub-area is marked as normal, warning, and danger. The mark value of each sub-area is changed if the occurrences of detected activities in the covered area change in the time by creating a time-dependent relationship between them, frequency and historical data of sub-area. To smooth the behavior of the system we use the following smoothed equation that permits to consider the past with a different weight by analyzing the application context and the environment conditions.

$$subarea_{i,j} = (1 - \alpha) * OV_k + (\alpha * HL_{i,j}) \tag{2}$$

In the Eq. 2 is shown how the framework evaluate the value for the the marking function made by the cloud system. Here, we have the index i and j that are used to find the sub-area of the ROI in a matrix based data structure. α is the smoothing factor that may change during the execution time because of the change of the environment conditions. Commonly the system starts with a value of 0.75 that has been evaluated by performing several simulation tests in a controlled environment.

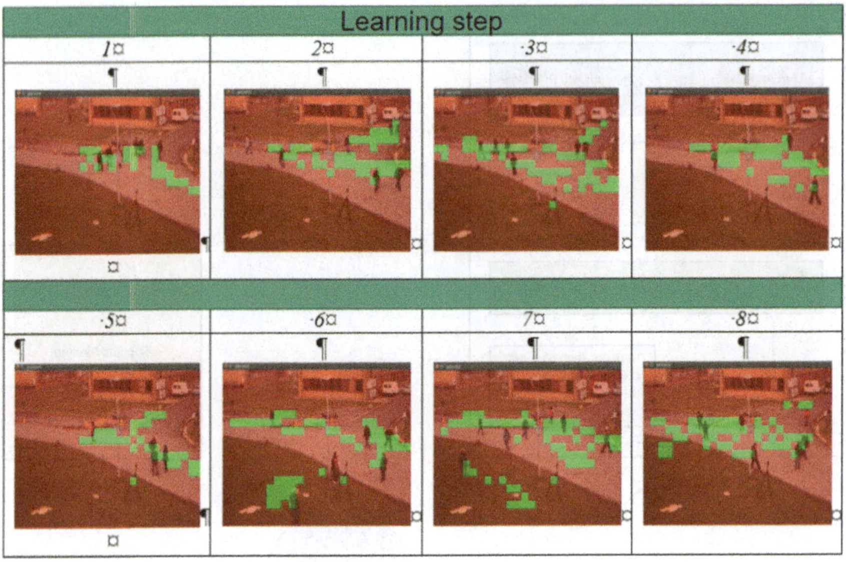

Fig. 12 Human activities in the *Region of Interest* (ROI) covered by a camera. Training step in an observation window with area mark update

OV_k is the $k-th$ observation window that measures the human occurrences in the monitored frames at the $k-th$ observation period. At the end $HL_{i,j}$ is the historical data of the occurrences for the sub-area in the position i, j. The results is shown in Fig. 13 where the final results obtained at the end of an observation window is shown. Here a different color mark is depicted for each area. If a human activity is confirmed in a warning or danger area, then an alarm is generated and it is sent to the monitoring system to perform notify events.

4.3.1 Cooperative Tracking of Detected Anomaly

A computer vision task consists in the identification of specific features of interest inside images. One major challenge is the translation of these simple features in a manner that a computer or a mathematical model can understand. A simple feature like an object color must be translated in a discrete representation to be elaborated by an algorithm. It is possible represent a single feature in different ways through different representations and each of these representation is more suitable to a specific detection technique. Moreover every representation is affected differently by the image characteristics that are analyzed, like image resolution, brightness, sharpness etc. The main components of which any computer vision algorithm is composed are 4:

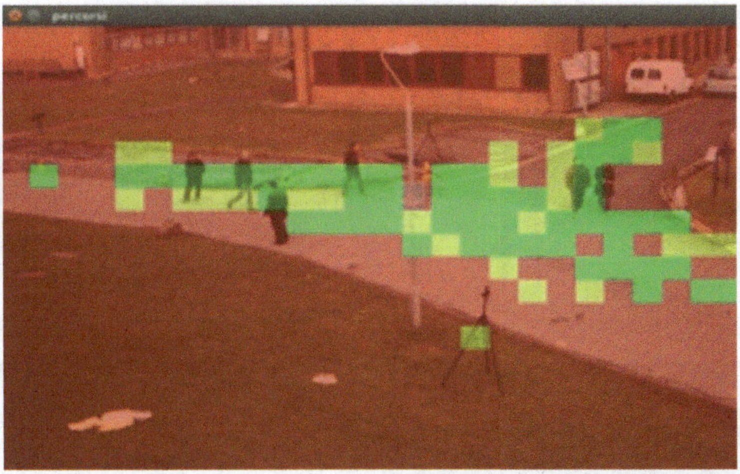

Fig. 13 Anomaly detection based on the depicted heating map related to the mark value of the ROI sub-areas

- Object Representation
- Tracking Features
- Object Detection
- Object Tracking.

Object Representation consist in the way of representing the object of interest inside the image. There are various representations in literature like using a cloud of points or some geometric shapes to highlight area of interest. Every of these representation highlight different characteristics depending on the algorithm that has been used. Tracking features are the characteristics or properties that have to be identified inside an image. They can be simple like a certain color or more complex like contours , movements or textures. The discovery and identification of these features is a very delicate phase and a lot of flexible models have been implemented in order to adapt to the specific case. After the identification of interesting features it's possible to continue with object detection. Object detection is the discovery and identification of an object of interest inside the image. The technique used for this kind of detection must be appropriate compared to the feature representation that has been used and compared to features selected before starting the search. Some classic examples are the detection of points of interest, background subtraction and image segmentation based on contours. Object tracking can be considered as the extension of detection techniques to a series of consecutive images that compose the video stream. Just like detection techniques also tracking algorithms include the analysis of point of interest, contours and objects which movement is detected through background subtraction or motion flow [21, 36, 37]. A complex scenario of whole system is shown in Figs.14 and 15 while in Fig.16 the switching procedure between cameras to keep the target in *LOS* is depicted.

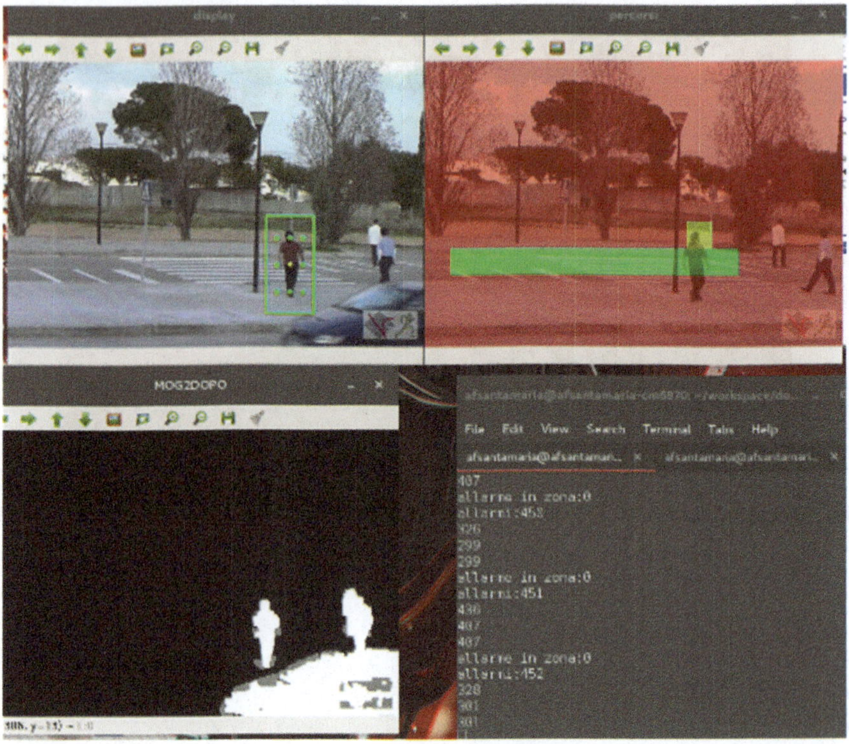

Fig. 14 Learning steps are performed in the cloud layer, human activities are recognized. Meanwhile, system is updating its internal data structures and tracking is performed locally in the edge layer to reduce latency on elaboration

Fig. 15 Once learning is going finish transitory, system can finally recognize anomalies sending an alarm notification by exploiting M2M communication protocols to interface other modules of the video-surveillance system. The result is visive messages on the central dashboard and snooze sounds on the registered end-user devices

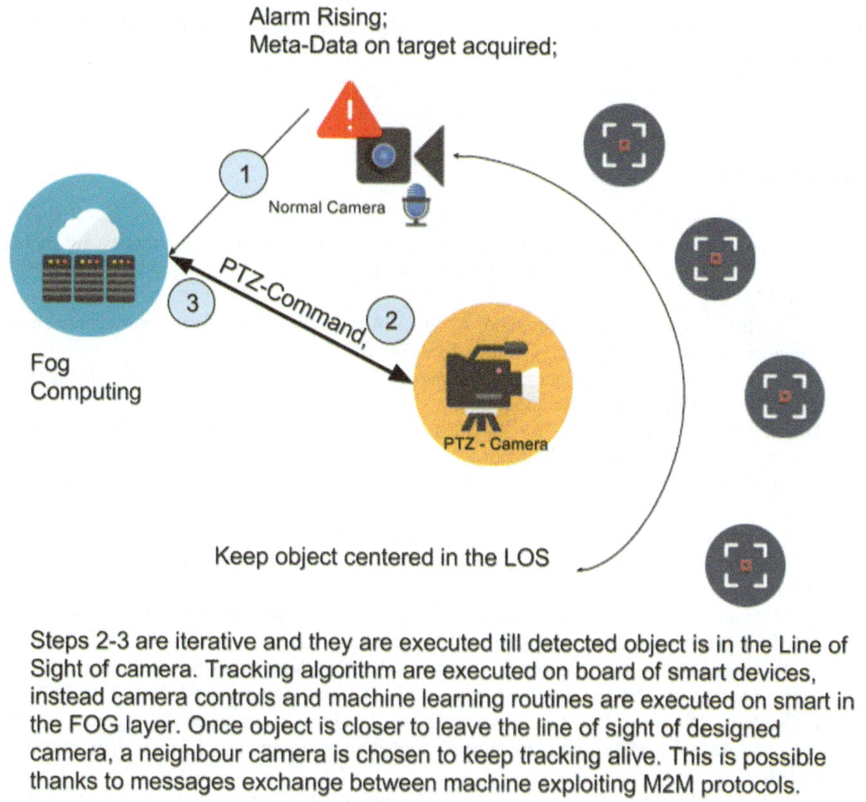

Fig. 16 Cooperation schema used by cameras to raise alarms, send notifications and activate tracking on neighbor camera for achieving the continuous tracking of detected anomaly

5 Conclusions

In this chapter we focus on a cooperative and distributed video-surveillance system able to detect and track anomalies along available cameras in the defined *Region of Interests* (ROIs). First of all we investigate current solutions and how new technologies and cooperative systems in the domain of IoT may help to improve overall performances. Distributed architectures can help to reduce local computation of the end devices for achieve better network performances avoiding to waste resources such as bandwidth and memory. This may help systems to work in a better way improving current results in terms of system response time to events. It can also help developers to introduce new services such as engine for anomaly detection based on continuous monitoring of the environments. Moreover, in this work we propose to introduce a smart framework able to allow camera cooperations to track anomalies during its travel in the monitored ROIs. This result can be obtained by exploiting M2M communication protocols and distributed architecture notifying alarm tracking message that will inform operators to follow detected anomaly in the available cameras covering space.

References

1. Z. Shao, J. Cai, Z. Wang, Smart monitoring cameras driven intelligent processing to big surveillance video data. IEEE Trans. Big Data **4**(1), 105–116 (2018). https://doi.org/10.1109/TBDATA.2017.2715815
2. F. Cicirelli, A. Guerrieri, G. Spezzano, A. Vinci, An edge-based platform for dynamic smart city applications. Future Gener. Comput. Syst. **76**, 106–118 (2017)
3. N.T. Siebel, *Design and Implementation of People Tracking Algorithms for Visual Surveilliance Applications* (University of Reading, 2003)
4. D. LI, *Low cost eye tracking for human computer interaction*, http://thirtysixthspan.com/openEyes/MS-Dongheng-Li-2006.pdf
5. S. Ali, M. Shah, Floor fields for tracking in high density crowd scenes, in *Proceedings of ECCV*, 2008
6. C. Liu, J. Yuen, A. Torralba, J. Sivic, W.T. Freeman, Sift flow, in *Proceedings of ECCV*, 2008
7. J. Yuen, A. Torralba, A data-driven approach for event prediction, in *Proceedings of ECCV*, 2010
8. A. Molinaro, F. De Rango, S. Marano, M. Tropea, A scalable framework for in IP-oriented terrestrial-GEO satellite networks. IEEE Commun. Mag. **43**(4), 130–137 (2005)
9. F. De Rango, M. Tropea, P. Fazio, S. Marano, Call admission control for aggregate MPEG-2 traffic over multimedia geo-satellite networks. IEEE Trans. Broadcast. **54**(3), 612–622 (2008)
10. F. De Rango, F. Veltri, P. Fazio, S. Marano, Two-level trajectory-based routing protocol for vehicular ad hoc networks in freeway and Manhattan environments. J. Netw. **4**(9), 866–880 (2009)
11. F. De Rango, M. Gerla, S. Marano, A scalable routing scheme with group motion support in large and dense wireless ad hoc networks. Comput. Electr. Eng. **32**(1–3), 224–240 (2006)
12. F. De Rango, M. Tropea, A.F. Santamaria, S. Marano, Multicast QoS core-based tree routing protocol and genetic algorithm over an HAP-satellite architecture. IEEE Trans. Veh. Technol. **58**(8), 4447–4461 (2009)
13. B. Zhou, Y.Z. Lee, M. Gerla, F. De Rango, GeoLANMAR: a scalable routing protocol for ad hoc networks with group motion. Wirel. Commun. Mob. Comput. **6**(7), 989–1002 (2006)
14. M. Satyanarayanan et al., The case for VM-Based cloudlets in mobile computing. IEEE Pervasive Comput. **8**(4), 14–23 (2009)
15. F. Bonomi et al., Fog computing and its role in the internet of things, in *Proceedings of 1st edn. MCC Workshop Mobile Cloud Computing (MCC 12)*, 2012, pp. 13–15
16. J. Gettys, K. Nichols, *Bufferbloat: Dark Buffers in the Internet*, vol. 9, no. 11 (ACM Queue, 2011)
17. M. Satyanarayanan, The emergence of edge computing. Computer **50**(1), 30–39 (2017). https://doi.org/10.1109/MC.2017.9
18. A.H.M. Amin, N.M. Ahmad, A.M.M. Ali, Decentralized face recognition scheme for distributed video surveillance in IoT-cloud infrastructure, in *IEEE Region 10 Symposium (TENSYMP)*, Bali, vol. 2016, 2016, pp. 119–124. https://doi.org/10.1109/TENCONSpring.2016.7519389
19. H.S. Parekh, D.G. Thakore, U.K. Jaliya, A survey on object detection and tracking methods. Int. J. Innov. Res. Comput. Commun. Eng. **2**(2) (2014)
20. N. Dalal, B. Triggs, Histograms of oriented gradients for human detection, in *2005 IEEE Computer Society Conference on Computer Vision and Pattern Recognition (CVPR'05)*, San Diego, CA, USA, vol. 1, 2005, pp. 886–893. https://doi.org/10.1109/CVPR.2005.177

21. S. Miguet A. Ilyas, M. Scuturici, Inter-camera color calibration for object re-identification and tracking, in *2010 International Conference of Soft Computing and Pattern Recognition*, 2010
22. https://www.learnopencv.com/handwritten-digits-classification-an-opencv-c-python-tutorial/
23. K. Balani, S. Deshpande, R. Nair, V. Rane, Human detection for autonomous vehicles, in *IEEE International Transportation Electrification Conference (ITEC)*, Chennai, vol. 2015, 2015, pp. 1–5. https://doi.org/10.1109/ITEC-India.2015.7386891
24. C.G. Saneem Ahmed, S. Saravanakumar, A. Vadivel, Multiple human object tracking using background subtraction and shadow removal techniques, in *Signal and Image Processing*, 2010
25. M.A. AlGhamdi, M.A. Khan, S.H. AlMotiri, Automatic motion tracking of a human in a surveillance video, in *IEEE First International Smart Cities Conference (ISC2)*, Guadalajara, vol. 2015, 2015, pp. 1–4. https://doi.org/10.1109/ISC2.2015.7366165
26. G. Sindhuja, S. Devi, M. Renuka, Comparative analysis of mean shift in object tracking, in *Power Control Communication and Computational Technologies for Sustainable Growth*, 2016, pp. 283–287
27. H. Ji, Real time robust 11 tracker using accelerated proximal gradient approach, in *IEEE Conference on Computer Vision and Pattern Recognition*, 2012, pp. 1830–1837
28. J.F. Henriques, C. Rui, P. Martins, J. Batista, High-speed tracking with kernelized correlation filters. IEEE Trans. Pattern Anal. Mach. Intell. **37**(3), 583–596 (2015)
29. S. Hare, S. Golodetz, A. Saffari, V. Vineet, M.M. Cheng, S.L. Hicks, P.H.S. Torr, Struck: structured output tracking with kernels. IEEE Trans. Pattern Anal. Mach. Intell. **38**(10), 2096 (2016)
30. F. Feng, X.J. Wu, T. Xu, Object tracking with kernel correlation filters based on mean shift, in *International Smart Cities Conference (ISC2)*, Wuxi, vol. 2017, 2017, pp. 1–7. https://doi.org/10.1109/ISC2.2017.8090863
31. R.T. Collins, A.J. Lipton, H. Fujiyoshi, T. Kanade, Algorithms for cooperative multisensory surveillance, in Proceedings of IEEE, Oct 2001
32. J. Mallett, V.M. Bove, Eye society, in Proceedings of IEEE ICME, 2003
33. C. Hong Lin, T. LV, W. Wolf, B. Ozer, A peer to peer architecture for distributed real time gesture recognition, in *Proceedings of International Conference on Multimedia and Exhibition* (IEEE 2004)
34. M. Bramberger, M. Quartisch, T. Winkler, B. Rinner, H. Scwabach, Integrating multicamera tracking into a dynamic task allocation system for smart cameras, in *Proceedings of the IEEE Conference on Advanced Video and Signal Based Surveillance*, 2005
35. L. McMillan, G. Bishop, Plenoptic modeling: and image-based rendering system, in *Proceedings of ACM SIGGRAPH*
36. B. Leibe, E. Koller-Meier, L. Van Gool, M. Breitenstein, F. Reichlin, Online Multi-Person Tracking-by-Detection from a Single, Uncalibrated Camera. Pattern Anal. Mach. Intell. (2010)
37. B. Schiele M. Andriluka, S. Roth, People-tracking-by-detection and peopledetection-by-tracking. Comput. Vis. Pattern Recognit. (2008)
38. E. Cassano, F. Florio, F. De Rango, S. Marano, A performance comparison between ROC-RSSI and trilateration localization techniques for WPAN sensor networks in a real outdoor testbed, in *Wireless Telecommunications Symposium, WTS 2009*, Prague, Czech Republic, 22–24 Apr 2009, pp. 1–8
39. P. Simoens et al., Scalable crowd-sourcing of video from mobile devices, in *Proceedings of 11th International Conference Mobile Systems Applications and Services (MobiSys 13)*, 2013, pp. 139-152

Personal Connected Devices for Healthcare

Adina Riposan-Taylor and Ian J. Taylor

Abstract Personal connected health technologies are fast becoming integral in a person's daily life to help improve their health and wellbeing. Aside from the ubiquitous personal fitness devices applications, there are numerous monitoring devices for personal healthcare, along with more proactive devices that enable users to engage in self-therapeutic techniques in order to improve their overall wellbeing. In this chapter, we provide a broad overview for a range of devices that allow people to monitor their own health and also those that provide self-therapeutic benefits using biofeedback or neurofeedback to learn to control their own physiologic functions. We discuss such devices within the context of the access to data and provide a taxonomy that covers the software for data integration, the type of connectivity and user interfaces for data collection, API authentication and authorization, and data access policies and formats.

1 Introduction

Urban ecosystems are facing some of the major challenges of our time, including global warming, pollution, waste management, energy efficiency, a rise in health issues, etc., and many smart city policies are evolving to address these issues. The Internet of Things (IoT) is driving these changes, by providing a paradigm where objects of everyday life are equipped with microcontrollers, transceivers and protocol stacks and are fast becoming an integral part of how users interact with the Internet. IoT has numerous healthcare applications from remote monitoring, diagnosis and treatment of patients using telemedicine, to smart sensors and medical device integration for the improvement of personal healthcare.

A. Riposan-Taylor
University of Notre Dame and Cardiff University, 341 N. Fletcher
Avenue Fernandina, Florida, FL 32034, USA

I. J. Taylor (✉)
Satya Sattva Consulting, 341 N. Fletcher Avenue Fernandina,
Florida, FL 32034, USA
e-mail: ian.j.taylor@nd.edu

F. Cicirelli et al. (eds.), *The Internet of Things for Smart Urban Ecosystems*,
Internet of Things, https://doi.org/10.1007/978-3-319-96550-5_14

As interconnectivity of data from such devices becomes more seamless, our aggregated view of our health information will become more and more commonplace in our daily lives. For example, technologies identified in previous books of this series have already identified techniques for representing devices in more intelligent ways, using smart objects (SOs) [1] or Cyber Physical Objects (CPOs) [2]. In such approaches, IoT is seen as a loosely coupled, decentralized system where SOs and CPOs can be used to discover, communicate, connect and optimize devices in a decentralized, distributed, and cooperative manner.

According to [3], the global connected health and wellness devices market is expected to reach $612.0 billion by 2024. Demand for wearable medical devices and remote patient monitoring systems is surpassing expectations because of the shifting consumer preferences towards healthier lifestyles. Furthermore, chronic diseases requiring continuous health monitoring have risen rapidly, and along with the availability of advanced products, such as remote sensors, adapters, and connected mobile communication, personal healthcare devices are fast becoming ubiquitous in households. For example, approximately 80 million people in the USA are hypertensive and at greater risk of suffering a heart attack or stroke. It is therefore becoming critically important to monitor blood pressure regularly and sharing that data with healthcare professions in order to avoid a catastrophic cardiac event.

Government initiatives and non-profit organizations such as the Healthcare Information and Management Systems Society (HIMSS) [4] are helping to raise awareness of such technologies, which is fueled by a rising emphasis on the development of advanced healthcare solutions in the connected health and wellness devices market. There are also a number of funding initiatives by government organizations, such as the Healthcare Research and Quality (AHRQ) USA, which provide funds to expand health information systems.

As such trends persist and technology continues to become a part of everyday life, the use of mobile devices to maintain a state of health and wellbeing and manage illness is becoming mainstream. However, the use of such devices for personalized use is complex, as it requires the integration of data from multiple sources and analysis to form a complete picture for the case of illness, care and treatment. Therefore, the promise of personal connected health will not be possible if data is only available in the devices used to record them.

The rest of the chapter is organized as follows. In Sect. 2, we provide an introduction to the current landscape of the two core themes of this chapter: Sect. 2.1 discusses personal connected devices and the standards and protocols that enable their integration; and in Sect. 2.2, we describe the use of biofeedback and neurofeedback devices for self-therapeutic monitoring to improve health and performance. In Sect. 3, we provide an overview of existing protocols that are used within this area in order to provide the reader with context for the metrics used in the evaluation taxonomy, described in Sect. 4. The taxonomy is first applied in Sect. 4.1 to devices, applications and data access to personal health data, and then in Sect. 4.2 to biofeedback and neurofeedback. Finally, in Sect. 5, we reflect on past experience using contemplative practices and discuss how lessons learned could lead to new future directions and techniques for human wellbeing.

2 The Landscape of Personal Connected and Wellness Devices

In this section, we introduce prior efforts that have led to standards and protocols that have paved the way to personal connected health devices being made available to patients for home monitoring. We then introduce self-therapeutic monitoring methods that employ the use of biofeedback and neurofeedback techniques and then finally, we discuss the recent advances that have shifted emphasis from clinical therapy to using personal biofeedback and neurofeedback devices in wellness.

2.1 The Evolution of Standards in Personal Health Monitoring

In an attempt to address the aforementioned issues, the Continua Health Alliance industry coalition was established in 2006. It was designed from the outset to address such issues as lifestyle, health management and demographic tendencies, which are some of the main factors causing a substantial rise in medical costs. The organization essentially promotes the establishment of technology standards and design guidelines that enable interconnection and operation of healthcare devices.

For manufacturers, standards decrease time-to-market, reduce development costs and increase efficiencies. They enable more rapid, less expensive integration to electronic health records (EHR) or health information exchange (HIE) platforms. Continua was a founding member of the Personal Connected Health Alliance (PCHAlliance), a non-profit organization formed by HIMSS. PCHAlliance members provide a vibrant ecosystem of technology and life sciences industry icons and innovative, early stage companies along with governments, academic institutions, and associations from around the world.

To support the vision, the PCHAlliance hosts an annual Connected Health Conference to exchange research, evidence, ideas, innovations and opportunities in personal connected health. The PCHAlliance publishes and promotes the global adoption of the Continua Design Guidelines, which enable an open framework for user-friendly, interoperable health data exchange in personal connected health. The Continua Design Guidelines are recognized by the International Telecommunication Union (ITU), an agency of the United Nations, as the international standard for safe, secure, and reliable exchange of data to and from personal health devices.

The high-level architecture of Continua is provided in [5]. The Personal Health Devices Interface is based on standards developed by the IEEE Personal Health Device working group [6] that define the data format and exchanges between the sensor and the personal health gateway. The Services Interface (WAN Interface) standardizes around the IHE Patient Care Device (PCD-01) [7] Transaction to move data between a Personal Health Gateway and the Health and Fitness Services. The Health Information Service Interface (using the Health Record Network (HRN) interface) standardizes around the HL7-based Personal Healthcare Monitoring Report (PHMR) [8] to move information between a Health & Fitness Service and the Healthcare Information Service provider (e.g. EHR). End-to-end security and privacy is addressed through a combination of identity management, consent management and enforcement, entity authentication, confidentiality, integrity and authentication, non-repudiation of origin, and auditing.

2.2 *Self-Therapeutic Monitoring Using Biofeedback and Neurofeedback*

Biofeedback (BFB) encompasses a number of treatments and scientifically based self-therapeutic techniques that allow patients to use instruments that help them become aware of their own physiologic functions [9, 10] (what is going on inside their bodies and their brains [11]), to monitor their responses to emotional, cognitive, or physical stimuli [12], in real time, and to gain more control over these responses, with practice and feedback [13–15]. As principles in cybernetics, one can control a variable only if information (or feedback) about the variable is made available, and that feedback makes learning possible [11]. The idea in biofeedback training is that the trainees receive direct feedback about their states and, based on this feedback, they learn to increase the control over their own biologic functions [11]. A consensus definition of biofeedback was given by the Association for Applied Psychophysiology and Biofeedback (AAPB) [16], Biofeedback Certification International Alliance (BCIA) [17], and the International Society for Neurofeedback and Research (ISNR) [18], in 2008: "biofeedback is a process that enables an individual to learn how to change physiological activity for the purposes of improving health and performance". The ultimate goal of biofeedback training is to being able to manipulate physiologic processes and responses at will [9, 10] and to establish permanent positive changes, so that, over time, these changes can endure without continued use of an instrument [10]. The Biofeedback Federation of Europe (BFE) [19] provides international education, training, and research activities in biofeedback and neurofeedback.

There are multiple types of biofeedback that include various types of physiologic measures, such as heart rate variability (HRV), surface electromyography (e.g., sEMG, muscle tension), galvanic skin response (GSR; i.e., skin conductance), surface temperature, breathing rates (via the use of a respiration belt), and/or electroencephalography (EEG) [12]. The devices are used to transmit data to the user (either as single measure or multiple measures) and the data is displayed by audio and/or visual means to provide information (or feedback) about an individual's biological functions. The user can experiment with different thoughts, feelings, and sensations and get immediate feedback on the physiologic changes [11], while controlling elements (such as the controls of a computer game) are used to learn to control these responses in order to reduce symptoms, improve performance, and/or increase health and well-being [13–15]. For example, if the biofeedback system provides a stress and relaxation assessment, the user will be immediately notified if there are abnormal or exaggerated responses [12]. Biofeedback may therefore be used to improve the physiological changes that naturally occur in conjunction with changes to thoughts, emotions, and behavior by following the real time visual/audio presentation of the measures associated with changes in thinking, emotions, and behavior, and learning to pursue the desired physiological changes.

Neurofeedback (NFB), also called neurotherapy or neurobiofeedback, on the other hand is a type of biofeedback that is specifically employed to teach self-regulation of brain function, and it typically makes use of electroencephalography (EEG) for real-time displays of brain activity (i.e., the brain waves). So, whereas Neurofeedback uses the electrical activity of the human brain (EEG) as feedback to the client, biofeedback uses various other physiological signals, except the EEG [11, 20]. In EEG-biofeedback, EEG sensors are placed on the scalp to measure the electrical activity of the brain and to produce a signal that is provided to the user using video displays or sound. In neurofeedback training, this display is outlined as positive feedback for desired brain activity and as negative feedback for undesirable brain activity that should be controlled, altered and ultimately changed. The subject becomes aware of the changes occurring during the training session, will consciously try to improve the brain patterns following the real-time changes occurring in the sound or video, and will have the possibility to keep track and assess the progress, which in time and with practice can lead to optimum performance. [21–23]. This change is thought to contribute to neural plasticity as a mechanism of training [12, 24].

Neurofeedback started in the 1960's with the Alpha activity experimentation in the human brain and the promising results inspired many new experiments; it is virtually undisputed today the fact that humans can learn to alter their brain function by the use of EEG-biofeedback [11]. EEG sensors are placed on the scalp on standard electrode placement locations to measure the electrical signals of the brain, i.e. on standard recording channels in the frontal, temporal, central, and occipital lobes, and neurofeedback systems provide audio and/or video feedback to the user (e.g., the power of a signal in a frequency band can be shown by a varying bar graph [21]). More advanced neurofeedback conducted in clinical and research settings may employ functional magnetic resonance imaging (fMRI) instead, or in combination with, electroencephalography (EEG), for real-time visualisation of brain activity. In user-oriented EEG-biofeedback, visualising the brainwaves in real-time teaches people to influence their brain activity consciously and adjust their state of mind, thereby teaching self-control of their brain functions. There are a series of neurofeedback hardware and software systems available, either for clinical/research or for personal use, and the neurofeedback treatment/training protocols typically include alpha, beta, alpha/theta, delta, gamma, and theta, or other ratio combinations [21–23].

2.3 From Clinical Therapy to Personal Devices in Wellness

The rise of health care costs has brought the need for more cost-effective, non-pharmacological treatments which converged with an increase in the population's interest in preventative medicine, holistic therapies and integrative medicine in the recent years. In clinical biofeedback and neurofeedback therapy, performed in medical and rehabilitation centres, the user is guided by a trained clinician [13–15] and there are international professional and scientific institutions and non-profit organisations that coordinate education, training, and research in biofeedback and neurofeedback [16–19]. However, due to the high cost of the treatment and equipment, and the fact that generally there is no insurance coverage for biofeedback training, the trend in industry and research is towards at-home use of biofeedback and neurofeedback portable devices. A series of personal wellness systems have been developed that appeal to people interested in relatively affordable, home-use, "wearable" devices that can be used for biofeedback self-therapy and training to monitor and improve well-being and performance, which is to develop the ability to generalise the control over one's physiologic processes to "in-vivo" settings [12] (taking the therapy out of the clinical settings and into one's home). Some of the processes that a person can learn to control include heart rate, brainwaves, breathing rates, muscle tone, skin conductance, and pain perception [25]. Recent advances in EEG technology and the ability to use "dry sensor" electrodes to pick up and digitize EEG signals [12] made neurofeedback accessible at home-use level and a variety of neurofeedback portable devices and headsets have been made available to the public.

Both biofeedback and neurofeedback may reduce the need of taking medication, and in some cases regular practice can replace treatment. Users can follow (see or hear) representations of data related to their own physiologic responses to stress or distraction, in real time, and they learn to alter these responses, being guided by the user-friendly applications to help increase attention or concentration, mindfulness, or to decrease activity in certain areas, by utilizing relaxation and breathing techniques, which correlate with patterns associated with healthy functioning, improved mood and welbeing [12, 24]. For example, biofeedback has been found to be effective for the treatment of headaches and migraines [26].

Neurofeedback—either referred to as neurotherapy, or as brain function training—is a safe and non-invasive procedure and it is considered an effective method to treat a variety of health problems at present. A list of different levels of efficacy of neurofeedback for various applications has been provided by The American Association of Applied Psychophysiology and Biofeedback (AAPB) [16] and The International Society for Neurofeedback and Research (ISNR) [18], which indicates that, for some applications, neurofeedback is an accepted treatment, while for other conditions it shows promise, and for some it should still be considered an experimental method [11]. With the recent advances in brain science, the use of "wearable" neurofeedback devices has been encouraged both to help patients self-regulate and reduce the antisocial behavior of individuals which has an undesirable impact on the society [27]. The neurofeedback training has been used in the treatment of depression; anxiety; insomnia; attention deficit/hyperactivity disorder (ADHD); autistism spectrum disorder (ASD); obsessive compulsive disorder (OCD); restless legs syndrome; learning disabilities, dyslexia and dyscalculia; drug addiction; eating disorders; migraines and headaches; pain management; fibromyalgia; epilepsy; schizophrenia; parkinson disease; recovery from an injury and stroke problems [21]. It has also been used for improving memory and for enhancing the performance of athletes, artists, and surgeons [21].

The effectiveness of neurofeedback therapy has been studied on many diseases and there are still methodological limitations and clinical ambiguities [21]. Neurofeedback has its own pros and cons as many other therapies, as its validity has been questioned in terms of conclusive scientific evidence of its effectiveness [21]. There are still questions such as how many sessions are needed for a participants to learn to exert an alert control over their own alpha waves (in the alpha treatment protocols), or how many sessions are needed to produce the expected therapy effect at the optimal performance level, or how long the positive effects may last without continuing the neurofeedback therapy, or if there are any long-term effects after the use of a device is discontinued [21, 28]. For example, neurofeedback has been recently studied for use in the treatment of traumatic brain injury (TBI) in both civilian and military populations [12]. In addition to treating direct symptoms such as headaches and migraines, mood changes, fatigue, and memory issues, the neurofeedback therapy also addressed comorbid diagnoses or symptoms that mimic those associated with traumatic brain injury (TBI)—e.g., attention deficit hyperactivity disorder (ADHD), post-traumatic stress disorder (PTSD), or substance abuse [12]. Preliminary data shows that TBI survivors who continue to suffer from longer-term symptoms and struggle with finding effective treatments to address their symptoms could greatly benefit from biofeedback and neurofeedback long-term therapy with affordable, home-use, portable devices that have the potential to improve their quality of life and life satisfaction [12]. Preliminary data also suggests that the use of personal user-friendly devices may additionally increase their motivation for treatment [12]. A review of research on the biofeedback efficacy for various medical conditions, that used an efficacy rating on the Association for Applied Psychophysiology and Biofeedback scale spanning from Level 1 ("Not empirically supported") to Level 5 ("Efficacious and specific"), has listed TBI as having a "Level 3; Probably Efficacious" rating [12].

At the clinical level, biofeedback and neurofeedback therapies are expensive procedures which are not covered by many insurance companies, they are time-consuming and their immediate benefits may not be long-lasting. Practice and training using biofeedback and neurofeedback may take a long time to reach the desired improvements and to potentially establish permanent positive changes. Therefore, the availability of affordable, home-use, portable devices as personal biofeedback and neurofeedback systems has the potential to increase the efficiency of such treatments as long-term training and self-therapies with cumulative training effect and longer-lasting benefits on health and well-being.

3 Technologies

This section provides a brief overview of some of the underlying technologies and protocols that are referenced in this chapter.

3.1 WiFi

WiFi[1] is a radio-waves technology for connecting devices to a local area network. Devices connect to a WiFi using a wireless access point, and the wireless access point, in turn, connects to the Internet WLAN. Such an access point (or hotspot) has a range of about 20 m (66 feet) indoors and a greater range outdoors.

Hotspot coverage can be as small as a single room with walls that block radio waves, or as large as many square kilometres achieved by using multiple overlapping access points. Once configured, WiFi provides wireless connectivity to your devices by emitting frequencies between 2.4 and 5 GHz, based on the amount of data on the network. Wireless networking is known as WiFi or 802.11 networking as it covers the IEEE 802.11 technologies. The major advantage of WiFi is that it is compatible with almost every operating system, game device, and advanced printer.

3.2 Bluetooth

Bluetooth [29] uses radio waves to transmit data from one device to another and it works in a similar way to other radio receivers, such as television. However, whereas televisions use radio waves that travel hundreds (possibly even thousands) of km/miles through the air, Bluetooth operates on a far shorter range (less than 10 m).

Each Bluetooth device contains a Bluetooth radio, which is a computer chip that contains software for pairing Bluetooth devices and streaming data between them. Bluetooth sends and receives radio waves in a band of 79 different frequencies (channels) centered on 2.45 GHz. The short range of Bluetooth transmitters mean that they use virtually no power and are theoretically more secure than wireless networks that operate over longer ranges, such as Wi-Fi.

Communication between Bluetooth devices happens over ad hoc networks known as piconets. A piconet is a network of devices connected using Bluetooth technology. When a network is established, one device takes the role of the master while all the other devices act as slaves. Piconets are established dynamically and automatically as Bluetooth devices enter and leave radio proximity.

---

[1] There are a number of theories about what the term means, but the most widely accepted definition is Wireless Fidelity.

3.3 *Bluetooth® Smart*

The Bluetooth Special Interest Group (SIG) released the Bluetooth Smart logo in 2011 to differentiate between the new low energy devices and the other Bluetooth devices. Bluetooth Smart Ready indicates a dual-mode device compatible with both Classic and low energy peripherals, whilst Bluetooth Smart indicates a low energy-only device which requires either a Smart Ready or another Smart device in order to function. However, in May 2016, the Bluetooth SIG phased out the Bluetooth Smart and Bluetooth Smart Ready logos and has reverted to using the Bluetooth logo, however, the logo uses a new blue color. Bluetooth Smart is also known as Bluetooth 4.0 and Bluetooth LE (BLE).

3.4 *ANT*

The Advanced and Adaptive Network Technology (ANT) is a protocol that uses the same frequencies as WiFi and Bluetooth (2.4 GHz) but the data transfer rate is much slower. It uses very low power so that it can run on a watch battery for a long time and the working distance is typically less than five feet.

There have been two versions: ANT and ANT+ [30]. ANT sensors were not designed to be interoperable but using ANT+, all manufacturers agreed to follow the same specification, so that heart rate monitors, heart rate chest straps, power meters, ect., are all interoperable. This allows the power meter designed by one manufacturer to work with the bike computer of another manufacturer.

3.5 *ANT FIT*

The Flexible and Interoperable Data Transfer (FIT) protocol [31] defines a set of data storage templates (i.e. FIT messages) that can be used to encapsulate information, such as user profiles and activity data as files. It is designed specifically for sharing sport, fitness and health devices data. The FIT protocol's main focuses are to be compact, interoperable and extensible.

The ANT File Share (ANT-FS) is an extension of ANT for transferring files wirelessly between two devices. ANT-FS is an optimized protocol that has ultra low power consumption and can be used in coin cell battery operated devices on both sides of the link. ANT-FS is typically used for fitness devices e.g. watches that collect real-time data from sensors and make this data available to other devices e.g. phones, PCs.

3.5.1 Google Fit

Google Fit [32] was originally a software framework allowing app developers and hardware makers to record, log and display health-related data with as little coding as possible. But after initial development, it has expanded to become a complete health-tracking platform developed for Android. It offers an API that blends data from multiple applications and devices, as well as collecting data from the activity tracker or mobile device, to record physical fitness activities (such as walking or cycling), which are measured against the user's fitness goals to provide a comprehensive view of their fitness. Users can choose who their fitness data is shared with as well as delete this information at any time.

The Google Fit API [33] allows developers to upload fitness data to a central repository where users can access their data from different devices and apps in one location. There are several APIs including: the Sensors API to provide access to raw sensor data streams; the Recording API that provides automated storage of fitness data using subscriptions; and the History API to provide access to the fitness history and it allows applications to perform bulk operations, like inserting, deleting, and reading fitness data. Applications can also import batch data into Google Fit.

Google Fit APIs cannot be used for non-fitness purposes, such as storing medical or biometric data, so the use is restricted.

3.6 Apple HealthKit

The Apple HealthKit [34] is similar to Google Fit. It collects in data from all various fitness apps and wearable companion apps on your iPhone and Watch, including Apple's own Watch workout app, and attempts to put the data in one, easy-to-read place. A user provides permission for a third party application to read and write health and activity data, and then this data is aggregated into a personal and central data collection point, which can be used by other applications to make more extensive uses of the data.

The HealthKit saves data in an encrypted database called the HealthKit store. Access to this store is through integration via an apple app, using Swift or Objective-C e.g.the database is accessed using the HKHealthStore class. Both the iPhone and the Apple Watch have their own HealthKit store, which is kept in sync, with old data being periodically purged from Apple Watch to save space.

4 Data Taxonomy for Personal Connected Health Devices

In a recent study comparing blood pressure devices for health monitoring [35], the authors compared eight devices that are currently in the American marketplace: Omron 10, Rating 4.3, Bluetooth connectivity and allows 2 users; Withings Wireless

BPM, Rating 3.7, Dual connectivity (Bluetooth + wired) and allows multiple users; Philips Upper Arm BPM, Rating 4.3, Bluetooth connectivity and allows 2 users; QuardioArm BPM, Rating 3.9, Bluetooth connectivity and allows up to 8 users; Blipcare Blip Wi-Fi, Rating 4.2, WIFI connectivity and allows 2 users; iHealth Feel BPM, Rating 3.8, Bluetooth connectivity and allows multiple users; iHealth View BPM, Rating 3.3, Bluetooth connectivity and allows multiple users; and Koogeek Smart Wrist BPM, Rating 3.8, Bluetooth connectivity and allows up to 16 users.

All of the above devices provide applications for the recording, interaction and visualization of the data. However, the way in which the data is stored and can be used varies significantly across manufacturers. Some companies provide an openly accessible APIs to enable third party authorized access to the data, whilst others keep the data completely private or provide access for specific approved business-to-business interactions. In this chapter, we take a look at several companies that operate in this space, the devices they provide and the type and means of access to the data. Specifically, the following metrics or criteria are used for the comparison:

1. *Type of Connectivity*. e.g. bluetooth, NFC, WiFi, wired
2. *Available Devices*. e.g. blood pressure monitor, heartbeat monitor, scales, etc.
3. *API Type*. How is the data access e.g. is access provided directly from the device or is it available through an API
4. *Authentication*. The type of authentication e.g. OAUTH 1, 2, etc.
5. *Data Access Policy*. Is the data available to third developers for making their own applications or is the data openly available to any developers that sign up to access it and are authorized to use it
6. *Data Format*. The data format for the API e.g. JSON, XML, FIT, etc.
7. *API Pricing Model*. Is the access to data free, is there a one-off license fee or a subscription model.
8. *Other*. Other comments about the availability or release state.

The focus of this comparison is on devices that can be used at home to monitor medical conditions and, therefore, personal fitness health monitoring devices are out of scope. There are also a number of companies that are not covered in this chapter, which offer Bluetooth (or other) connectivity to applications on Android, Apple or Windows devices, but those that do not offer an API or direct device integration are not covered here e.g. 1byone [36] is an example that offers a blood pressure monitor and an app but does not provide developer access to such data.

Using this taxonomy metrics, devices are covered in detail and are divided into the following two broad sections: "Personal Health Monitoring" and "Personal Self-Therapeutic Biofeedback for Wellbeing".

4.1 Personal Health Monitoring

We apply this taxonomy to the following companies that provide devices, applications and data access to personal health data:

1. Philips
2. Garmin
3. Omron
4. Beets Blu
5. Withings
6. iHealth
7. Quardio.

4.1.1 Philips HealthSuite

The Philips HealthSuite [37] is built for complex healthcare challenges and provides clinical databases, patient privacy, industry standards and protocols, and both personal and population data visualizations. It uses the cloud to store health data securely and provides analysis tools that can be used to provide insight into individuals for healthcare professionals.

Philips offers multiple connected consumer and health systems products, such as Philips: Lumify tablet-based ultrasound; Health Watch; Upper Arm Blood Pressure Monitor; Wrist Blood Pressure Monitor; Body Analysis Scale; Ear Thermometer; and Avent uGrow smart baby monitor. The devices are connected using the Bluetooth or Bluetooth Smart technology.

The data is HIPAA-compliant and developers have access to a variety of sources, including electronic medical records (EMRs), personal devices, diagnostic, imaging and monitoring equipment. A developer also has access to the complex analytics algorithms that the platform offers, which can deliver predictive, personalized insights e.g. to help motivate healthy behavior through digital coaching, or to alert medical teams to potential problems.

At the time of writing the HealthSuite SDK and API has not yet been released [38] but they say it will be based on open standards, such as FHIR, REST, and oDATA. The API will be offered as a Platform-as-a-Service (pay per use), as part of the HealthSuite Digital Platform to avoid the cost and effort of setting up, maintaining and operating a backend. Instead, the online service communications, such as firmware updates, user registration and data collection, are handled through the platform.

Philips does have an experimental platform called "Connect to Healthy", which supports different tools, devices and wearables [39] including: Withings, MyBasis, Fitbit, iHealth and Jawbone. This platform had a lot of potential for providing a common API to multiple types of devices but it is currently not part of their production environment and therefore only contains prototypes of the data for experimentation. As of writing, the API provides access to the following information: blood glucose, blood pressure, burned calories, energy intake, fat percentage, hba1c (glycohemoglobin), heart rate, resting heart rate, skin temperature, sleep duration, steps and weight.

Fig. 1 Garmin index smart scale

4.1.2 Garmin

Garmin is an American multinational technology company founded in Lenexa, Kansas and is now based in Schaffhausen, Switzerland. The company is primarily known for its GPS technology development and its use in automotive, aviation, marine, outdoor, and sport activities and utilities. However, more recently they have moved into the health and fitness marketplace, developing wearable technology and devices for health monitoring.

Garmin mostly focuses on fitness monitoring but they do make some health monitoring products also: the Garmin Index Smart Scale (Fig. 1 and wearable devices (e.g. vívosmart and forerunner), which monitors heart rate, amongst other things [40]. Furthermore, Garmin offers a full suite of data APIs which can be used to interact with the data.

The Garmin Connect API [41] is an online API designed to work with Garmin's devices and it analyzes end-user data and supports more than 30 different activity types. The Garmin Connect API provides authorized third-party developers access to data with end-user consent and provides the original data file (FIT, GPX or TCX, depending on the device) using PUSH notifications. Near real-time data is pushed directly to third party servers that a developer can set up to collect user data. They also have a suite of REST APIs that can be called to extract specific information e.g. workout service.[2]

Garmin offers another API (Wellness API), which is primarily used as a corporate, research/clinical solution and provides only summary data (including steps/sleep/all day monitoring). The Wellness API has no licensing fee.

4.1.3 Omron

Omron is a market leader in heart health and a distributor of personal wellness products. Omron's products include home blood pressure monitors and electrotherapy devices. Omron invented its first blood pressure monitors nearly 40 years ago and

---

[2]https://connect.garmin.com/proxy/workout-service-1.0/.

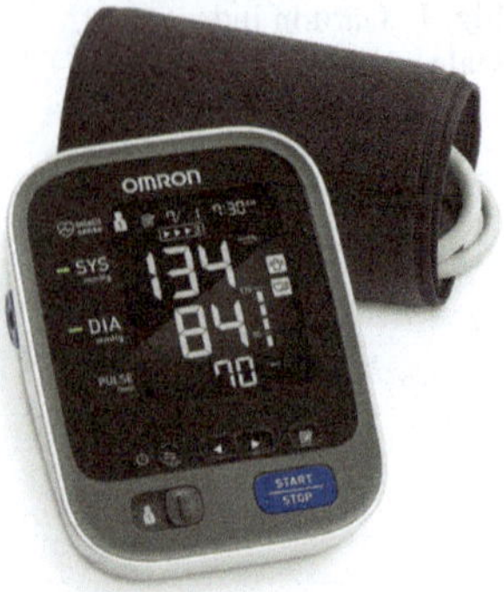

Fig. 2 Omron blood pressure monitor

since then, has been passionate about empowering people to take charge of their personal health.

Omron has a number of bluetooth enabled devices world wide but in the states, only blood pressure monitors and activity trackers allow such connectivity, for example, the Omron Health Care BP786 Digital Blood Pressure Monitor (Fig. 2) and the Omron HJ-327T Alvita Wireless Activity Tracker.

Omron's API allows the integration of Omron end user's blood pressure and/or fitness activity data into applications. Interaction with the API can be achieved in any programming language, as long as it supports HTTPS-based request/response scenarios. The API uses the OAuth 2.0 and OpenID Connect protocols for authentication. There is a free staging environment for development and the user data flows from the devices to Omron's backend site to the client applications. User must authorize access to any client application and once enabled the application can perform on-demand polling and pulling of data. There is also an Omron Wellness app offered for monitoring blood pressure but there are mixed reviews on Android [42].

Omron also has a Telehealth initiative that provides three further ways of integrating devices into applications [43] but their uses is restricted to certain geographical areas (as of Mar 2017):

1. Device integration via OMRON connect App (USA, Canada, India, Korea, Japan, Thailand, Taiwan, China, Australia, Indonesia, Philippines, Singapore and Vietnam, China, Japan, Taiwan, Australia, China, Malaysia, Vietnam, Thailand, Singapore, Hong Kong, Philippine, Japan, New Zealand and Indonesia). This option allows direct integration with the device itself using an Omron Telehealth device and adopting standard Services and Profiles designated by Bluetooth SIG.
2. Integration via the server (China and Japan), where data is recorded on the server and the applications query that server.
3. Integration via OMRON connect App provided by OMRON (Taiwan, Australia, China, Malaysia, Vietnam, Thailand, Singapore, Hong Kong, Philippine, Japan, New Zealand and Indonesia) where data is collected on the servers and also OMRON connect App can be used to interact with the data.

Fig. 3 Beets BLU digital scale

4.1.4 Beets BLU

Beets BLU was founded in 2012 by a group of specialists in the fields of mobile applications, fitness technologies and electronics, specializing in wireless sensors with Bluetooth Smart technology. Beets Blu [44] offers two bluetooth devices that are designed to use Bluetooth Smart technology, which is a protocol designed to connect low energy devices with mobile phones.

They offer an easy-to-use bathroom scales (see Fig. 3) and a heart rate monitor. Beets do not offer a Web API, instead they offer direct integration to Android, Windows or Apple devices and provide example applications for developers to use [45]. The heart rate monitor uses the standard protocol from Bluetooth SIG but the smart scale uses a proprietary protocol.

4.1.5 Withings

Withings[3] [46] is a French consumer electronics company that is known for design and innovation in connected health devices. Withings was purchased by the Finnish company Nokia on 26 April 2016 and since then has began the transitioning and by year end 2017 all products with carry the Nokia brand name.

Withings have been in the connected health space for some time, releasing the first ever Wi-Fi scale in 2009. They have a large suite of products including a: WiFi Body Scale (Fig. 4), FDA approved Blood Pressure Monitor, Smart Baby Monitor, Smart Kid Scale, Smart Body Analyzer and Wireless Scale, Pulse Ox (heartbeat, steps taken, distance, etc.), Steel (activity and sleep), and Thermo (thermometer).

They have an extensive open API [47] to access the device data, and a developer can sign up and get access instantly. It uses OAuth V1 for request signing with a hash-based message authentication code (SHA1) [48]. This algorithm creates a signature by hashing the request information, mixed with a secret composed of a consumer key and the access token.

---

[3]pronounced "with-things".

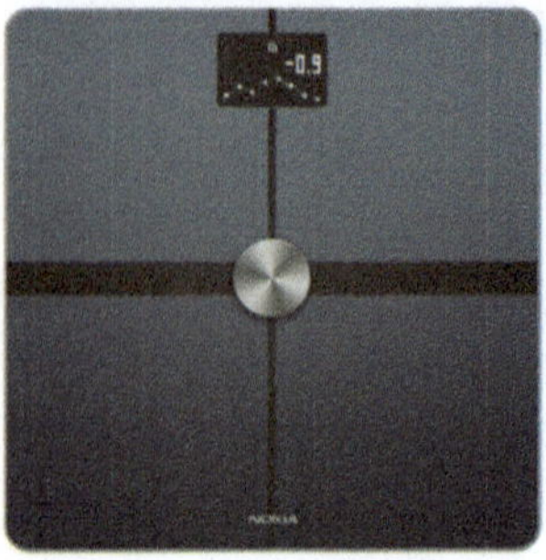

Fig. 4 Nokia (Withings) digital scale

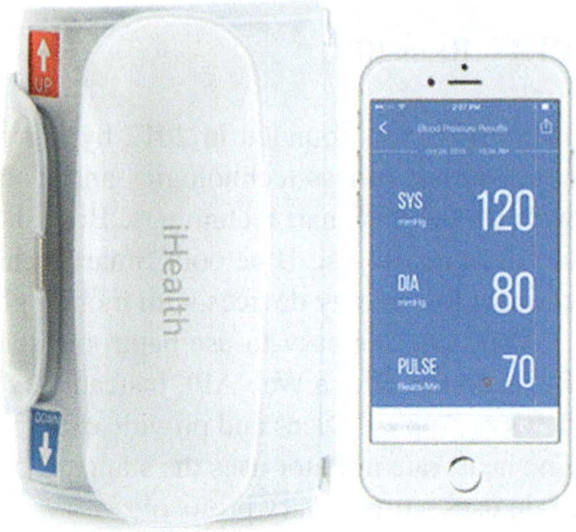

Fig. 5 iHealth feel wireless blood pressure monitor

The documentation is very clear and they also provide a push and pull model, using web hooks within their notification system. So, for example, a developer can subscribe to specific data from specific devices and a web method is called with updates as and when data from a particular user is updated in the database [49]. The callback provides a start time, end time and data type, the application can then use the query method to collect the actual data.

4.1.6 iHealth

iHealth Labs Inc [50] is based in Mountain View, California and is a subsidiary of Andon Health, a Chinese manufacturing company and one of the largest OEM health technology manufacturers in China. The company is known for its wireless devices (Bluetooth or WiFi enabled) that can measure vital signs and are compatible with Apple and Android devices.

iHealth devices include blood pressure monitors (e.g. iHealth Feel in Fig. 5), blood glucose monitors, body analysis scales, pulse oximeters and activity and sleep trackers. All of the devices sync directly with a free mobile app that makes monitoring, viewing, storing and sharing of health vitals simple and comprehensive.

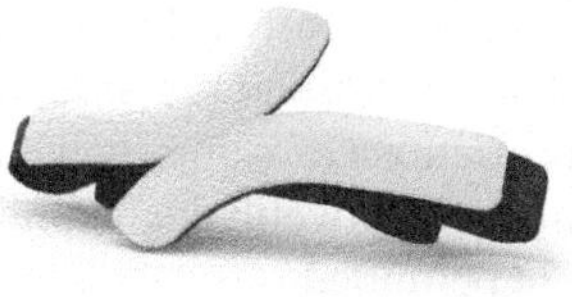

Fig. 6 QardioCore wireless continuous ECG monitor

iHealth's open API [51] enables an application to interface with user data collected using iHealth products i.e. the data travels from the iHealth product to the iHealth cloud, and then to the application. Developers can develop applications in the iHealth Sandbox [51], which allows experimentation and testing before migrating to active user data.

The iHealth API authenticates and authorizes using the OAuth 2.0 protocol. The API provides the ability for developers to set unit profiles (e.g. cm or in, kg or lbs), which can either be based on the type of language (English US, UK or the default, which uses the European measurements) or a user can provide their own profile. Data is requested via https via the API located at https://api.iHealthLabs.com:8443. The API supports XML or JSON payloads and supports pagination making it particularly suitable for front end Web applications.

Currently, the API provides access to the following data: blood pressure, weight, blood glucose (BG), blood oxygen, activity report or sleep report. The API also supports subscription notifications for users that authorize the partner app for data access through OAuth. When one of these events are created or deleted, a notification is pushed to the registered application webhook, which is set using the application configuration page when it is registered.

4.1.7 Quardio

Quardio [52] is an American technology company based in San Francisco that focuses on wireless wearable technology for heart health monitoring, including a blood pressure monitor, smart weighing scale and body analyzer, multi-sensor electrocardiography (ECG) monitor (See Fig. 6).

Their QuardioMD [53] product is a platform that allows Quardio devices connect to the QuardioMD platform to help doctors manage chronic care patients. Patients measure their blood pressure, ECG, weight at home with Quardio's medically accurate devices and doctors use the QuardioMD web portal to monitor patients' measurements. QuardioMD automatically analyzes and prioritizes patients based on individually set criteria so that the doctor knows which patients are doing well and which needs more attention.

Quardio take a somewhat different approach to making data available by using mobile device manufacturers's APIs. Quardio have integrated their devices with both Google Fit [32] and Apple Health [34] to allow the sharing of data with those platforms. Users with Android devices can connect their Quardio accounts and automatically send all of the QuardioBase weight measurements to Google Fit [54]. It has also recently (March 2017) been integrated into Apple's HealthKit, enabling the sharing of blood pressure, heart rate and other health data [55]. However, there are restrictions of using such an approach due to license agreements, as generally only fitness and wellbeing applications are allowed to use the data and medical uses are not permitted.

The Quardio app [56] has one of the highest ratings of any Apple store health monitoring application (4.8 out of 5 at the time of writing).

4.2 Personal Self-Therapeutic Biofeedback for Wellbeing

For biofeedback and neurofeedback we cover the following devices:

1. Emotiv
2. InteraXon
3. Spire
4. Electrical Geodesics (EGI).

4.2.1 Emotiv

Emotiv is a bioinformatics company founded in 2011. The company has its headquarters in San Francisco, USA, and facilities in Sydney, Hanoi and Ho Chi Minh. Emotiv develops BCIs (Brain Computer Interface) technology also referred to as MMI (Mind Machine Interface), DNI (Direct Neural Interface), BMI (Brain Machine Interface). While the applications for the EMOTIV technology and interface can span a large variety of industries, the company mainly focuses on empowering individuals to understand the human brain and the use of electroencephalography (EEG) to track cognitive performance and monitor emotions, and to employ machine learning of trained mental commands to control both virtual and physical objects. Providing technology and knowledge to accelerate brain research globally, Emotiv's community of developers and researchers spans over 100 countries.

MyEmotiv [57], the mobile app recently made available by Emotiv, is compatible with both their wearable EEG neuroheadsets Insight and Epoc+, uses Bluetooth wireless connectivity, and is supported by Apple iOS and Android (See Fig. 7). The users can choose either the Insight 5 Channel EEG headset, or the Epoc+ 14 Channel EEG headset, to capture electrical signals from their brain activity and measure their mental performance and fitness. Performance metrics provide real-time detection of cognitive and emotional states, and the 3D BrainViz viewer provides a real-time display of the brain's activity patterns, which feed information to the users to track and optimize their brain performance. Saving the recordings and playback make it possible to compare the results to previous sessions and participate in the Emotiv on-line community.

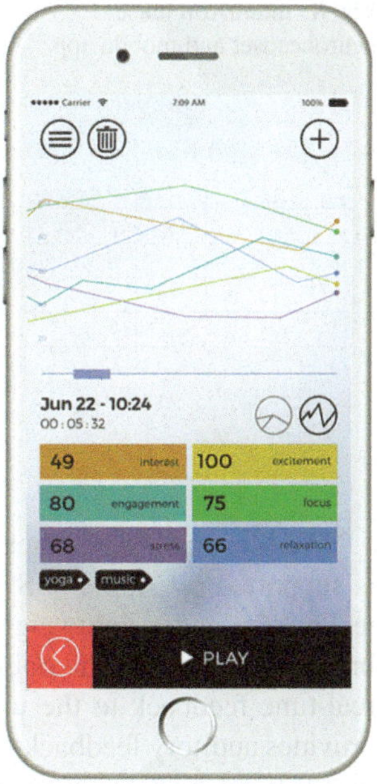

Fig. 7 Emotiv MyEmotiv neurofeedback mobile App

Emotiv has an SDK Community Edition [58] called Cortex, which allows interaction with the Emotiv Insight, Epoc and Epoc+ products. The Cortex API is a locally accessible API and corresponding SDK that is built on JSON and WebSockets to allow access from different programming languages and platforms. You can use Cortex to create games and applications, record data for experiments, etc. Cortex uses WebSockets, JSON, and JSON-RPC to provide a real-time connection to the underlying Cortex service for use in both desktop and web-based applications. By default, a developer can work with Cortex without client id/ secret but if access to raw EEG data and high-resolution performance metrics is required, a client ID and secret needs to be supplied to the authorize method to verify the subscription level grants access. They offer 3 types of license: basic (free), Pro, or commercial. The basic API does not allow access to the EEG data or performance metrics, the Pro has EEG but not performance metrics and the commercial allows everything. Cortex support for iOS and Android mobile application development is not available at the time of writing but is expected to be available soon.

4.2.2 InteraXon

InteraXon is a brain-sensing technology innovator based in Toronto, Canada. InteraXon has worked with a variety of industries to create brain-sensing customer programs. The company launched their first product, Muse ("the brain sensing headband"), in 2014.

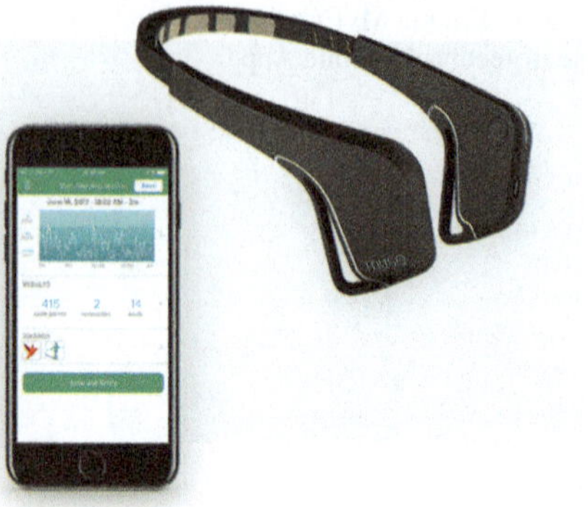

Fig. 8 InteraXon muse neuroheadset and mobile app

The Muse neurofeedback device [59] uses Bluetooth wireless connectivity and is supported by Apple iOS and Android (See Fig. 8). Muse utilises brain-sensing technology and a lightweight headband with 7 dry sensors to detect and measure brain activity at specific points across the forehead and behind the ears and provide real-time feedback to the user for improving the meditation practice. The device provides auditory feedback to the users to ascertain states of calmness and alertness in real-time and the apps record and track the progress with practice so that the users can assess their scores and earn "rewards" for attaining longer periods of calmness, as measured by the device. Recordings and real-time experiences can be shared with multiple accounts in the on-line community.

Muse has an sdk [60] that enables integration with a number of platforms (iOS, Android, Windows and Unity) and Tools for other platforms (Windows, Mac, Linux). The data is collected using MuseDirect or MuseIO and then data is sent around using Open Sound Control (OSC), originally intended as a successor to the control protocol for electronic instruments (MIDI), but it became far more general. Data can be downloaded into an app using an OSC server, which is bundled with the software.

4.2.3 Spire

Spire is a small group of professionals from four continents, specialising in hardware and software engineering, algorithms, marketing, healthcare, and operations, that teamed up to develop solutions for empowering individuals to control their own mental and physical health by the use of continuous respiration sensing technology, real-time interventions, and actionable feedback. The product design utilised research work in respiratory psychophysiology at Stanford University's Calming Technology Lab.

The Spire Stone biofeedback device [61] is a personal stress management and activity tracker that uses Bluetooth wireless connectivity and the Spire App for iOS and Android (See Fig. 9). Spire measures the user's breath via the expansion and contraction of the torso, and provides permanent feedback to the user to guide breath

Fig. 9 Spire stone biofeedback device

adjustment. The Spire App algorithms classify respiratory patterns to correlate with cognitive and emotional states and the app categorises breathing as calm, tense, or focused. Following a real-time view of their own breathing patterns on the device's screen, in-app guidance and notifications of sudden changes, the users can learn to reset their breath, establish new patterns, and watch their progress metrics over time.

Spire exposes a Web API for interfacing with your personal data [62] and there are third part SDKs available, such an a NodeJS package [63]. The API requires a token for a user, which is obtained by emailing the customer support. Third party integration is not currently supported at tis time. Since Spire monitors breathing to help reduce stress and improve your sleep, the API provide access to three parameters: "br"—breath rate in breaths per second; "steps"—number of steps; "calories"—number of calories burned.

4.2.4 Electrical Geodesics (EGI)

EGI (Electrical Geodesics, Inc.) is a neurodiagnostic medical technology company founded in 1992. The company is based in Eugene, Oregon, and it operates in the United States, Canada, Europe, Asia, and internationally. As of July 21, 2017, Electrical Geodesics, Inc. operates as a subsidiary of Philips Holding USA Inc. EGI provides non-invasive dense array electroencephalography (dEEG) devices for monitoring and interpreting brain activity and the HydroCel Geodesic sensor net, the brain acquisition and visualisation software environment, electrical source imaging and head modelling, and the neuroscience online laboratory high-performance computing environment. The company also offers wireless portable EEG systems for research and personal monitoring.

The Avatar EEG is a portable EEG device [64] that uses Bluetooth wireless connectivity and a mobile app for Android smart phones and tablets (See Fig. 10). Avatar is an 8-channel recorder device and the channels can be used for recording EEG, ECG, EOG, and EMG data. Avatar can be used as a biofeedback and neurofeedback personal device.

The Avatar recorder is designed for direct integration and measures EEG, ECG, EOG, and EMG. There is a version of the data collection software for Android

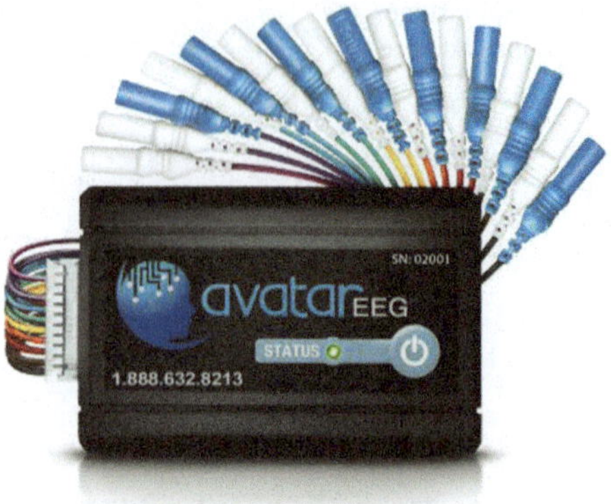

Fig. 10 Electrical geodesics (EGI) avatar EEG biofeedback and neurofeedback device

[65] and provides a distribution with the device in Python [66] to: receive data via Bluetooth from one or more Avatar EEG recorders; convert an Avatar EEG recorder native binary file to comma separated format (CSV); convert an Avatar EEG recorder native binary file to European Data Format (EDF); and to convert an Avatar EEG recorder native binary file to BDF+ format—the 24 bit version of a European Data Format (EDF+) file. Since this is written in Python it supports all major platforms.

4.2.5 Other Biofeedback and Neurofeedback Initiatives

This section describes several other companies that make devices in this space, but do not have interfaces to the data at this time.

- The HeartMath institute [67] provides the Inner Balance [68] device for personal use that uses a Bluetooth HRV Sensor for iPhone and Android, and Lightning sensor for iPhone. The Inner Balance Bluetooth Sensor attaches to an iPhone, iPad or Android device and an earlobe and feeds the user's heart rhythms recordings into the app. emWave 2 [69] is a standalone handheld device used to observe the heart rhythm during a session and provide real-time coherence feedback. These devices provide real-time feedback to train the users to shift and replace emotional stress with emotional balance and coherence, by analyzing and displaying the heart rhythm, measured by heart rate variability (HRV), which indicates how emotional states affect the users' nervous system during live sessions. "Quick Coherence" training techniques and real-time coaching tips are used to guide the users to reach and maintain the state of coherence and track progress over time.
- The Biofeedback company [70] offers the Flow [71] biofeedback personal device for Android. The biofeedback heart rate sensor attaches to the finger or the ear; the HRV wearable device analyzes data from the sensor and sends it wirelessly to the smart phone. The mobile apps display real-time feedback in user friendly graphics and are particularly recommended for training in ADHD and ADD.

- WellBe [72] offers the WellBe Stress Balancing Bracelet [73] with mobile apps for iPhone and Android. The wireless bracelet includes biofeedback sensors that monitor the heart rate variability (HRV) and uses a patent pending stress analysis algorithm to determine stress and calmness levels based on personalized stress triggers.
- The Pip Personal Stress Trainer [74], with mobile apps for iPhone and Android, combines electrodermal activity (EDA) and biofeedback based relaxation training (BART). My Pip [75] HIPPA compliant cloud platform syncs seamlessly with all Pip Apps.
- Stens company offers a DP200M Complete Home System called the DynaPulse Home Monitor [76], which is a home-use electronic blood pressure monitor. DynaPulse displays the waveform allowing the user to monitor irregular heartbeats (arrhythmia) and it can be used as a biofeedback device showing each individual heartbeat during a measurement.

4.3 Taxonomy Comparison

This section provides a table containing each company that offers personal connected health devices using the taxonomy identified in Sect. 4, for the "Personal Health Monitoring" and the "Personal Self-Therapeutic Biofeedback for Wellbeing" areas. The information is presented as a table with the companies representing the columns and the criteria or metrics being integrated into the rows.

In Fig. 11, it can be seen that each company has its own niche or individual offering on features and focuses. For the Web-based APIs the competitors of the free offerings are Withings and iHealth. For paid platform as a service models, Omron and Philps are direct competitors with Garmin offering one time fee. In terms of comprehensiveness, Withings probably has the edge at the time of writing but things can change very quickly in this space.

In terms of the number of available devices supported the clear winners are Philips, Withings and to a lesser extent iHealth in the USA, but in the worldwide market, especially in Asia, Omron has an excellent extensive portfolio. It remains to be seen however, whether these devices make it to Europe or the USA.

In Fig. 12, the four companies: Emotiv, InteraXon, Spire and EGI are compared for self-therapeutic techniques and wellbeing applications. For these types of applications, the general interface preference is using an SDK because typically, the devices are directly integrated using a mainstream OS e.g. Windows, Mac or Linux, or mobile platforms. Spire is the only company in this category that collects its data via a cloud server and provides a Web API to it. The other three, provide data via Bluetooth or WiFi to a mobile or desktop application and then provide SDKs for those devices to interact with the data directly. All allow third party applications to interface with the device to access the data for the user that is using the device at that point in time.

Metric \ Company	Philips	Garmin	Omron	Beets Blue	Withings	iHealth	Quardio
Type Of Connectivity	WIFI and Bluetooth	WIFI, Bluetooth and ANT+	WIFI and Bluetooth	Bluetooth	WIFI and Bluetooth	WIFI and Bluetooth	Bluetooth
Available Devices	Multiple	Scales and Hearthbeat	Blood Pressure and Activity (USA). Multiple worldwide	Scales and Hearthbeat	Multiple	Multiple	Blood pressure, Scale,Body Analyzer, ECG
API Type	Web	Web	Web and Direct Integration	Direct Integration	Web	Web	Apple HealthKit and Google Fit
Authentication	Not Disclosed	Oauth 1	Oauth 2 and Open ID	In App	Oauth V1	Oauth V2	In App
Data Access Policy	Open	Open	Open	In App	Open	Open	In App
Data Format	JSON	JSON with File payloads (FIT, GX and TCX)	JSON	In app	JSON	JSON	Apple HealthKit and Google Fit
API Pricing Model	Platform as a service (PAAS)	One Time Fee ($5k)	Paid Subscription	Free	Free	Free	Free
Other	Not Released Yet		Access to staging is free for development				

Fig. 11 Personal health monitoring taxonomy

Metric \ Company	Emotiv	InteraXon	Spire	EGI
Type Of Connectivity	WIFI and Bluetooth	Bluetooth	Bluetooth	Bluetooth
Available Devices	INSIGHT and EPOC+ (Neurofeedback EEG headsets 5-channel, 14-channel) MyEmotiv Mobile App	Muse Brain-Sensing Headband (Neurofeedback EEG headset 7 dry sensors) Mobile App	Spire Stone (Biofeedback Breathing patterns) Spire App	Avatar Wireless (Biofeedback Neurofeedback 8-channel recorder EEG, ECG, EOG, EMG) Mobile App
API Type	SDK & Web Sockets	SDK	Web API	SDK
Authentication	In App, API Client/secret	NA	API Client/Secret	In App
Data Access Policy	Depends on Subscription Level	In App	In App	In App
Data Format	JSON	Muse data over OSC	JSON	RAW, CSV, EDF
API Pricing Model	Paid Subscription	Commercial Licenses for Third-Party Applications	Free for Personal Use	Free
Other	Mobile SDK Not Released Yet			

Fig. 12 Personal self-therapeutic biofeedback for wellbeing taxonomy

5 Contemplative Practices: Learning from the Past and Future Directions in Wellbeing

In this section and as a conclusion to this paper, we reflect on some of the traditional contemplative disciplines and their transition into today's IoT space for personal wellbeing. We see this fusion of "East meets West" being an increasing focal point as IoT in this space further personalizes health and wellbeing.

Applied biofeedback was originally developed based on cultural factors that contributed to merging Eastern and Western traditions, with the rise in popularity of meditation and mindfulness practices and the so-called "expansion of consciousness" [11]. As Yoga and Zen masters were thought to be able to alter their physiological states through meditation, some biofeedback experiences were developed based on related phenomena, and biofeedback received its nicknames of "the Yoga of the West" and "electronic Zen" [11, 12] Psychotherapies can be used in conjunction with biofeedback and neurofeedback; cognitive–behavioral therapy (CBT) or mindfulness-based therapies have shown to help improve the patients' coping skills and to affect thought patterns that lead to learning how to alter the behaviors, habits, patterns, and responses that contribute to the persistance of symptoms and reduce well-being, performance or function [13–15]. Contemplative practices that help establishing coherence between the heart and the brain, increase parasympathetic modulation and improve heart rate variability (HRV), can be implemented as protocols in biofeedback therapy for internal self-regulation (cognitive, affective, and autonomic regulation) as an integrated function of the heart and the brain [77]. Meditation techniques that bring awareness to the sinoatrial (SA) node (the heart's natural pacemaker that controls the heart rate and helps adjust abnormal heart rhythms), such as Ramana Maharshi's self-inquiry methods and Hridaya meditation of the heart [78], may be used as such guided methods to enable internal self-regulation.

As mindfulness meditation has been scientifically proven to reduce symptoms associated with stress, anxiety, and depression and to improve attention, performance, self-efficacy, calmness, mindfulness, and quality of life [12, 59], some neurofeedback devices use mindfulness-based guided meditation exercises (provided as electronic "personal meditation assistants") to help the user reach states of calmness and/or focus while adjusting from undesirable states of distress and/or distraction. Harvard Medical School and Spaulding Rehabilitation Hospital are currently conducting a pilot study of 20 patients with chronic mild-to-moderate TBIs (traumatic brain injury) [12] comparing a group using the Muse neurofeedback device [59] and a group receiving the same mindfulness meditation intervention via the app without the use of the device, over a 6-week intervention of daily meditation practice [12]. The purpose of the study is to assess if there is a benefit in this population when using the neuroheadset device on measures of attention, mood, and mindfulness. The preliminary results suggest increased motivation and consistency in the daily practice with the Muse neuroheadset groups, potentially due to the feedback provided by the device, as these groups have shown longer time spent practicing on their own and above the required daily intervention [12]; such results suggest that survivors of TBI could benefit from improved motivation for treatment as well as decrease of symptoms related to attention, mood, and mindfulness with the addition of neurofeedback therapy to mindfulness meditation training [12].

Other types of biofeedback devices use breathing function measurements as indicators of the user's mindset and the respiratory patterns are correlated with cognitive/emotional state [61], which relates to the Eastern self-cultivation practices of "pranayama" (breathing exercises, breath awareness and control) that were recently translated into protocols for clinical interventions shown to alleviate anxiety and pain, increase heart rate variability, and reduce blood pressure.

Biofeedback devices provided by the HeartMath institute [67] use real-time heart rate variability (HRV) feedback to train the users to shift and replace emotional stress with emotional balance and coherence. Coherence is an optimal physiological state in which the heart, mind and emotions are operating in sync and in balance, and the immune, hormonal and nervous systems function in a state of harmonious coordination; physiological entrainment may be established during coherence (i.e., rhythmic physiologic patterns becoming harmonious and synchronized) [67]. Quick training techniques and real-time coaching tips are used during the biofeedback sessions to guide the users to reach and maintain the state of coherence and track their progress over time.

Contemplative practices are associated with positive trends in heart rate variability, blood pressure, reduced stress, anxiety and depression, increased mindfulness, harmonization of the emotional and psychological functions, improved self-acceptance, reduced catastrophizing, increased coping mechanisms and accommodative coping skills, and positive effects in chronic pain patients. The integrated innate system for internal self-regulation works through a series of feed-back and feed-forward loops, and these disciplines have the potential to enhance the balance of the autonomic nervous system (ANS) by adjusting sympathetic and parasympathetic activity, improving functions of the immune and autonomic nervous systems.

It will be extremely interesting to see if, how and when such techniques are integrated using IoT to provide the next generation of personalized practice and wellbeing applications—bringing the old wisdom to widespread adoption and usage.

References

1. G. Fortino, P. Trunfio, *Internet of Things Based on Smart Objects: Technology, Middleware and Applications* (Springer, 2014)
2. A. Guerrieri, L. Valeria, R. Anna, F. Giancarlo, *Management of Cyber Physical Objects in the Future Internet of Things* (Springer, 2016)
3. Grand View Research on Personal Connection Healthcare, http://www.grandviewresearch.com/industry-analysis/connected-health-wellness-devices-market. Accessed 10 May 2017
4. Healthcare Information and Management Systems Society (HIMSS), http://www.himss.org/. Accessed 10 May 2017
5. Fundamentals of Data Exchange—Personal Connected Health Alliance, https://cw.continuaalliance.org/document/dl/13473. Accessed 10 May 2017

6. PHD—Personal Health Device Working Group, https://standards.ieee.org/develop/wg/PHD.html. Accessed 10 May 2017
7. Patient Care Devices, https://www.ihe.net/Patient_Care_Devices/. Accessed 10 May 2017
8. Personal Healthcare Monitoring Report, http://wiki.hl7.org/index.php?title=Personal_Healthcare_Monitoring_Report. Accessed 10 May 2017
9. D.H. Barlow, V.M. Durand, Abnormal psychology: an integrative approach. *Nelson Education*, 2011
10. What is Biofeedback—at bottom of page in Association for Applied Psychophysiology and Biofeedback, https://en.wikipedia.org/wiki/Association_for_Applied_Psychophysiology_and_Biofeedback
11. Stens Corporation, What is Biofeedback? What is Neurofeedback? https://stens-biofeedback.com/pages/what-is-neurofeedback
12. S.N. Gray, an overview of the use of neurofeedback biofeedback for the treatment of symptoms of traumatic brain injury in military and civilian populations. Med. Acupunct. **29**(4), 215–219 (2017)
13. M.S. Schwartz, F. Andrasik, *Biofeedback: A Practitioner's Guide* (Guilford Publications, 2017)
14. I.Z. Khazan, *The Clinical Handbook of Biofeedback: A Step-by-Step Guide for Training and Practice with Mindfulness* (Wiley, 2013)
15. D.L. Frank, L. Khorshid, J.F. Kiffer, C.S. Moravec, M.G. McKee, Biofeedback in medicine: who, when, why and how? Ment. Health Fam. Med. **7**(2), 85 (2010)
16. Association for Applied Psychophysiology and Biofeedback, https://www.aapb.org/
17. Biofeedback Certification International Alliance (BCIA), http://www.bcia.org
18. International Society for Neurofeedback and Research (ISNR), https://www.isnr.org/
19. Biofeedback Federation of Europe (BFE), http://www.bfe.org/
20. Napoletano, Margo. Neurofeedback: Frequently Asked Questions.Child and Family Counseling Center, http://www.childandfamilycenter.com/services-specialized-neurofeedback.htm
21. H. Marzbani, H.R. Marateb, M. Mansourian, Neurofeedback: a comprehensive review on system design, methodology and clinical applications. Basic Clin. Neurosci. **7**(2), 143 (2016)
22. T. Dempster, An investigation into the optimum training paradigm for alpha electroencephalographic biofeedback. Ph.D. thesis, Canterbury Christ Church University, 2012
23. D.J. Vernon, Can neurofeedback training enhance performance? an evaluation of the evidence with implications for future research. Appl. psychophys. Biofeedback **30**(4), 347 (2005)
24. S. Larsen, L. Sherlin, Neurofeedback: an emerging technology for treating central nervous system dysregulation. Psychiatr. Clin. **36**(1), 163–168 (2013)
25. R. Christopher deCharms, F. Maeda, G.H. Glover, D. Ludlow, J.M. Pauly, D. Soneji, J.D. Gabrieli, S.C. Mackey, Control over brain activation and pain learned by using real-time functional mri, in *Proceedings of the National Academy of Sciences of the United States of America*, vol. 102, no. 51, 2005, pp. 18626–18631
26. Y. Nestoriuc, A. Martin, Efficacy of biofeedback for migraine: a meta-analysis. Pain **128**(1–2), 111–127 (2007)
27. Y. Gil, G. Li, J. Lee, Integrated real-time neurofeedback system to raise the frontal lobe activity: design and implementation, in *Annual International Conference of the IEEE Engineering in Medicine and Biology Society, 2009. EMBC 2009* (IEEE, 2009), pp. 845–848
28. T. Mauro, S.A. Cermak, The Everything Parent's Guide to Sensory Integration Disorder: Get the Right Diagnosis, Understand Treatments, and Advocate for Your Child. *Simon and Schuster*, 2006
29. Bluetooth, https://www.bluetooth.com/. Accessed 10 May 2017
30. ANT+, https://www.thisisant.com/consumer/ant-101/what-is-ant/. Accessed 10 May 2017
31. FIT: Flexible and Interoperable Data Transfer, https://www.thisisant.com/developer/ant/ant-fs-and-fit1. Accessed 10 May 2017
32. Google Fit, https://www.google.com/fit/. Accessed 15 May 2017
33. Google Fit API, https://developers.google.com/fit/. Accessed 15 May 2017
34. Apple HealthKit, https://developer.apple.com/healthkit/. Accessed 15 May 2017

35. Health and Fitness, Gadgets and Wearables: Top Smart Blood Pressure Monitors, http://gadgetsandwearables.com/2017/03/06/leading-smart-blood-pressure-monitors/. Accessed 10 May 2017
36. 1byone Website, https://www.1byone.com/. Accessed 14 May 2017
37. Philips Healthsuite Application, http://www.usa.philips.com/healthcare/innovation/about-health-suite. Accessed 10 May 2017
38. Philips Health Suite API, http://www.usa.philips.com/healthcare/innovation/about-health-suite/developers-healthsuite/philips-healthsuite-digital-platform-developers-portal. Accessed 15 May 2017
39. Philips Connected Life hackathon, http://www.philips.com/a-w/en_connectedlife-hackathon/api.html. Accessed 15 May 2017
40. Garmin Wellness Devices, https://explore.garmin.com/en-US/wellness/. Accessed 10 May 2017
41. Garmin Data APIs, https://developer.garmin.com/. Accessed 10 May 2017
42. Omron Wellness App, https://play.google.com/store/apps/details?id=com.omron.wellness. Accessed 15 May 2017
43. Omron Telehealth Integration Comparison, http://www.telehealth.omron.com/home/difference/. Accessed 15 May 2017
44. Beets Blu, http://beetsblu.com/. Accessed 14 May 2017
45. Beets Blu SDK, http://beetsblu.com/sdk.html. Accessed 14 May 2017
46. Withings, https://www.withings.com/. Accessed 15 May 2017
47. Withings API Reference, https://oauth.withings.com/api/. Accessed 15 May 2017
48. H. Krawczyk, M. Bellare, R. Canetti, *HMAC: Keyed-Hashing for Message Authentication, RFC 2104 (Informational)*, Feb 1997
49. Withings Notiication API, https://oauth.withings.com/api/doc#api-Notification-notify_subscribe. Accessed 15 May 2017
50. iHealth Labs, https://ihealthlabs.com. Accessed 15 May 2017
51. iHealth API and Sandbox, http://sandbox.ihealthlabs.com. Accessed 15 May 2017
52. Quardio, https://www.getqardio.com/. Accessed 15 May 2017
53. QuardioMD, https://www.getqardio.com/qardiomd-heart-health/. Accessed 15 May 2017
54. Qardio Google Fit Integration, https://support.getqardio.com/hc/en-us/articles/207901675-Sharing-data-via-Google-Fit. Accessed 15 May 2017
55. Quardio Apple Health Integration, https://www.getqardio.com/news/qardioarm-is-the-first-medical-device-with-health-support-available-on-the-app-store/. Accessed 15 May 2017
56. Qardio Blood Pressure Monitor and Weight Tracker, https://itunes.apple.com/us/app/qardio-blood-pressure-monitor-and-weight-tracker/id855275752. Accessed 15 May 2017
57. Emotiv and MyEmotiv, https://www.emotiv.com/myemotiv/
58. Emotiv SDK Community Edition, https://github.com/Emotiv/community-sdk/
59. InteraXon, Muse, The Brain Sensing Headband, http://www.choosemuse.com/
60. MUSE SDK, http://developer.choosemuse.com/sdk
61. Spire Stone, https://spire.io/pages/stone
62. SPIRE Developers API, https://developer.spire.io/
63. SPIRE NPM Tracker Package, https://www.npmjs.com/package/spire-tracker
64. Electrical Geodesics (EGI)—Avatar Wireless, https://www.egi.com/research-division/eeg-systems/wireless-eeg-system
65. Avatar Android App, http://play.google.com/store/apps/details?id=com.avatareeg.smarteeg
66. Avatar Software Overview, https://drive.google.com/file/d/0B388xdH0Vxl2R1ZYV2NHRlgwVGc/view
67. HeartMath Institute, https://www.heartmath.com/
68. HeartMath, Inner Balance, https://store.heartmath.com/innerbalance
69. HeartMath, emWave 2, https://store.heartmath.com/emwave2
70. The Biofeedback company, http://thebiofeedback.com/
71. The Biofeedback, Flow, http://thebiofeedback.com/flow-hrv-monitor-biofeedback
72. WellBe, https://thewellbe.com/

73. WellBe, WellBe Stress Balancing Bracelet, https://thewellbe.com/pages/features
74. The Pip Personal Stress Trainer, https://thepip.com/en-us/
75. The Pip, My Pip HIPPA compliant cloud platform, https://thepip.com/en-us/my-pip/
76. Stens Corporation, DP200M Complete Home System, https://stens-biofeedback.com/collections/computerized-biofeedback-systems/products/dp200m-complete-home-system
77. R. McCraty, F. Shaffer, Heart rate variability: new perspectives on physiological mechanisms, assessment of self-regulatory capacity, and health risk. Glob. Adv. Health Med. **4**(1), 46–61 (2015)
78. Hridaya Meditation of the Heart, http://www.hridaya-yoga.com/

Evacuation and Smart Exit Sign System

V. Ferraro and J. Settino

Abstract Smart systems based on the Internet of things are being successfully applied in multiple sectors. In this work, the application of the Internet of things to emergency evacuation systems will be tackled. Fixed direction signs represent the current practice in identifying evacuation paths, but they can be misleading or lead evacuees to unsafe areas. Smart evacuation systems can overcome all these issues and allow the shortest and safest path to be determined based on the real time data gathered from sensors. A description of the proposed algorithm is presented. The communication between the different devices and exchange of information are of crucial importance in real-time operations. A reliable evacuation algorithm has to consider if an escape route is no longer usable and promptly provide an alternative path. The evacuees are guided by "*smart exit signs*" which can change their direction dynamically. Simultaneously, more precise and accurate data is available for the rescuers to ensure a more effective intervention.

1 Introduction

The intertwine between information and communication technologies and physical devices has led to the development of smart systems based on the Internet of things technology and smart objects (SO). Through sensors, actuators and a connection module, smart objects are able to communicate between each other, with the surrounding environment and interface with the user. An insight into new methods, protocols and architecture for the management of SO is given by Guerrieri et al. [8]. The emergence of the internet of things and smart objects coincided with the development of sensor technologies whose advancements have further enhanced the benefit

V. Ferraro (✉)
University of Calabria, Via Pietro Bucci, 87036 Arcavacata di Rende, CS, Italy
e-mail: vittorio.ferraro@unical.it

J. Settino
University of Malta, Msida 2080, MSD, Malta
e-mail: jessica.settino@um.edu.mt

F. Cicirelli et al. (eds.), *The Internet of Things for Smart Urban Ecosystems*,
Internet of Things, https://doi.org/10.1007/978-3-319-96550-5_15

of smart systems. They allow to meet the users expectations of higher comfort level and quality of the services provided while ensuring a more efficient management of the resources. This novel approach has significant implications in multiple sectors. Fortino and Trunfio [5] provide an extensive description of the many possible application of smart objects (SO) in a wide variety of smart environment: Smart Cities, Smart Health, Smart Buildings and Advanced IoT Projects.

This work focuses on the application of the Internet of things technology to emergency evacuation systems. The need of readily available information, gathered from different sources, and of a prompt response is never so impelling as in case of emergency. Unlike traditional systems, smart systems allow to fulfill these requirements thanks to an efficient communication network between the different devices. An Emergency Evacuation Service is proposed by Gokceli et al. [7]. The authors describe in detail the system and its real-time implementation for Building Automation Systems. A smart building evacuation system has also been proposed and implemented by Chen [2] using a fuzzy logic method.

In this work, a smart evacuation system based on Dijkstra's shortest path algorithm is described.

The chapter is structured as follows. In Sect. 2, traditional evacuation systems and their shortcomings have been analyzed. An overview of a smart guidance evacuation system is provided in Sect. 3. Each physical and software component is analyzed in detail: sensors, actuators and shortest path algorithm. In Sect. 4, a shortest path algorithm suitable for a smart guidance system is proposed. Results and possible applications are discussed in Sect. 5. Final remarks are provided in Sect. 6.

2 Evacuation Exit and Direction Signs

Fire regulations provide clear guidance on fire prevention and fire-fighting systems. Nevertheless, scientific literature points out that in case of emergency, the statutory evacuation procedures could be insufficient to adequately support occupants in the first moments of an evacuation alarm, when they can rely only on themselves. According to Kobes et al. [10], the factors influencing the fire response performance are essentially three:

- Fire characteristics: propagation speed, generated heat, presence of smoke and/or toxic gases;
- Features of the building: building layout, ease of way-finding, number and location of emergency exits;
- Human behavior: individual characteristics, familiarity with the building, relationship between occupants.

Fire characteristics strongly affect the possibility to escape. Extremely dangerous are fires with a high growing rate. In these cases, detect the fire at its early stage and promptly evacuate the building are of crucial importance. Nevertheless, as observed by Proulx [14], the reaction to a fire alarm is not immediate, often the danger of a certain situation is not clearly perceived. Presence of smoke is a clearer indicator of a fire; hence the pre-movement time is reduced [9]. The pre-movement time is the time interval between when the alarm is given and the first action undertaken by people to escape, while the movement time refers to the actual time necessary to escape. In reducing the movement time, a key role is played by the location and number of the emergency exits. The regulation prescribes a maximum distance of 30 m of each area of the building from an evacuation exit, while the number of exits depends on the congestion (number of people per square meter). The location of the closest exit has to be highlighted by signs. According to the regulations, the signage consists of both maps of the building, indicating the current position of the evacuee and the location of the exits, and permanent signs, pointing towards the closest exit or safety areas. These are fixed signs and their direction cannot be modified. Signs have to be durable and resistant to impact and adverse conditions. A good visibility of the signs has to be ensured even under poor light conditions and in presence of smoke; hence either electric signs have to be used or radio and photo-luminescencent exit signs. The former relies on electrical energy provided by a centralized back-up system or locally by batteries. On the other hand, radio and photo-luminescencent signs do not require an electrical source, luminescence is induced by radiation of a radioactive material or it is due to phosphorescence, i.e. the absorbed light is later re-emitted. These signs are also known as "self-luminous" and "glow in the dark" respectively. According to Cho et al. [3] the signals currently used increase the risk of ambiguity and disorientation, adding more stress in a situation already characterized by low levels of lucidity. Moreover, the occupants are more likely to incur in dangerous situations since with permanent signs is not possible to take into account the location of the fire and re-direct the evacuees accordingly.

Among the factors influencing the fire response, human behavior is the most difficult to predict. People have different reactions depending on the situation and circumstances [16]; hence, the mutual influence between human behavior and environment has to be taken into account. As stated by Paulsen [12], the early studies on human behavior in case of fire have been carried out by Wood in Great Britain and by Bryan in the United States, in the second half of the 20th century. People involved in fire incidents have been interviewed to understand their reactions and the reasons behind them. An element highlighted by both works is the importance of relationship between people, that re-entered into the burning building to search for family members [12] or waited for them before escaping. Moreover, in case of residential fire, people would try to extinguish the fire themselves rather than escape. Personal features have also an influence on the undertaken decisions. The experimental study carried out by Kobes et al. [11] shows that people staying overnight in a hotel, not used to this situation, tend to escape trough the main exit, thus through more familiar routes. Moreover, the authors compared the behavior of people which had received a building evacuation training to those who had no. Experiments under different conditions of fire and smoke have been carried out. The results show that when no smoke is perceived the majority of trained participants (55.6%) escape through the nearest exit compared to the 36.4% of not trained evacuees. In the smoke scenario the nearest exit was used by 88.9% of trained and 60.7% by not trained participants.

3 Smart Evacuation Guidance System

The problems pointed out by many experimental studies, lack of awareness of the presence of exit signs, people hesitation and uncertainty of the right direction to take, can be overcome with a smart guidance system. The use of smart evacuation guidance systems allow to guide people step-by-step providing clear and unequivocal instructions. It is composed of three main subsystems: sensors, algorithm to determine the escape routes and actuator systems.

3.1 Sensors and Data gathering

Sensors have a crucial importance in ensuring a real-time response of the guidance system. They provide all the necessary information, ranging from the location of a fire and its extend to the number of occupants in a given environment. Fire detectors are based on a constant or short intervals monitoring of at least one of the physical and/or chemical phenomena associated with a fire: heat, flame and smoke. Thus, they can be divided into three categories:

- Heat detectors
- Flame detectors
- Smoke detectors

Fire detectors can be further divided into point-to-point detectors, which monitor what is happening at a given point in the environment where the sensor itself is placed, or linear detectors, which allows to control the parameters on a continuous line. Heat detectors are sensitive to temperature rise. They can be fixed-temperature or rate-to-rise, also known as thermo-velocimeters. In the first case, the sensor determines if the temperature exceeds the set threshold value. This value is usually 10–35 °C higher than the highest expected temperature of the ambient. Typical values are between 50 and 60 °C. Thermo-velocimeters or rate-to-rise detectors are able to detect sudden changes in temperature. Such a detection mechanism is effective if fire develops very rapidly a large amount of heat.

Flame detectors are sensitive to the infrared and ultraviolet radiation emitted by the flames. Thus, the fire can be detected from its very beginning, ensuring the possibility of timely intervention. An important factor which has to be taken into account is that the system is ineffective in presence of dense smoke.

Smoke detectors can detect the presence of combustion products. Depending on the physical principle exploited, they can be categorized into optical and ionizing smoke detectors. The former are based on the so-called Tyndall effect. They have a source of light in the infrared field and a sensitive photodiode. An optical labyrinth prevent the transmission of the signal to the receiver. Smoke entering the detector reflects the light emitted by the infrared LED, making it a path to the receiver, which emits the alarm signal. This type of detector is unable to detect dark smoke, for which the Tyndall effect is not appreciable. Unlike optical sensors, ionization detectors are suitable for detecting all types of smoke, however they are less used because of maintenance requirements and radioactive contamination due to the use of Americio 241.

Considering the advantages and shortcomings of each fire detector, it could be necessary to instal more than one type to ensure the detection of any type of fire since its initial stage, avoiding false alarms. The location and number of sensors should also be carefully determined based on the characteristics of the environment (surface, height, shape), identifying the most critical areas. All the information gathered from the different sensors should be analyzed and crosschecked to avoid false alarm and have a clear and accurate picture of the situation.

Another important aspect which should be monitored is the presence of people [17]. This can be done in different ways, using: infrared detectors, vision based systems, which analyze the images registered by a video camera [1], or indirectly, by detecting the presence of a device through Bluetooth and Wifi signals.

3.2 Shortest Path Algorithms

Shortest-Path algorithms represent the heart of smart evacuation guidance systems. A proper translation of the building layout into a graph is the basis of all evacuation algorithms. A graph (G) is a set of objects called nodes (N) connected through edges (E). Information on the connection between two nodes can be stored in adjacency matrices (A) or lists. The adjacency matrix is a $N \times N$ Boolean matrix: the a_{ij} element is equal to 1 if there is an edge connecting node N_i to node N_j or it is equal to 0 if there is no connection. The representation of a graph through adjacency lists is instead based on the association of each vertex with a list of neighboring nodes. The use of adjacency lists is usually preferred in sparse graphs, with a limited number of edges, since the storing space required is lower. Nevertheless, check whether two nodes are connected is straightforward when adjacency matrices are used, whilst for adjacency lists the procedure is slower. In the following discussion we will refer to the representation by adjacency matrix.

It is also important to consider that information on the distance between two nodes is also necessary. Graphs can be weighted or unweighted, depending if a value is assigned to each edge or not. For the application under analysis, graphs are weighted and oriented (directed), i.e. edges are represented by arrows to indicate the direction of movement.

Each physical area of the building (rooms, corridors, stairs, exits etc...) is associated to a node and the weight of the edge, connecting two nodes, is equal to the distance between the two corresponding locations in the real building.

This information is typically stored in square matrices called distance matrices. The only difference with adjacency matrix is that the a_{ij} element is equal to the distance d_{ij} from node i to node j and it is equal to infinite if there is no connection between the two node. Thus, in undirected graphs the distance matrix is symmetric,

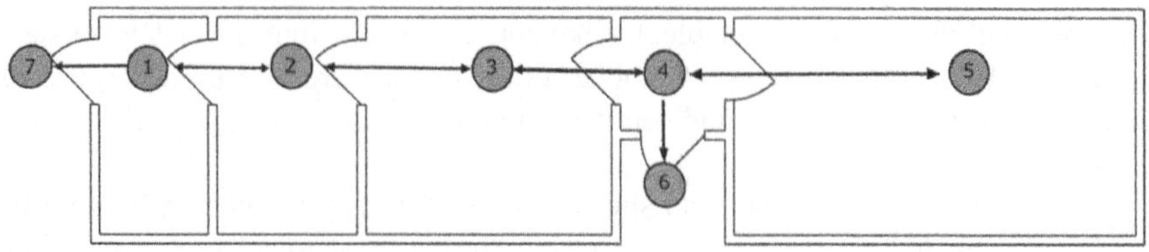

Fig. 1 Building layout and the corresponding graph

while this does not necessary occur for directed graphs where there could be the possibility to move from node i to j but not vice versa, i.e. $a_{ij} = d_{ij}$ but $a_{ji} = Inf$. Figure 1 shows a simple building layout and the corresponding graph and distance matrix. Scaled value of the distance can also be used without influencing the algorithm results.

$$A = \begin{bmatrix} - & 1.2 & - & - & - & - & 0.6 \\ 1.2 & - & 1.8 & - & - & - & - \\ - & 1.8 & - & 1.7 & - & - & - \\ - & - & 1.7 & - & 2.8 & 0.4 & - \\ - & - & - & 2.8 & - & - & - \\ - & - & - & - & - & - & - \\ - & - & - & - & - & - & - \end{bmatrix}$$

Once the distance matrix is known, the shortest path between each source node and the exit nodes, representing the evacuation exits, has to be determined. Cho et al. [3] classified the shortest path algorithms in four categories:

- Single-pair (SPSP), based on a single start point and a single destination point;
- Single-source (SSSP) with a single start point and multiple destination points;
- All-pairs (APSP) which search for the shortest path between all the possible combinations of single start points and destination points;
- Single-destination (SDSP) that search for the shortest path between multiple start points and a single destination point.

Single-Pair (SPSP) and Single-Destination (SDSP) algorithms are unsuitable for identifying the shortest path, considering that it is necessary to select among various exits the closest one. Among single-source shortest path (SSSP), Dijkstra's and Bellman-Ford algorithm are the main examples. Dijkstra's algorithm can only be used with positively weighted graphs, while the Bellman-Ford algorithm is suitable also for negative weights. However the complexity of the Dijkstra's algorithm is O((N + E)logN) while for the Bellman-Ford algorithm it is O(NE). Thus, it is evident the advantage of using the Dijkstra's algorithm, moreover if it is considered that the weights cannot be negative since they represent a distance.

An alternative is the FloydWarshall algorithm, which allows to determine the shortest path between each pairs of vertex, with a computation time of $O(N^3)$. This strategy is equivalent to apply the Dijkstra algorithm to all vertex of the graph; thus with a computational time of O(N(N + E)logN) or O(EN + N^2logN) with an optimized code based on the fibonacci heap. The benefit of the Dijkstra algorithm, compared to the FloydWarshall's one, is evident for sparse matrices with a low number of edges (E) and considering that determine the path between all pairs would result in unnecessary calculations.

It is also worth noting that there are substantial differences between a generic shortest path algorithm and the one which is required to control a smart exit system. In the latter, identify the shortest path is not enough, all the information, gathered by the different sensors, have to be elaborated in order to exclude unsafe paths and provide a real-time evacuation management strategy. A modified version of the Dijkstra's algorithm suitable for a smart evacuation system will be described in Sect. 4.

3.3 *Smart Actuators*

Once the dangerous situation has been detected, it is necessary to promptly alert the occupants. Even if bell alarms are usually used, the studies of Kobes et al. [10] suggest that vocal messages or directives would sort a better effect on people. The authors observed that rather than escape as soon as the traditional alarm sound is heard, the occupants were uncertain and waited for other people action, while a shorter delay has been observed in case of voice messages. This simple expedient can have an important role in reducing the pre-movement time and consequently increase the survival rate.

An important component of the smart guidance system is represented by the smart directional signals. Unlike conventional signs which have a permanent and fixed direction, 'smart exit signs' are designed to modify the pointing direction according to the information provided by the shortest path algorithm. To identify the escape routes, a dynamic guidance system has been developed by the University of Greenwich and its partners in the GETAWAY project. It is based on the use of the traditional exit signs properly modified and a red cross is used to indicate that a certain direction is not viable any more, as indicated in Fig. 2.

In the DOMUS project [4], an alternative solution has been taken into account, i.e. the use of electrical light paths. The experiments carried out by Kobes et al. [9] highlight the benefit of using light paths, even more if they are placed at floor level, since they can be more visible even when smoke is present. The authors show that when no smoke is visible only a minority of the occupants would escape through the nearest fire exit, ignoring the traditional exit signs, but the percentage of people escaping via the closest exit increases using green light paths located at floor level. Figure 3 shows a typical element of a light path suitable in the application of a smart

Fig. 2 Dynamic guidance systems. The signs show that only one exit way is accessible. *Source* Galea et al. [6]

Fig. 3 Element of a light path for smart guidance systems

guidance system. One of the two arrows is highlighted to dynamically change the direction and the evacuees are guided in each step without any risk of confusion.

4 Direction Setting Algorithm

Various versions of the original Dijkstra's algorithm have been proposed by different authors and adapted for being used in a smart evacuation algorithm. Cho et al. [3] assumed as source node the exits and calculated backwards the different paths from the exit towards all the remaining nodes. The reason of this choice is due to the unknown starting position of the evacuee, which could be anywhere.

Even though still based on the Dijkstra's algorithm, a different approach has been used by Shekhar et al. [15], that proposed a Capacity Constrained Route Planner (CCRP), which takes into account the maximum amount of people which can transit through a certain area at the same time. The proposed algorithm determines the optimal paths to reduce the evacuation time rather than to minimize the distance. The authors applied the algorithm to an area with a radius of 10 miles, proving the suitability of this approach also for large areas.

A problem which could arise in a capacity constrained system for building evacuation is that evacuees from the same location would be required to leave the building through different routes. Even if this strategy aims at reducing the evacuation time, it can sort the opposite effect, creating confusion and disorienting the occupants. Thus, people behavior, in a real situation of emergency, should once more be carefully taken into account.

In this section, a detailed description of the Dijkstra's algorithm for evacuation systems is provided. The algorithm aims at minimizing the distance the evacuees have to travel before reaching the exit. Four are the main steps required:

1. input the distance matrix, obtained as described in Sect. 3.2,
2. identify the source nodes and the exit nodes,
3. update the matrix according to fire information provided by sensors,
4. determine the shortest paths.

The algorithm can be applied to a wide extend of situations, due to its flexibility. It is sufficient to properly represent the building layout with an oriented graph and provide the corresponding distance matrix.

Based on the particular structure of the matrix, the source nodes and exit nodes can be easily identified. The exit nodes N_e, which represent the evacuation exits, only have arrows pointing towards the node, the a_{ej} elements from node N_e to any other node N_j is equal to infinite, therefore the e row of the matrix contains only elements with a value of infinite. All the remaining nodes are classified as source nodes.

The minimum distance paths are then determined applying the Dijkstra's algorithm to each source node. Nevertheless, if the minimum path for a certain node N_x has still been not determined, but N_x occurs in the minimum path of the N_i node, then the minimum path of N_x is automatically assigned and N_x together with N_i are added to the list of visited nodes. Applying the Dijkstra's algorithm for the N_x node would result in unnecessary calculation, whilst, computational time can be reduced, through simply verify whether the nodes have already been visited or not. Reduce the computational time, providing a prompt response to the data collected by sensors is a key parameter for a reliable and efficient evacuation guidance system. In case of emergency, if a fire is detected the involved areas are determined, based on the information gathered from sensors. The graph is updated and a new graph is generated, which reflects the real-time situation inside the building. Dangerous areas are excluded, the corresponding N_i node connections are deleted setting all the a_{ij} and a_{ji} elements of the distance matrix equal to infinite. The evacuation paths are then recalculated based on the new graph. A flowchart of the process is provided in Fig. 4.

A step by step explanation is provided by the following example for the building layout shown in Fig. 5.

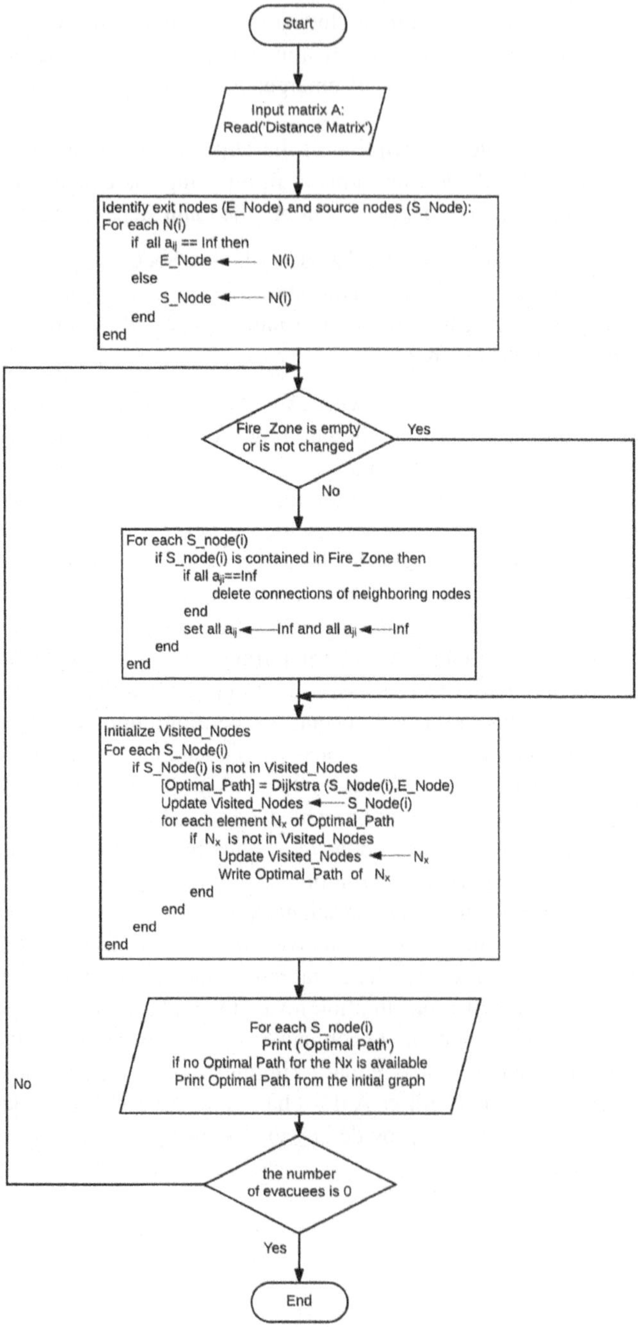

Fig. 4 Flowchart highlighting each step of the algorithm

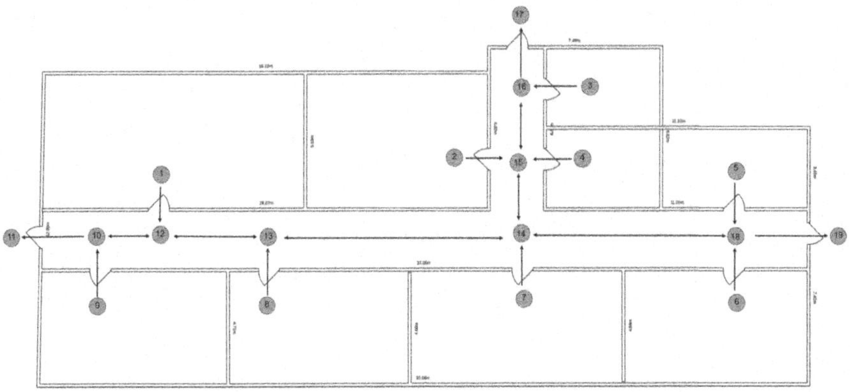

Fig. 5 Building layout and graph

<u>*Step* 1 : *Input the distance matrix*</u>

$$A = \begin{bmatrix}
- & - & - & - & - & - & - & - & - & - & - & 0.6 & - & - & - & - & - & - & - \\
- & - & - & - & - & - & - & - & - & - & - & - & - & - & 0.6 & - & - & - & - \\
- & - & - & - & - & - & - & - & - & - & - & - & - & - & - & 0.6 & - & - & - \\
- & - & - & - & - & - & - & - & - & - & - & - & - & - & 0.6 & - & - & - & - \\
- & - & - & - & - & - & - & - & - & - & - & - & - & - & - & - & - & 0.6 & - \\
- & - & - & - & - & - & - & - & - & - & - & - & - & - & - & - & - & 0.6 & - \\
- & - & - & - & - & - & - & - & - & - & - & - & - & 0.6 & - & - & - & - & - \\
- & - & - & - & - & - & - & - & - & - & - & - & 0.6 & - & - & - & - & - & - \\
- & - & - & - & - & - & - & - & - & 0.6 & - & - & - & - & - & - & - & - & - \\
- & - & - & - & - & - & - & - & - & - & 0.6 & - & - & - & - & - & - & - & - \\
- & - & - & - & - & - & - & - & - & - & - & - & - & - & - & - & - & - & - \\
- & - & - & - & - & - & - & - & - & 0.6 & - & - & 1.5 & - & - & - & - & - & - \\
- & - & - & - & - & - & - & - & - & - & - & 1.5 & - & 3.5 & - & - & - & - & - \\
- & - & - & - & - & - & - & - & - & - & - & - & 3.5 & - & 1 & - & - & 3.1 & - \\
- & - & - & - & - & - & - & - & - & - & - & - & - & 1 & - & 1 & - & - & - \\
- & - & - & - & - & - & - & - & - & - & - & - & - & - & 1 & - & 1 & - & - \\
- & - & - & - & - & - & - & - & - & - & - & - & - & - & - & - & - & - & - \\
- & - & - & - & - & - & - & - & - & - & - & - & - & 3.1 & - & - & - & - & 0.8 \\
- & - & - & - & - & - & - & - & - & - & - & - & - & - & - & - & - & - & -
\end{bmatrix}$$

<u>*Step* 2 : *Define the exit nodes and the source nodes*</u>

The exit nodes can be identified considering that they only have edges pointing towards them, all the remaining nodes are classified as source nodes. For the example under analysis:

ExitNodes = [11, 17, 19]

SourceNodes = [1, 2, 3, 4, 5, 6, 7, 8, 9, 10, 12, 13, 14, 15, 16, 18]

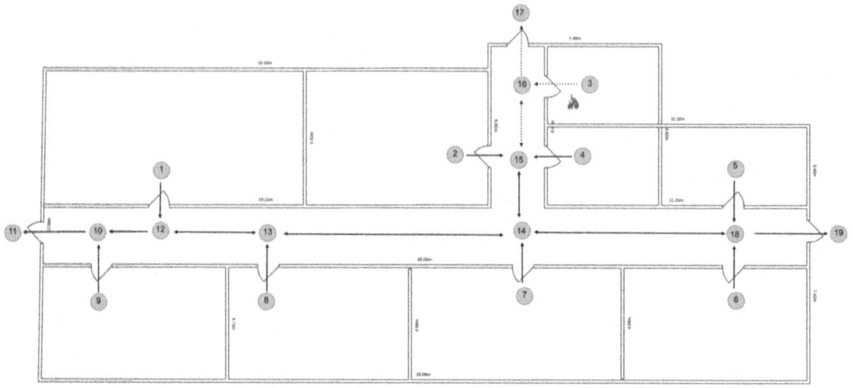

Fig. 6 Graph modification in case of fire

Step 3 : *Update the graph*

In case of fire, the graph is updated and the distance matrix is modified accordingly. For sake of security, if the node N_i, where the fire is located, has only arrows originating from the node, not only the a_{ij} and a_{ji} elements are set to infinite but also the connections of the neighboring nodes are deleted, as Fig. 6 shows for node 3.

In this case, not only $a_{3,16} = Inf$ but also the connection of node 16 are deleted: $a_{15,16}$, $a_{16,15}$ and $a_{16,17}$ are set to infinite.

Step 4 : *Determine the shortest path for each source node*

For each source node the Dijkstra's algorithm is applied. For the example under analysis, at the first iteration the algorithm is applied to node 1 and the calculated path is:

$OptimalPath(1) = [1, 12, 10, 11]$

Thus, the shortest paths for nodes 10 and 12 are already known and *OptimalPath* and *VisitedNodes* are updated.

$$OptimalPath(12) = [12, 10, 11]$$

$$OptimalPath(10) = [10, 11]$$

$$VisitedNodes = [1, 10, 12]$$

The process is iterate until the best evacuation routes for all the source nodes are known. Step 3 to 4 are repeated and if necessary new evacuation routes are determined until there are no people in the building.

Figure 7 shows the shortest paths determined when all routes are still viable, whilst Fig. 8 and Fig. 9 show the new paths recalculated when the presence of fire is detected.

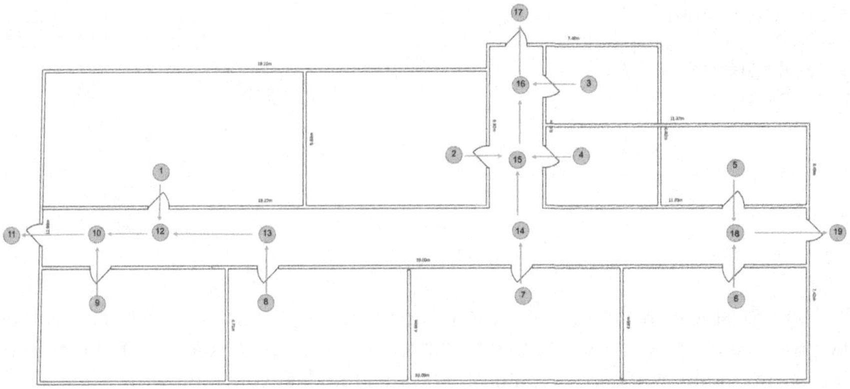

Fig. 7 Shortest Paths when no fire is detected

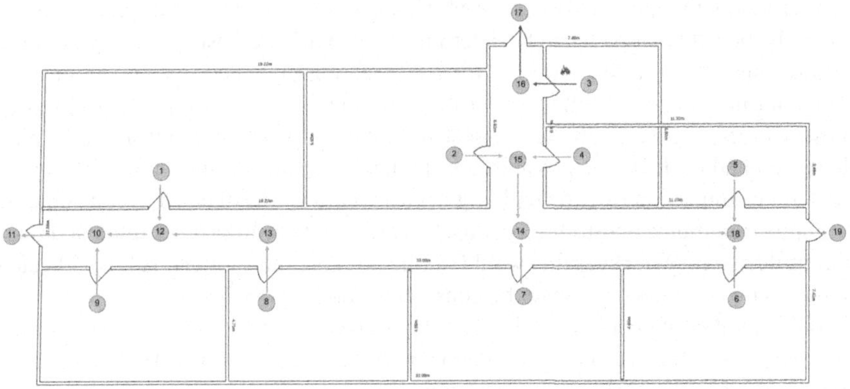

Fig. 8 Shortest Paths obtained under the hypothesis of fire in node 3

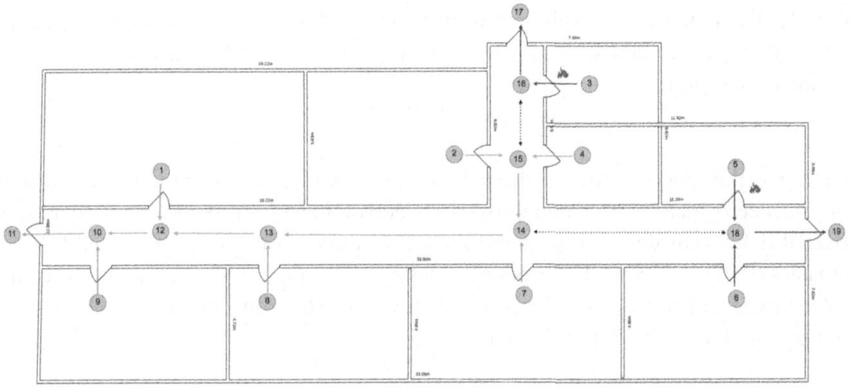

Fig. 9 Shortest Paths obtained under the hypothesis of fire in nodes 3 and 5

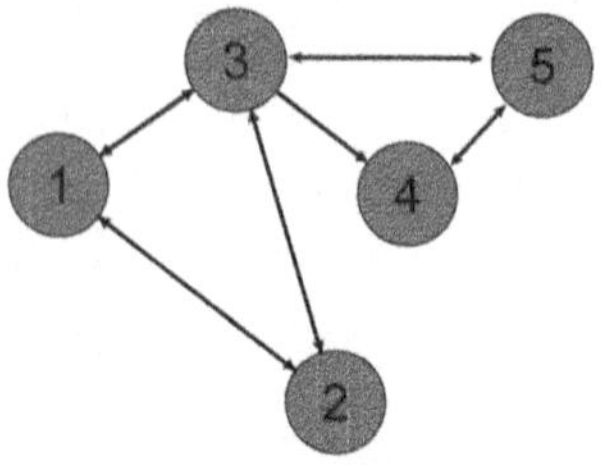

Fig. 10 Cyclic structure: 1-3-5-4-3-2-1 is an example of closed walk while 1-2-3-1 is a simple cycle

The robustness of the algorithm is verified even in particular situations, such as the presence of cycles, as confirmed by the tests performed. Cycles occur when from the starting node N_i there is a path of edges and nodes which allow to reach again N_i, without any repetition of the edges, while nodes can repeat or not, depending whether it is a closed walk or a simple cycle (Fig. 10).

A critical situation may occur, in which some nodes have no exit path, or more precisely, no safe exit path can be determined since the fire obstructs every possible escape route. Even in this case the smart evacuation guidance system can provide important information on the most critical areas and on the number of people still present in those areas. This is of great importance for rescuers to promptly plan the action and evaluate the possibility of finding alternative routes to help the evacuees, for example entering the building from a window. The planned action could be promptly communicated to the occupants avoiding them to take extreme and dangerous decision. Whenever this would not be possible the shortest path is indicated to the user even though it cannot be considered safe.

If the number of people in each area is know, based on the evacuation paths determined, it is possible to evaluate whether there is the risk of overcrowded interconnection points. This is a potentially dangerous situation, a bottleneck in the evacuation process, which can increase the risk of injuries because of people pushing each other and panic. People behavior in crowded area have been analyzed by Pelechano et al. [13]. The density of people has an important influence on the movement speed. According to Pelechano et al. [13], an estimation can be obtained through the following correlation:

$$v = k - akD \tag{1}$$

where v is the speed (m/s), k is the velocity factor which depends on the area where the evacuee is located. It is equal to 1.4 for corridors and doorways but it decreases to 1.19–1 for stair depending on the riser. D is the density of occupants and a is a constant (0.266 m^2/pers). If the people density is less than 0.55, the evacuees would not be expected to interfere between each other. Thus, the maximum velocity is assumed and it can be determined as follow:

$$v = 0.85 \cdot k \tag{2}$$

Based on this considerations, it is possible to estimate the time required to escape. Nevertheless, other aspects should also be considered, such as the individual mobility, presence of people with disabilities but also the presence of smoke in the environment which can slow down the moving speed.

5 Results Discussion and Applications

The algorithm described in Sect. 4 can be applied to any type of building, very complex structures and scenarios with a large number of exit routes, but it can also find interesting application in large evacuation plan of a whole geographical area.

It is also worth noting that the smart guidance system can be an extremely powerful tool to help rescuers, too. The current best practice for firefighters consists in identifying the location of the fire once arrived on the scene, its extend, presence of people trapped into the building and then plan the intervention accordingly. A smart guidance system can provide to the rescues all necessary information allowing them to plan their actions even before arriving on-site, therefore saving precious time.

Examples of application of an evacuation guidance system are provided. Figure 11 shows the layout of a building and its corresponding graph. The ID of each node and the connections among them are highlighted. Node 1, 7, 16, 27, 33, 38 and 42 represent the exit doors. The shortest path for each source node towards the nearest exit is provided in Fig. 12 and highlighted with green arrows. These paths, obtained assuming that no fire alarm is detected, coincide with the ones used in traditional evacuation system to guide people towards the closest exit.

To assess the reliability of the proposed evacuation guidance system, different scenarios have been hypothesized. Figure 13 shows the recalculated shortest path when fire is detected in node 26. Due to its proximity to node 26, node 24 is considered as a potentially dangerous area. Hence, evacuees, located in node 20, 21, 22 and 23, which originally where transiting through node 24, are now guided to the nearest safe exit, represented by node 33. In case the fire continues to expand involving node 23, the new shortest paths are identified among all the possible safe alternatives. As shown in Fig. 14, people located in node 22, 21 and 20 are guided towards the exit node 38. The lack of safe path is underlined in Fig. 13 and Fig. 14 with red arrows.

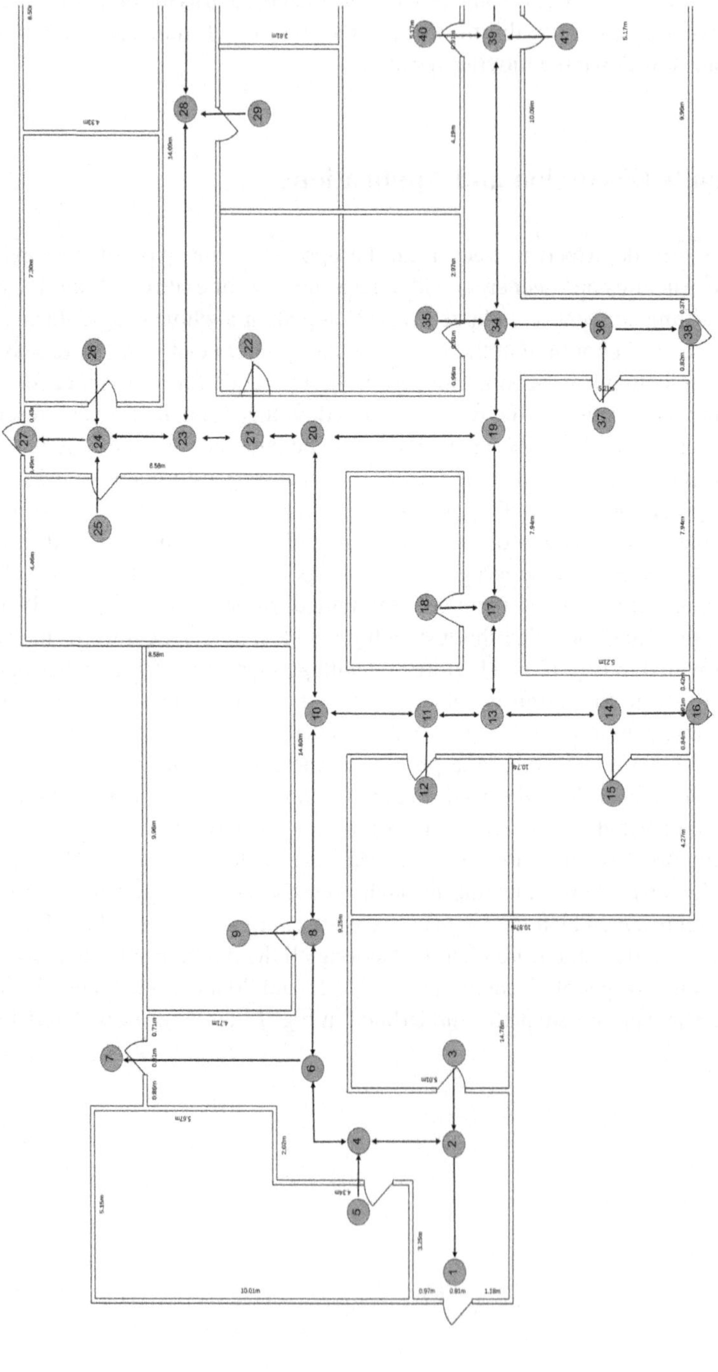

Fig. 11 Graph representing the building layout

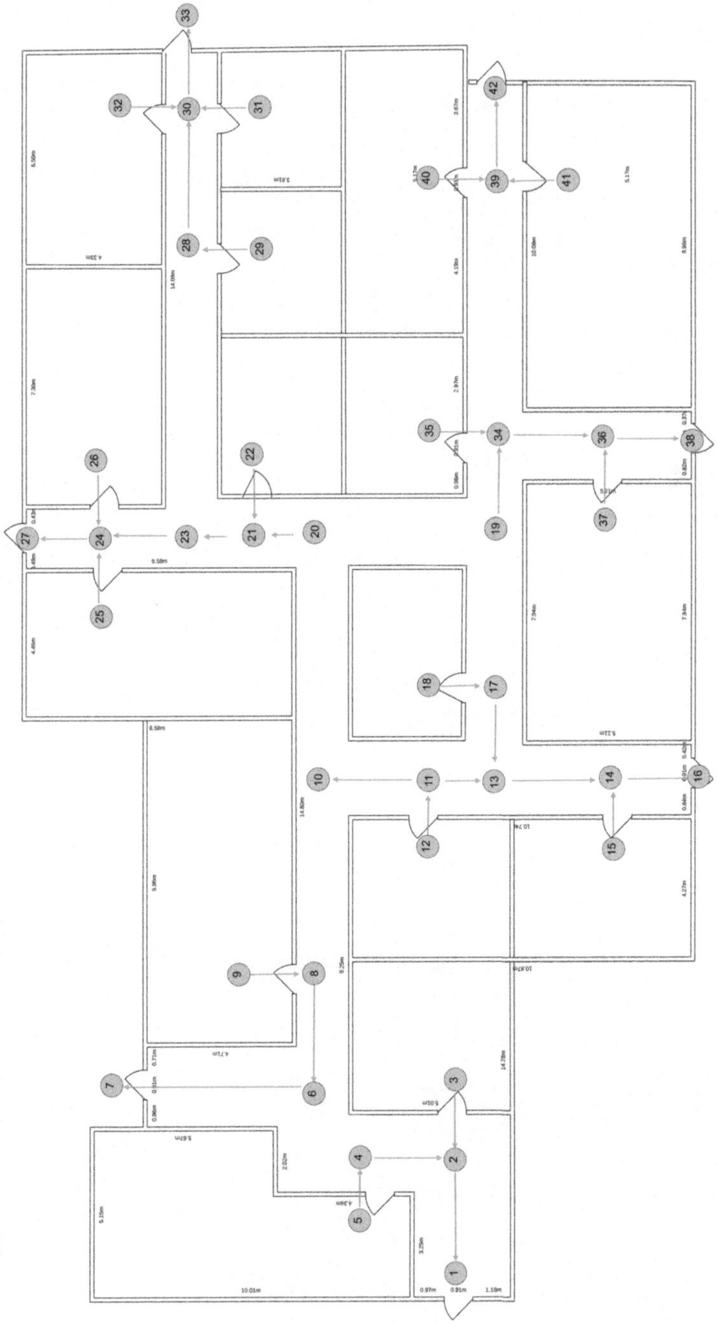

Fig. 12 The shortest evacuation paths are highlighted

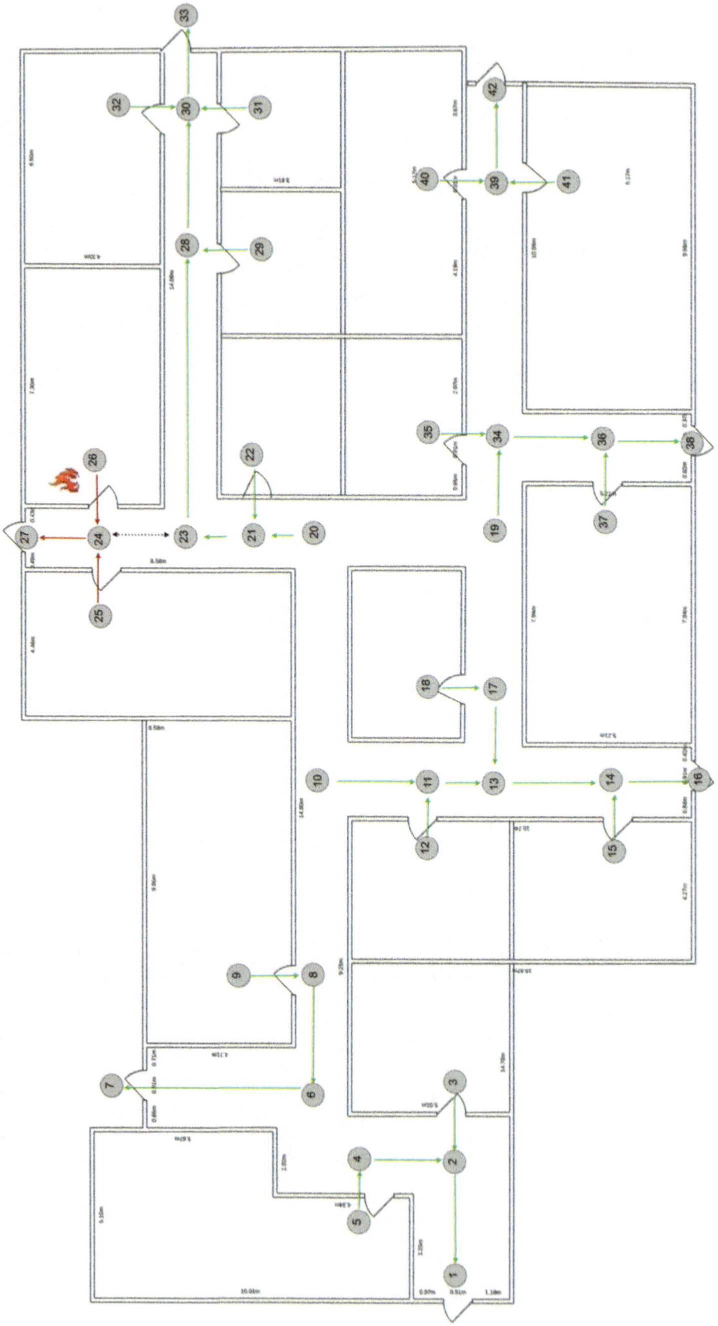

Fig. 13 Shortest evacuation paths in case of fire in node 26

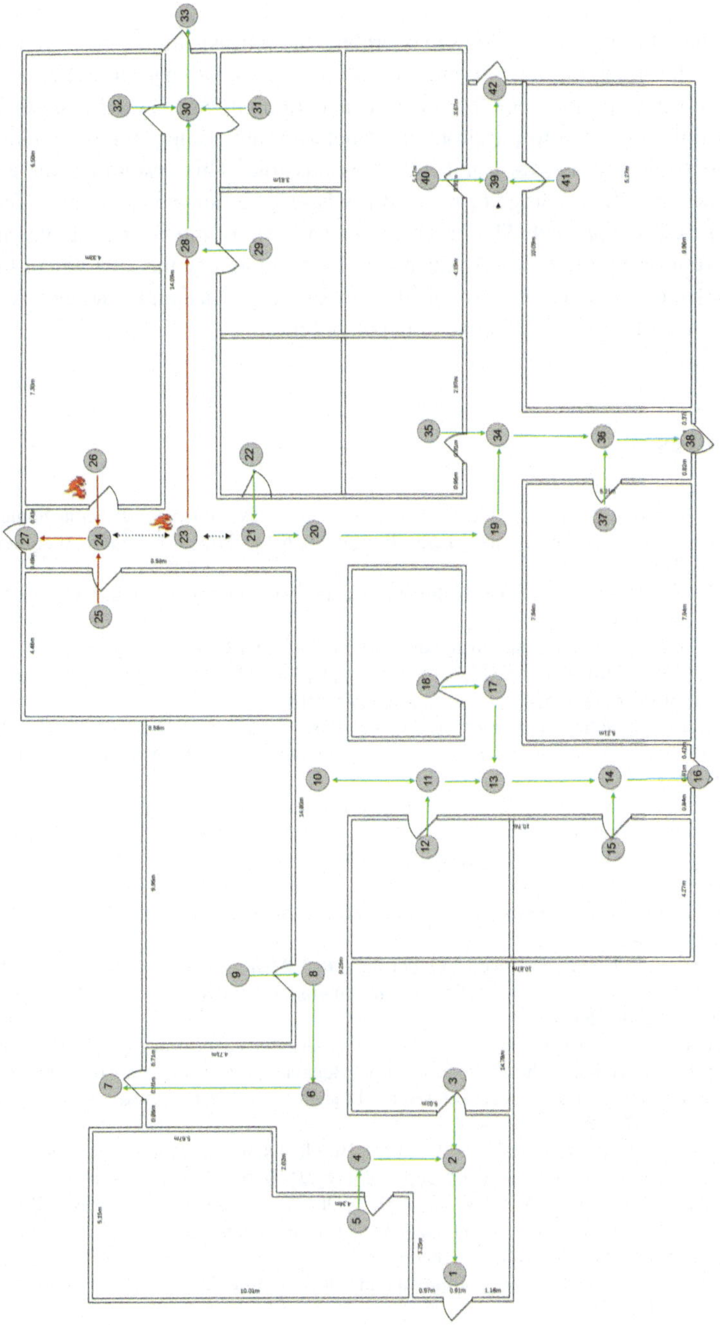

Fig. 14 Shortest evacuation paths in case of fire in nodes 26 and 23

6 Conclusions

The Internet of things (IoT) allows to overcome the intrinsic limit of traditional technologies, addressing the safety issues of traditional evacuation systems that multiple experiments have pointed out. The smart evacuation guidance system allows a general real-time approach to the problem of determining escape routes, enabling the occupants to escape as safest and fastest as possible. This system can be applied to various types of buildings or even to a whole geographical area, provided that they are suitably represented by graphs. It also allows for more accurate planning of the intervention by the rescuers, since it is easy to identify the area where the fire developed and its extend, but also information on people in danger and their location can be provided for more effective emergency relief.

References

1. Y. Benezeth, H. Laurent, B. Emile, C. Rosenberger, Towards a sensor for detecting human presence and characterizing activity. Energy Buildings (2011). https://doi.org/10.1016/j.enbuild.2010.09.014
2. C.Y. Chen, The design of smart building evacuation system. Int. J. Control Theory Appl. **5**, 73–80 (2012)
3. J. Cho, G. Lee, S. Lee, An automated direction setting algorithm for a smart exit sign. Autom. Constr. (2015). https://doi.org/10.1016/j.autcon.2015.05.004
4. http://www.distrettodomus.it/it/. Accessed 3 Jan 2018
5. G. Fortino, P. Trunfio (eds.), *Internet of Things Based on Smart Objects: Technology, Middleware and Applications* (Springer International Publishing, 2014). https://doi.org/10.1007/978-3-319-00491-4
6. E.R. Galea, H. Xie, P. Lawrence, Intelligent active dynamic signage system: bringing the humble emergency exit sign into the 21st century (2018). http://www.sfpe.org/general/custom.asp?page=Issue3Feature1. Accessed 3 Jan 2018
7. S. Gokceli, N. Zhmurov, G.K. Kurt, B. Ors, IoT in action: design and implementation of a building evacuation service. J. Comput. Netw. Commun. (2017). https://doi.org/10.1155/2017/8595404
8. A. Guerrieri, V. Loscri, A. Rovella, G. Fortino (eds.), *Management of Cyber Physical Objects in the Future Internet of Things* (Springer International Publishing, 2016). https://doi.org/10.1007/978-3-319-26869-9
9. M. Kobes, I. Helsloot, B. de Vries, J.G. Post, Exit choice, (pre-)movement time and (pre-)evacuation behaviour in hotel fire evacuation. Behavioural analysis and validation of the use of serious gaming in experimental research. Procedia Eng. (2010). https://doi.org/10.1016/j.proeng.2010.07.006
10. M. Kobes, I. Helsloot, B. de Vries, J.G. Post, Building safety and human behaviour in fire: a literature review. Fire Saf. J. (2010). https://doi.org/10.1016/j.firesaf.2009.08.005
11. M. Kobes, I. Helsloot, B. de Vries, J.G. Post, N. Oberije, K. Groenewegen, Way finding during fire evacuation; an analysis of unannounced fire drills in a hotel at night. Build. Environ. (2010). https://doi.org/10.1016/j.buildenv.2009.07.004
12. R.L. Paulsen, Human behavior and fire: an introduction. Fire Technol. (1984). https://doi.org/10.1007/BF02384147
13. N. Pelechano, J.M. Allbeck, N.I. Badler, *Virtual Crowds: Methods, Simulation, and Control* (Morgan & Claypool Publishers, 2008)

14. G. Proulx (2001) Occupant behaviour and evacuation, in *Proceedings of the 9th International Fire Protection Symposium*, pp. 219–232
15. S. Shekhar, K. Yang, V.M.V. Gunturi, L. Manikonda, D. Oliver, X. Zhou, B. George, S. Kim, J.M.R. Wolff, Q. Lu, Experiences with evacuation route planning algorithms. J. Geogr. Inf. Sci. Int. (2012). https://doi.org/10.1018/13658816.2012.719624
16. J.D. Sime, An occupant response shelter escape time (ORSET) model. Saf. Sci. (2001). https://doi.org/10.1016/S0925-7535(00)00062-X
17. T. Teixeira, G. Dublon, A. Savvides, A survey of human sensing: methods for detecting presence, count, location, track, and identity. ACM Comput. Surv. **5**, 427–450 (2010)

The manufacturer's authorised representative in the EU is Springer Nature Customer Service Centre GmbH, Europaplatz 3, 69115 Heidelberg, Germany. If you have any concerns regarding our products, please contact ProductSafety@springernature.com

Printed and bound by CPI Group (UK) Ltd, Croydon, CR0 4YY
30/09/2026
02225643-0012